Telecommunications in Africa

Other volumes in the Global Communications series of the Columbia Institute for Tele-Information

TELEVISION IN EUROPE
By Eli M. Noam

TELECOMMUNICATIONS IN EUROPE
By Eli M. Noam

TELECOMMUNICATIONS IN THE PACIFIC BASIN
An Evolutionary Approach
Edited by Eli M. Noam, Seisuke Komatsuzaki, and Douglas A. Conn

TELECOMMUNICATIONS IN LATIN AMERICA
Edited by Eli M. Noam

TELECOMMUNICATIONS IN WESTERN ASIA AND THE MIDDLE EAST
Edited by Eli M. Noam

Telecommunications in Africa

EDITED BY

ELI M. NOAM

New York Oxford
OXFORD UNIVERSITY PRESS
1999

Oxford University Press

Oxford New York
Athens Auckland Bangkok Bogotá Buenos Aires Calcutta
Cape Town Chennai Dar es Salaam Delhi Florence Hong Kong Istanbul
Karachi Kuala Lumpur Madras Madrid Melbourne Mexico City Mumbai
Nairobi Paris São Paolo Singapore Taipei Tokyo Toronto Warsaw

and associated companies in
Berlin Ibadan

Published by Oxford University Press, Inc.
198 Madison Avenue, New York, New York 10016

Library of Congress Cataloging-in-Publication Data
Telecommunications in Africa / edited by Eli M. Noam.
p. cm. — (Global communications series)
Includes bibliographical references and index.
ISBN 0-19-510201-0
1. Telecommunication—Africa. 2. Telecommunication policy—Africa. I. Noam, Eli M. II. Series.
HE8464.T45 1999
384'.096—dc20 96-36158

1 3 5 7 9 8 6 4 2

Printed in the United States of America
on acid-free paper

Contents

Contributors

G. O. Ajayi, Obafemi Awolowo University, Ile-Ife, Nigeria

Felix K. Akorli, University of Science and Technology, Kumasi, Ghana

Francis K. A. Allotey, University of Science and Technology, Kumasi, Ghana

Klaus Dierks, Deputy Minister of Works, Transportation, and Communications, Namibia

Girma Feyissa, Department of Transportation and Communications, Addis Ababa, Ethiopia

E. Bisimwa Ganywa, University of Bukavu, Zaire

Robert B. Horwitz, University of California, San Diego

Heather E. Hudson, University of San Francisco, California

Janice Hughes, Booz, Allen & Hamilton

Alphonse Kabayiza, Ministry of Communications, Kigali, Rwanda

David Kaplan, University of Cape Town, South Africa

Andrea L. Kavanaugh, Virginia Polytechnic Institute

Hugues Koné, CERCOM, University of Abidjan, Ivory Coast

M. L. Luhanga, University of Dar es Salaam, Tanzania

Eli M. Noam, Professor of Finance and Economics, Columbia Business School, and Director, Columbia Institute for Tele-Information

Mansur M. Nuruddin, Columbia Institute for Tele-Information

Gehan Rachty, Dean of Faculty of Mass Communications, Cairo University

T. I. Raji, Oyo State University of Technology, Ogbomoso

Helena Renfrew, Booz, Allen & Hamilton

R. I. Salawu, University of Lagos, Lagos

Bukasa Tshilombo, University of Bukavu, Zaire

Abii Tsigie, Department of Transportation and Communications, Addis Ababa, Ethiopia

Michael Tyler, Booz, Allen & Hamilton

Telecommunications in Africa

Introduction

ELI M. NOAM

Africa comprises 20 percent of the world's land mass and contains 12 percent of its population. But it accounts for only 2 percent of the world's telecommunications. While telephone penetration in the United States in 1994 was almost 65 main lines per 100 population and 47 for all OECD countries,[1] it was 1.5 in Africa (see Table I.1). In 1994, only one African country had more than 5 main lines per 100. Almost 40 percent of Africa's telephone lines were concentrated in a single country, South Africa. The other sub-Saharan countries that account for 80 percent of the continent's population were served by only 10 percent of its phone lines. That region of 600 million people had fewer phones than Manhattan south of 59th Street. In the entire continent of Africa there were fewer phone lines than in New York State.

With few exceptions, such as Ethiopia, telecommunications was introduced to Africa by the colonial nations as a mechanism of control and governance. The era of political independence in the 1960s brought with it a growing recognition of the linkages between telecommunications and economic development. Telecommunications was refocused to serve as domestic tools of national coherence and development.

Unfortunately, the European colonial powers left not only inadequate telecommunications hardware infrastructure but also an obsolete organizational structure—the state monopoly PTT (Post, Telephone, and Telegraph) system. This system was ineffective even in the advanced European countries in meeting the needs of the emerging information-based economy and high-technology industries. In time, it collapsed in Europe in favor of increasingly privatized and competitive structures. It was even less effective in poorer countries without significant indigenous technological resources and funds for redistribution. Yet the state monopoly system was in tune with the more general economic organizations of many of the emerging countries, which emphasized state control, centralized planning, and national sovereignty. Thus, while low telephone penetration can be partly ascribed to poverty and political instability, this is only part of the explanation. In the field of telecommunications, investment resources are not as scarce as they are elsewhere. First, each country has people and businesses who desire a telephone badly enough to be willing to pay the investment necessary in advance.

Table I.1 Telephone Penetration for Selected African Countries

Country	Main Lines, 1994 (thousands)	Main lines per 100 Inhabitants
Algeria	1,068	3.97
Egypt	2,374	4.26
Libya	240	4.76
Morocco	820	3.07
Tunisia	421	4.89
North Africa	**4,923**	**4.19**
South Africa	**3,659**	**9.00**
Benin	20	0.39
Botswana	43	3.10
Burkina Faso	21	0.22
Cameroon	57	0.45
Central African Republic	6	0.21
Congo	36	0.09
Congo-Brazzaville	19	0.76
Côte d'Ivoire	20	0.59
Djibouti	7	1.28
Ethiopia	132	0.25
Gambia	16	1.60
Ghana	48	0.30
Kenya	214	0.85
Malawi	32	0.35
Namibia	69	4.46
Rwanda	11	0.16
Senegal	64	0.80
Zambia	78	0.91
Zimbabwe	128	1.20
Sub-Sahara (excluding South Africa)	**1,021**	**0.45**
Total Africa	**9,605**	**1.52**

Source: ITU/BDT/INFS database.

And second, foreign investors are eager to enter most countries' telecom markets. Thus, much of network construction can be self-financing. And once networks are operational, profitability is normally assured.

However, governments interfered heavily in telecommunications. They used the network as a funding source for their general budgets by taxing away a significant part of the operator's surplus domestically and by pocketing most of its foreign currency earned internationally. Furthermore, governments forced the operators to provide for the state and its officials below cost, causing a significant revenue drain. The operating monopoly also became a place for patronage employment, its procurement contracts were often embroiled in international diplomacy and personal gain, and its management was politicized. The emphasis on national, centralized control usually prevented other public solutions, such as multi-country operators, municipal systems, or cooperatives. Indeed, in most countries the success of such local arrangements in Finland and rural America is quite unknown. Privatization and foreign participation were politically unthinkable to governments, and competition was unattractive to many foreign investors.

One cannot, of course, deny the burden of overcoming low levels of development and the shortages of specialized human and capital resources. But even here other approaches might have worked. As the industrialized world introduced modern digital equipment, it junked perfectly functional analog equipment in great quantities. Such equipment could have been available at the price of transportation or less. But it was rarely taken up, because a combination of Western equipment manufacturing interests and African national pride stood in the way of accepting second-hand equipment.

The deficiencies existing within the African telecommunications sector are compounded by the absence of an engineering infrastructure for the production of spares and components and by the scarcity of an indigenous technically trained workforce. Less recognized is the shortage of qualified staff to deal with legal and economic matters of telecom reform. Nonetheless, a number of African countries including Egypt, Ethiopia, Gambia, Ghana, Ivory Coast, Kenya, Namibia, Senegal, South Africa, Tunisia, Zambia, and Zimbabwe have gradually introduced new pro-competition and pro-privatization policies and a commercially reoriented management structure.

By the 1980s, following the lead of the United States, Britain, and Japan, many developed countries reluctantly began to dismantle their own inefficient monopoly systems. These concepts also found their way to developing countries. Thus, the 1980s also saw the first telecom reforms instituted in Africa. The pace of government policy reforms was initially slow, as many chapters in this volume show. But this situation is changing rapidly. In financing restructuring programs, international development organizations, such as the World Bank, have pushed corporatization and privatization schemes for public telecommunications operators. By linking loans to the telecommunications sector on real progress in reforming PTTs, the World Bank has goaded several African governments, such as Guinea and Cameroon, to restructure telecommunications. Following corporatization, the next step is often a partial privatization and the introduction of competition into the sector.

International development banks have acted as major forces of change because of the absence of the large business users that have agitated for telecommunications reform elsewhere. Both a low use of communications and a frequent lack of a democratic framework render users' groups nonexistent or, at best, ineffective as forces of change. PTT trade unions likewise had only marginal impact on the reform process. There are cases where users' groups have had an impact on the restructuring process, such as in Egypt, South Africa, and Tunisia. In the case of both Algeria and Morocco, however, telecommunications labor unions, fearing unemployment, played an active role in protesting and postponing corporatization plans for the PTT. And in South Africa they helped obtain a five-year exclusivity period for the incumbent, monopolist.

African policy makers were cautious about privatization of the national telecommunications administration due to political sensitivities about foreign penetration into this strategic sector. While there is now widespread agreement among policy makers on the need for reforming the sector, there is also a strong reluctance by governments to relinquish control. Even in major countries where a privatization of the telecommunications sector has been initiated, as in South Africa and Egypt, the major shareholder of the corporate entity is the state, and the corporation is accountable to government ministries beyond the regulatory authorities.

Despite the difficulties associated with privatization, several governments concluded it was a viable option. Some governments were prompted by a desire to reduce national debt. Others sought foreign expertise and capital. In most cases, privatization programs sought to sell a minority stake of the PTT to a strategic investor, that is, to a foreign telecommunications carrier.

Are these hopes realistic? In contrast to other regions of the world, Africa has experienced a reluctant interest in foreign investment in its telecommunications infrastructure. Generally, the trend of privatization in developed countries has resulted in international expansion. Some telephone companies sought to enter foreign markets to escape residual domestic restrictions. And importantly, the globalization of large firms has created a worldwide demand for advanced and seamless global business telecommunications services. This has led to numerous joint ventures, alliances, acquisitions, and entries. In 1995, 29 percent of international ventures included partners from Western Europe, and 17 percent included partners from North America. Their extra-territorial investments were partly in their home regions but mostly in the developing world. And it was remarkable how few of them were in Africa. In 1996, while 32 percent of all international telecom ventures were in Eastern Europe and 26 percent in South America, only 2 percent of such venturing activity was in Africa.[2] Even those ventures were largely accounted for by interest in post-apartheid South Africa and by supportive activities of semi-official French organizations in politically closely linked Francophone countries.

Thus, France Télécom and its subsidiary France Câble et Radio have been active in the Central African Republic (40 percent of Socatel), Equatorial Guinea (DGCT), Ivory Coast (51 percent of CI-Telecom), Senegal (Société National de Télécoms), and in Congo-Brazzaville for mobile communications. Meanwhile, Investcom, a close partner of France Télécom in Lebanon, holds mobile networks in Guinea and in the Congo through its subsidiary Spacetel. The Société Française de Radio Téléphonie, the second official French telecom organization, was a cobidder in Cameroon's digital mobile network. It also has an interest in Tunisia's mobile communications, as does Canada's Teleglobe and the United Kingdom's Mobitel and Cellnet. Britain's Cable & Wireless, another former colonial telecommunications administration, is active in South African mobile service and in the Seychelles and Botswana. Telekom Malaysia is involved in South Africa (with the American Bell company SBC) and in several other countries. But these are the exceptions. Many major international players are still absent from Africa.

Why this lack of interest in African telecommunications? Official protestations notwithstanding, Africa was slow to restructure its basic telecommunications market. African countries have to compete with investment opportunities in many other regions. The frequent reluctance among African countries to allow international investment and the perception of Africa as a troubled region were significant liabilities. But as investment opportunities in other regions get snapped up and as Africa now undergoes accelerated restructuring and liberalization, we may see an upturn in Africa's share of international telecommunications investment.

As an alternative to direct investments in national networks, some companies have initiated regional projects. AT&T wants to link the continent through Africa One, a network based on a 30,000-mile underwater fiber optic cable. Other projects include Afrilink, a plan by Siemens to lay underwater cable; Atlantis-2, a plan to lay submarine fiber optic cable between Brazil, Senegal, Spain, and Portugal; and Afrinetwork, a regional plan put forth by NTT.

The impediments to telecommunications development in Africa have had their greatest negative impact on rural areas. Although comprising the major share of population, rural areas have severely undeveloped telecommunications systems. Opponents to telecom reform often fear that development of the rural infrastructure would be neglected by commercial operators. They argue that a state monopoly would be more ready to cross-subsidize services to the approximately 70 percent of the population residing in rural areas. This argument, made similarly in the developed world, is analytically incorrect, although it may be politically realistic. There are alternative approaches to state ownership to generate cross-subsidies to achieve universal connectivity. For example, when a country wants to subsidize the connectivity of its rural population, it can do so by taxing all communications services and creating a universal service fund. There is no need to implement subsidies through monopoly market structure and its extension to vastly uneconomic international rates. Indeed, the fact that rural areas have not much benefitted from decades of monopoly suggests that new approaches should be explored.

Pressing needs to extend basic telephony countrywide are one part of the equation. The other part is that in order to expand African economic competitiveness, its businesses must have access to advanced telecommunications capabilities and information-related activities. This has been the motivation behind the liberalization of value-added services, along with the belief that liberalization of these services would face the least opposition from the state monopolies. The Internet, in particular, has been a focus for change, starting with the connectivity of universities to the rest of the world. Many groups and organizations are using the Internet to achieve global connections for economic and personal communications. Political information flows become harder to control unless a government is willing to sacrifice the economic benefits of access to global information sources. And high international telephone rates can be undercut by Internet telephony. For these reasons governments often try to control access and usage of the Internet. Even in South Africa, the traditional operator, Telekom SA, has asserted a monopoly over Internet service provision.

Generally speaking, nations' telecommunication networks around the world have evolved along the following stages:

1. *Cost-sharing stage networks.* Growth is based on the sharing of costs and increasing the value of interconnectivity. The countries in this stage rely heavily upon government intervention, through subsidies and long-term

planning, to develop their basic telecommunications infrastructure. However, lack of capital and sometimes political instability have resulted in inadequate investment and slow expansion of service.

2. *Redistributory stage networks.* These grow through politically directed expansion and through transfers from some users to others. The countries in this stage have gone well beyond the early developmental stages and have expanded their networks through a redistributory process, controlled by a monopoly carrier.
3. *Pluralistic stage networks.* These are not necessarily more advanced technologically than redistributory stage networks, but they have progressed institutionally. They are not a centralized system but a federation of sub-networks. The United States, Japan, and Europe are in transition to this stage.

Developing nations do not need progress in the same slow sequence as developed countries did in the past. They can leapfrog a lengthy redistributory stage and enter the pluralistic network stage more quickly, combining it with redistributory policies. Most of Africa is still in the cost-sharing stage, but several countries are entering a pluralistic stage of networks. Examples are Ghana and Nigeria. On the other hand, for South Africa and Namibia, the redistribution process has been more central and is intertwined with these countries' overcoming of racial divisions. They believe that a move to a pluralistic system needs to wait for the redistributory stage to have run its course.

In the 1990s, as new political and economic developments have been shaping the policy environment, the pace of change has been as diverse as the countries that comprise the continent. It is this diverse and changing environment that we present in this book. The studies presented in this volume focus not on technology but on the institutional structure and the process of change.

Andrea Kavanaugh of the Virginia Polytechnic Institute analyzes telecommunications in the three North African countries of Algeria, Morocco, and Tunisia. These countries have significantly outpaced their sub-Saharan neighbors in telecommunications development. In fact, in the late 1980s, these countries have been ahead of many industrialized nations in telecommunications investment as a percentage of gross fixed capital formation.

Gehan Rachty, dean of the faculty of Mass Communications at Cairo University, discusses the evolution of Egypt's telecommunications network from the nineteenth century through the part-privatization of the national carrier ARENTO and its transformation into Telecom Egypt. Egypt has been moving rapidly toward an information economy, with Egyptnet, the country's packet-switched network, growing rapidly. However, progress in restructuring the sector has been reluctant. Rachty argues that only through continued deregulation and privatization will Egypt be able to take advantage of the global communications revolution.

In a similar vein, Abii Tsigie of Ethiopia's Department of Transportation and Communications notes that although telecommunications were indigenously introduced to Ethiopia from the top through the initiative of Emperor Menelik II, today's needs include non-state solutions and the development of a local telecommunications manufacturing industry.

In their chapter on Kenya, Michael Tyler, Janice Hughes and Helena Renfrew, a team of consultants then from Booz, Allen and Hamilton, Inc., explore the foreign exchange cost to the Kenyan economy incurred by the country's inability to meet the demands for telecommunications services of twenty of its most profitable businesses. The study lends credence to the view that expanding the telecommunications sector is vital to national economic strategy.

M. L. Luhanga's analysis of Tanzania indicates that the country, despite its long-standing statist tradition, is moving toward a pluralistic network. The challenge for Tanzania is to balance the need for basic telecommunications for the majority of its citizens with the push for specialized networks and value-added services by the commercial sector. This requires a balancing of the cost sharing and the pluralistic phases of network development.

In discussing telecommunications development in the Congo (formerly Zaire), E. Bisimwa Ganywa and Bukasa Tshilombo of the University of Bukavu assert that among the many obstacles that have hindered network expansion, the major problem has been the failure to view telecommunications as a public good; instead, telecommunications service has been primarily limited to a privileged elite. It was part of the unhappy state of affairs in Zaire's history.

A different problem exists in Rwanda. Alphonse Kabayiza explicates the effects of political instability on African telecommunications development. As Kabayiza notes, prior to its civil war, Rwanda had made significant strides in the development of its telecommunications network, i.e., in the redistributory stage. Prior to the civil war, the government-owned carrier Rwandatel was slated to be the first African telecommunications carrier to be privatized, but these plans had to be shelved.

In discussing the Ivory Coast (Côte d'Ivoire), Hugues Koné of CERCOM at the University of Abidjan notes that the country has one of the best telecommunications infrastructures in sub-Saharan Africa. In order to improve performance in the sector, the government decided to privatize CI-Telecom and allow competition in cellular telephony. Koné concludes that despite this change, policy must be focused on addressing fighting fraud, developing rural telecommunications, and meeting the backlog for basic value-added services.

In their chapter on Nigeria, G. O. Ajayi of Obafemi Awolowo University, R.I. Salawu of the University of Lagos, and T. I. Raji of Oyo State University of Technology in Nigeria describe the evolution of Nigeria's telecommunications network from a primitive system to one that employs technology. Despite this progress, Nigeria has a long way to go. In 1993, in a move to improve service, Nigeria's telecommunications sector was restructured, a regulatory commission was created, and several telecommunications markets were opened to competition.

In Ghana, as described by Francis Allotey and Felix Akorli of the University of Science and Technology in Ghana, the evolution of a telecommunications network progressed from the installation of the first telegraph line in 1881 to a radical restructuring in the mid-1990s. Ghana's telecommunications initiations were hampered by numerous obstacles, and it had one of the lowest penetration rates in the world. The government embarked on an ambitious restructuring program, created a separate telecommunication regulatory agency, and sold a majority stock of Ghana Telecom to a strategic international investor. It also licensed a second carrier.

South Africa is the economic giant of the continent. It merits two chapters, one on its equipment industry, the other on its network. David Kaplan of the University of Capetown examines the South African telecommunications equipment industry. In an attempt to ensure a national high-tech industry, the South African Post Office entered in the late 1950s into long-term agreements with a few favored foreign companies, such as Siemens and Alcatel. In exchange for exclusive contracts these companies agreed to establish extensive local production of telecommunications equipment. This apartheid-era policy introduced significant distortions into the sector and was not effective in achieving true technology transfer. Today, with economic sanctions lifted, the South African telecommunications equipment sector is entering export markets, especially in Africa.

Robert Horwitz of the University of California at San Diego provides an overview of the historical development and present state of telecommunications in South Africa. He describes a transition from the apartheid-era structure that combined posts and telecommunications as well as regulatory roles in one entity, the South African Post & Telecommunications (SAPT). Today's regime includes a separate and part-privatized telecommunications entity, Telekom SA, engaged in providing service to all citizens. As in all aspects of South African society, the system of apartheid was the framework under which SAPT's activities were carried out. This led to a disparate distribution of the network. The major challenges facing the new Telekom SA is rectifying the historical inequities in the distribution of telecommunications services, while meeting the needs of South Africa's private sector for advanced and inexpensive telecommunications.

Dr. Klaus Dierks, deputy minister of Works, Transportation and Communications in Namibia, notes that with a teledensity of about 5 percent, Namibia has more than the average number of telephones for Africa. Nonetheless, the newly commercialized national carrier Telecom Namibia still cannot satisfy the demand for basic or value-added telecommunications services. A major goal in Namibia is bridging the telecommunications gap between the two Namibias, the modern urban sector and the former homelands where the majority of Namibians live. Namibia is embarked on a process of regulatory and institutional reform. Although Telecom Namibia has a monopoly on telecommunications services, Dierks foresees increased competition in the future from cellular carriers and other entities such as banks and mining companies.

In an effort to address the sector's deficiencies, African leaders are increasingly looking to regional cooperation. They have attempted to join efforts and resources in the establishment of pan-African telecom bodies. One regional attempt was the Pan-African Telecommunications Network (Panaftel). Mansur Nuruddin of Northern Business Information examines the forces behind the evolution of this regional effort. Nuruddin also discusses several other models for regional telecommunications networks in Africa.

Heather Hudson of the University of San Francisco analyzes the role of new communications technologies in African socioeconomic development. A major impediment to the sector's development is that the telecommunications administration acts as a bureaucratic bottleneck. Such administrative inefficiencies could be remedied by allowing greater autonomy for providers, with independent management and commercial goals. African countries are increasingly considering the advantages of converting telecommunications entities into private or semi-private companies, often with the participation of foreign companies.

The information contained in this book is the result of the cooperation and collaboration of more than 20 authors, each of whom contributed research, knowledge, and expertise of particular countries. Inevitably, it proved extraordinarily complex to coordinate the completion of chapters by so many authors in a fast-moving environment. In several cases where countries underwent political upheavals we lost contact with authors for many months. All this makes us even more grateful to those who helped us in bringing this volume to conclusion. It is the final volume in a six-volume series on global telecommunications, published by Oxford University Press.

Besides the contributing authors, many others were involved in this volume. We wish to thank especially Dr. Ray Akwule, who contributed much to the selection of authors of this volume at its initial stages. Credit goes also to Lamiaa E. Elfar, who contributed to the summaries of this chapter, and to Sabi Muteshi, who began the task of reviewing the original submissions. Mansur Nuruddin coordinated the updates and contributed his own research to several chapters in the book, including this introduction. Further informational assistance was provided by Elizabeth Ridley. Paul Bodine completed the developmental editing of the project, and Lisa Domonkos and Caterina Alvarez coordinated the editing and updates. We also wish to acknowledge a grant by the World Bank that helped us in the editorial process. And we would like to thank Oxford University Press, in particular Herb Addison and Tamara Destine, for their faith in this global project and support. We are also grateful to Françoise Bartlett of G&H Soho for a splendid editorial job.

Today, African telecommunications are at a crossroads. There are considerable obstacles and hidden opportunities. Africa must overcome years of telecom inertia and establish a vibrant sector for its societies and economies.

African governments will have to deal with the following issues:

- How to provide telecommunications services across society and geography.
- Whether to allow foreign ownership of national telecommunications carriers and how to reconcile the goals of investors and government.
- Where to allow competition.
- How to effectively regulate a privatized telecommunications sector.
- How to balance the need for basic communications and the need for advanced business services.
- How to collaborate internationally outside traditional government-to-government frameworks.

It is our hope that the present volume will contribute to the development of innovative, effective, and timely solutions to the challenges that face African telecommunications as they enter a period of historic transformation.

Notes

Thanks go to Lamiaa E. Elfar and Mansur Nuruddin, who contributed to the summaries of this chapter.

1. International Telecommunication Union, *Challenges to the Network: Telecommunications and the Internet* (Geneva: 1997). ITU Table 4, p. A-17.
2. Federal Communications Commission, International Bureau, "Global Communication Alliances: Forms and Characteristics of Emerging Organizations," prepared by Douglas Balbi and Chris Keating, February 8, 1996, as cited in Eli M. Noam and Anjali Singhal, "Supra-national Regulation for Supra-national Telecommunications Carriers?" *Telecommunications Policy,* Vol. 21, No. 10, pp. 769–815 (London, 1996).

1
Algeria, Morocco, and Tunisia

ANDREA L. KAVANAUGH

Algeria, Morocco, and Tunisia are all former dependencies of France: Algeria from 1830 to 1962, Morocco from 1912 to 1956, Tunisia from 1881 to 1956. All three Maghrebi states share common cultural, linguistic, and religious characteristics, and all are classified by the World Bank as middle-income countries (see Table 1.1). In 1992, per capita gross national product (GNP) was U.S.$1,840 in Algeria, U.S.$1,720 in Tunisia, and U.S.$1,030 in Morocco. Sixty percent of the population of all three states were under twenty-five years of age in the mid-1980s, and in 1992 their adult literacy rates were 43 percent (Algeria), 51 percent (Morocco), and 35 percent (Tunisia).

Algeria is the second largest country in Africa, covering an area of slightly less than 2.4 million square kilometers with a population in 1992 of 26.3 million. Ninety-five percent of its inhabitants live along the narrow northern coastal zone lying between the Mediterranean and the Atlas mountains. Fifty-four percent of its inhabitants live in urban areas, with 12 percent concentrated in the capital city of Algiers. Algeria is an oil exporting nation. The economy of the southern region is based on hydrocarbons (petroleum and liquefied gas), while virtually all the non-hydrocarbon-related activities of its economy—including agricultural production, governmental units, services, and industry—are concentrated in the northern coastal zone.

Morocco covers an area of 447,000 square kilometers, and in 1992 its inhabitants numbered 26.2 million (almost the same as Algeria). In 1990, 47 percent of the total population lived in urban areas, and 4 percent lived in the capital city of Rabat. Unlike Algeria, Morocco is an oil importer whose major export is phosphate, a major source of its resources for socioeconomic growth. Morocco is a constitutional monarchy, in which the king rules with a Parliament. Although during the 1960s and 70s, Morocco's King Hassan dissolved Parliament and ruled by decree several times during "national emergencies," he has permitted a kind of limited democracy since 1977 and has emphasized institutional continuity and system maintenance. In general, Morocco benefits from greater freedom of the press than Tunisia or Algeria.

Table 1.1. Basic Indicators (1992)

	Algeria	Morocco	Tunisia
Population (million)	26.3	26.2	8.4
Population growth (%)			
(1965–80)	3.1	2.5	2.1
(1980–92)	2.8	2.5	2.3
(1993–2000)	2.2	1.8	2.2
Life expectancy (years)	67	63	67
Area (sq km 10^3)	2,382	447	164
GNP per capita ($U.S.)	$1,840	$1,030	$1,720
GNP per capita growth (%)			
(1980–92)	–0.5	1.4	1.3
GDP growth (%)			
(1970–80)	4.6	5.6	6.8
(1980–92)	2.6	4.0	3.8
Inflation rates (%)			
(1970–80)	14.5	8.3	8.7
(1980–92)	11.4	6.9	7.2
Investment rates *			
(1970–80)	7.2	9.9	6.1
(1980–92)	12.4	2.6	–0.3
Adult literacy (%)			
female 1990	55	62	44
total 1990	43	51	35

*Investment rate = gross domestic investment/gross national product.
Source: World Bank Development Report (1994).

Tunisia covers an area of 164,000 square kilometers, with a population of 8.4 million in 1992. Fifty-seven percent of its total population live in urban areas, and 20 percent live in the capital city, Tunis. Although the petroleum industry used to be Tunisia's leading foreign exchange earner, growth in Tunisia's economy now depends largely on the results of its agriculture sector, textile exports, and tourism. Tourism and textiles provide the greatest proportion of its foreign exchange revenues (about U.S.$1 billion and U.S.$1.25 billion, respectively, in 1990).

1.1 The Past

After independence in 1962, Algeria followed a centrally planned or "command" economy approach to development. The Algerian government's guidelines for economic development have been based on the goal of national self-sufficiency in important sectors, including telecommunications. Throughout the 1970s, Algeria's national plans emphasized investment in heavy industry, which has been supported extensively by the export of energy resources (liquefied natural gas and petroleum). The major increase in oil revenues after 1973 made possible large investments in many sectors of Algeria's economy, including telecommunications, although a drop in oil prices in the mid-1970s brought lower than expected government revenues. By 1980, almost 95 percent of Algeria's total export earnings (which amounted to approximately 65 percent of total revenues) came from petroleum and liquefied natural gas production. Algeria's burden of debt in 1980 was high (38.7 percent of GNP), but because oil prices were also high its balance of payments was positive.

In 1961, five years after Morocco's independence, King Hassan II, who is both secular and religious head of state, inherited the throne from his father. Historically, Morocco's economic decision making and other broad powers have been centralized in the person of the king, and under Hassan's rule there has been a large measure of government participation in economic development. The Moroccan government has kept close control of transportation and communications, along with many other productive services. Morocco's development strategy has been based on a mixed economy with private ownership of property, private enterprise, and a capitalist form of monetary apparatus.

Morocco's economic development has historically relied heavily on phosphate mining. Phosphates—and to a lesser extent other primary products—have accounted for 90 percent of Morocco's merchandise exports between 1956 and 1976. In 1974, Morocco offset the general increase in oil prices by raising the price of its major export, phosphate, and was therefore able to launch an ambitious public investment program. In the late 1970s, Morocco suffered an economic slowdown as cheaper phosphate sources, such as mines in Florida, took some business away. Between 1973 and 1975, Morocco's gross domestic product (GDP) growth rose from a two-decade average of 4 percent annually to 7.5 percent annually. But the large public investment expenditures combined with increases in oil import prices and increases in defense expenditures in the western Sahara strained the public treasury. Morocco resorted to considerable foreign borrowings to finance treasury deficits.

Tunisia has oscillated between "reconciliatory" and "mobilization" methods of governance and development since independence in 1956 (Hermassi 1972). During the 1960s, Tunisia attempted to mobilize society with rapid and radical social and economic change through a centrally planned economy. Throughout the 1970s and 80s, however, the government sought reconciliation and compromise among groups and emphasized moderate programs and a mixed economy. On November 7, 1987, the newly appointed prime minister and former minister of the interior, Zine El Abidine Ben Ali, declared long-time president Bourghiba incompetent to rule on medical grounds and proclaimed himself the new president of Tunisia. Most of the people of Tunisia, as well as other national governments, greeted the coup with some relief insofar as Bourghiba's failing health shed doubt upon his good judgment in the final years of his long reign (1956–87), which began with Tunisia's independence. While Ben Ali promised democratic reforms, he has been criticized for censoring the press, outlawing the militant fundamental Islamic party, limiting other opposition parties, and rigging elections at the municipal and parliamentary levels.

1.1.1 Telecommunications

The telecommunications operating entities of Algeria, Morocco, and Tunisia were all established under colonialism and modeled after the French Ministry of Post, Telegraph, and Telephone (PTT), a government-owned department. The majority of telephone service in all three states was provided by the PTT. During colonialism and the years immediately following independence, all international telephone calls from North Africa were routed through Paris. Internal communications networks followed only a few main arteries of information exchange, as telephones were concentrated in urban areas. Most of the radio equipment used before independence in each of the three states came from French manufacturers, most notably the Compagnie Française Thomson-Houston and Société Française Radioélectrique-Afrique, which manufactured standard telephone sets at its subsidiary company in Algiers.

Prior to the late 1950s, a maximum of only nine domestic calls could be placed simultaneously along Algeria's north-south axis. To overcome these limitations, the foreign oil companies operating in Algeria began to maintain an extensive local radio system of their own to aid in the search for oil in the Sahara (Hermassi 1972). The public telephone lines operated by the PTT were transmission wires strung on poles. During the national uprising for independence beginning in 1954, these lines were easy targets for Algerians trying to disrupt the long distance communications of the French. In order to overcome this vulnerability, the French developed a new long distance network based on radio communications centrally controlled through the PTT in Algiers.

Algeria's 1976 National Charter declared the associated development of an adequate telecommunications infrastructure to be a prerequisite for achieving the country's goal of national self-sufficiency. Among the industries established to promote Algerian self-sufficiency in the early 1970s (when analog technology was state of the art) was telecommunications equipment manufacturing.

Although Morocco had a private telephone concession (ending in 1964) in the former Spanish zone, and another in Tangier (ending in 1967) that served about 7,000 subscribers, most of Morocco's telephone service has historically been provided by the state-run PTT. Most of Morocco's 160,000 subscribers in 1969 were concentrated in Casablanca and a few other large cities.

Telecommunications technology played an important role in Tunisia's struggle for statehood—during the revolutionary unrest that led to independence in 1956, Tunisian nationalists used telegrams to organize demonstrations and campaigns throughout the country (Anderson 1987). By 1968, 83 percent of Tunisia's telephones were in Tunis, the capital city, where 8 percent of the total population resided. Tunisia's rural areas have historically been served by radiotelephone, where they have been served at all.

1.2 The Present: Economic Trends

1.2.1 Algeria

Algeria's rapid industrialization and overdependence on energy revenues have resulted in a number of problems, most notably, shortages in agriculture and consumer goods and economic instability owing to changes in oil and gas prices. These economic problems, along with the death of long-time president Boumedienne by natural causes in 1978, inspired a reevaluation of development strategies and a shift toward greater privatization of some economic activities in the 1980s—a trend that continued into the 1990s. National plans since 1980 have encouraged the private sector and called for the decentralization of many government operations in an attempt to stimulate the economy.

1.2.2 Morocco

Financial assistance from the World Bank, together with considerable economic reforms and an outward-oriented development strategy, has helped Morocco reduce budget deficits and improve economic performance since 1983. Morocco's debt-to-GDP ratio dropped from 117 percent in 1985 to about 80 percent in 1991. The debt service ratio over the same period dropped from 58 percent to about 45 percent of GDP.[1] Morocco's economy in the early 1990s was an active, expanding, free enterprise system in which state intervention was steadily declining and private investment predominated in most areas of economic and commercial activity.

1.2.3 Tunisia

The decline of Tunisia's petroleum industry continued in the 1990s. External debt was less than 52 percent of GDP in 1990 (down from 57 percent in 1989). Tunisia began a series of World Bank–International Monetary Fund structural adjustment reforms. Among these reforms was the liberalization of laws and administrative procedures regulating foreign investment. The computer market has benefited greatly from the reforms, which have lowered import taxes and simplified bureaucratic procedures that hampered the industry.

1.2.4 Economic Growth

A factor influencing the three states' ability to sustain telecommunications development is economic growth. On the whole, Algeria, Morocco, and Tunisia experienced economic growth during the 1970s, followed by economic slowdown in the 1980s and some recovery in the 1990s. Algeria has seen another downturn, however. The GNP per capita growth (in real terms) between 1980 and 1992 was –0.5 percent in Algeria, 1.4 percent in Morocco, and 1.3 percent in Tunisia. In Algeria, growth in GDP averaged 4.6 percent between 1970 and 1980 then fell to 2.6 percent between 1980 and 1992. In Morocco, GDP grew at an average annual rate of

5.6 percent between 1970 and 1980 and dropped to 4 percent between 1980 and 1992. In Tunisia, GDP growth fell from 6.8 percent (1970–1980) to 3.8 percent (1980–1992); in 1994 real growth in GDP was down to 2.1 percent.

1.3 The Present: Telecommunications

1.3.1 Ownership and Regulation of Telecommunications

Post and telecommunications services have been combined in the same organization in all three Maghrebi states throughout the colonial and postcolonial periods. In Algeria and Tunisia, however, the personnel and financial affairs of post and telecommunications were separated. In Morocco, these were not separate functions until after the reorganization in 1984. Algeria and Tunisia have had similar experiences with their telecommunications labor unions opposing any restructuring. In both countries, the labor unions have opposed any change, claiming it would threaten the guaranteed employment clause that exists under current civil service statutes. The fear of unemployment is a reasonable concern in these countries. For example, in Tunisia the official unemployment rate was 14 percent and the unofficial rate was as high as 25 percent.

1.3.1.1 Algeria

In Algeria, government ownership of the telecommunications entity continued after Algeria's independence from France in 1962. Nonetheless, in 1976, the agency was reorganized on a commercial basis with daily administrative autonomy. Hereafter referring to the Ministry of PTT as the Ministry of Post and Telecommunications (P&T), the decree was intended to improve efficiency within the administrative structure.[2] According to the reorganization, the services of equipment and exploitation (which had been separate directorates) were integrated under the directorate general of telecommunications. Postal and financial services were integrated under a director general of posts. The director of finance serving under the director general of posts was responsible for computing activities, as well as the definition and design of the accounting system, including cost accounting for both post and telecommunications services. The directorate of accounting ("Agence Comptable"), which was responsible for post and telecommunications accounts, remained independent as before. The director of budget was a part of the Directorate of General Administration.

The reorganization did not produce all the desired effects, however, and in 1978, the general director of P&T, Abdelkader Bairi, expressed his support for studies to further improve the organization. Additional organizational changes were introduced with effect from January 1983. Three presidential decrees were promulgated to redefine the functions, responsibilities, and the organization of P&T at the level of the ministry's central administration. The 1983 organization aimed at further improving P&T's efficiency, as well as its programming and planning capability, and at streamlining the ministry's various branches of activities. It specified that P&T would be headed by a minister, assisted by a secretary general, inspector general, and several advisers. There were five directorates general for: (1) telecommunications; (2) postal services; (3) common services (buildings, transport, and procurement); (4) financial and human resources and training; and (5) planning and information systems. Each directorate had subdirectors and bureaus.

Under the authority of the minister, the Ministry of P&T was organized into a central administration in Algiers and directorates in the "wilayas" (provinces). The directorates were responsible for service to the public in the wilayas and to the "walis" (prefects) who represented the government at the regional level. There was a directorate for post and telecommunications in each of the thirty-one wilayas headed by a director belonging either to posts or telecommunications assisted by a subdirector of the other branch.

In 1985, the position of director general for telecommunications was eliminated, and all directorates relating to telecommunications, post, financial, and postal services reported directly to the secretary general. In October 1989 the Ministry of P&T was scheduled to be transformed into a public corporation, the National Office of Post and Telecommunications (ONPT), with government-owned shares. Due largely to resistance from labor, however, the reorganization was postponed. The World Bank, which encouraged the reorganization, decided to discontinue loan activities for the sector in 1992.

1.3.1.2 Morocco

From the time of national independence in 1956 until 1984, all public telecommunications, post, and postal/financial services in Morocco were provided by the government department, the Ministry of P&T, which had been established by the French during colonization. The organization of the P&T was of a simple administrative type, with directors of the external services reporting to the minister, either directly or through the secretary general of the ministry. The central administration was organized into three directorates for general administration (including budget, personnel, social affairs, buildings, and transport services), postal/financial services, and telecommunications; and three divisions for data processing, international and public relations, and training. Services to the public were provided by fourteen regional directorates, which corresponded to the country's administrative organization.

In 1984, a legally, financially, and administratively autonomous public corporation, called the National Office of Post and Telecommunications (ONPT) was created by law.[3] This corporation was created, partly as a result of the World Bank's recommendations, with the express purpose of operating on a commercial basis. The overall aim was to reduce central government intervention in day-to-day operations. All operational and contractual responsibilities for telecommunications and postal services, including staff, assets, and liabilities, were to be transferred from the Ministry of P&T to ONPT in January 1984. But two years later sector assets and liabilities had still not been transferred to ONPT, and ONPT remained basically a government department in terms of its organization and management. By 1988, four years after the decree, the separation of responsibilities between the Ministry of P&T and ONPT, its semi-independent branch, was clearer. Staff and budgets of the two were separated: the budget of the ministry was part of the government budget while ONPT's budget consisted of its earned revenues plus commercial borrowing.

As before the reorganization in 1984, the Moroccan telecommunications entity was governed by an eight-member board of directors, chaired by the prime minister. Its members included the ministers of Finance, Defense, Interior, Planning, Economic Affairs, Industry and Equipment, and Transport. In 1985, the World Bank reiterated that the membership of Morocco's public enterprise boards should become more professional and effective by opening them up to people with technical and managerial experience. The board of directors determines ONPT's internal organization decree, organization chart, and major appointments (director, deputy director, and senior executives). It had organized ONPT into a general secretariat and six departments: telecommunications, posts and financial services, finance, general administration (personnel, social, and legal affairs), common services (buildings and transport), and procurement. In 1988, the French management consulting firms SEMA-MATRA and SOFRECOM recommended further changes in ONPT's organizational structure and management. These proposals provided for greater autonomy within ONPT's principal branches of activity (telecommunications, post, and financial services), while increasing responsibility and accountability at all levels. Regional delegations were also accorded greater operational responsibilities, and policy control at headquarters, in Rabat, was strengthened.

1.3.1.3 Tunisia

During the colonial period, the Tunisian Ministry of Post, Telephone, and Telegraph (PTT), along with various other technical bureaucracies ("directions") including Finance, Agriculture, and Public Works, was formally under the control of the prime minister. But in actual practice, the PTT was a separate ministry, controlled and staffed exclusively by French nationals (Anderson 1987). The exclusion of Tunisians was justified on the grounds that they lacked technical education. Since independence in 1956, the telecommunications entity has remained a conventional government department ("secretariat d'état") supervised and regulated by the Ministry of Communications and Transportation. In 1987, the PTT gained the right to retain earned revenues for direct reinvestment in the sector; in every other way, however, it remained a conventional government department. In the early 1990s, proposals for restructuring the entity into a public enterprise or other semiautonomous organization were met with hostility from the PTT labor union. A demonstration in 1992 by the labor unions opposing greater autonomy measures was met with assurances from the Ministry of Communications that no changes were imminent. Nonetheless, the issue was prominent in the discussions leading to the VIIIth National Plan (1992–96). An International Telecommunication Union (ITU) study conducted in 1990 strongly recommended institutional reforms, specifically restructuring the entity in order to provide legal access to greater financial borrowing sources and the authority to offer competitive salaries to its staff. On April 10, 1995, the Chamber of Deputies (the Tunisian Parliament) adopted a law specifying the creation of a national office of telecommunications. This restructuring changed the entity from an administrative structure to a public enterprise with industrial and commercial orientation.

In 1996, the Tunisian government established Tunisie Télécom as a corporatized entity to succeed the General Directorate of Telecommunications. The Ministry of Communications still retained the authority for the development of infrastructure. For any telecom entity, foreign ownership levels can be up to 49 percent, with the exception of Tunisie Télécom whose cap is 10 percent.

1.3.2 Financial Health and Independence

Throughout the 1970s and 1980s, the greatest portion of the budget of the telecommunications agencies in Algeria and Tunisia came from the central government (until 1987 in Tunisia). In Algeria, for example, 96 percent of the financial resources came from the public treasury between 1977 and 1986. In Morocco, on the other hand, the greater part of sector financing before the 1984 reorganization came from the private sector (80 percent during the early 1980s). During the years in which each of the telecommunications agencies were government departments, they were required to turn over their revenues to the public treasury and reapply for funds during each national plan period (until 1984 in Morocco, 1987 in Tunisia).

In Algeria, where the agency is still a government department, P&T submits a budget proposal (operating budget and investment budget) for each national plan period. The operating budget is prepared at the wilaya level and reviewed by the budget division of the general administration department and is required to show an overall surplus. The investment budget, on the other hand, is prepared by the central departments of P&T. The first draft budget goes to the secretariat of plan and requires the approval of the Ministry of Planning. The Council of Ministries then discusses the proposed operating and investment budgets, along with the budgets of all the other ministries. Both operating and investment budgets are submitted to the National Assembly for approval in October and November of each year; final decisions are made in June. The National Assembly adopts the budget through vote and incorporates it into finance law, which details expenditures and authorizes commitments and borrowing. Algeria financed only about 26 percent of construction requirements from earned revenues or net internal cash generation (NICG) in the early 1970s (1974–76). But increases in tariffs brought this proportion up to over half (57 percent on average) of the investment being financed by NICG between 1983 and 1988.

Like that of Algeria, the Moroccan P&T budget (until 1984) was annexed to the general budget of the state. Annexed budgets were submitted by all government agencies in Morocco that resembled commercial institutions insofar as they had some independent sources of operating revenue (e.g., the private sector). The annexed budgets were linked to the general budget of the state through the transfer of their current surpluses to the general budget resources and through the funding of their current deficits and investment outlays from the general budget expenditures. A draft of the finance law was prepared annually by the Ministry of Finance in cooperation with the planning division of the Five-Year Plan organization. The draft was presented not later than November 1 to the Council of Ministers for their consideration, and approval was then sent to the legislature to be discussed and voted into law. Since the level of sector investment was so low prior to 1984, P&T was able to finance 100 percent of investment from NICG.

During the period 1980–84, 50 percent of earned revenues covered all investment costs, another 45 percent was transferred to the Moroccan government, and the remaining 5 percent covered the debt service. After 1984, when investment levels rose dramatically, ONPT was able to finance about 57 percent of total costs and continued to transfer money to the government. In 1991 and 1992, ONPT financed 50 percent of investments from internal revenues. Between 1987 and 1994, ONPT was expected to collect for the government a 12 percent surcharge on billings (amounting to DH3.4 billion between 1987 and 1994). It was subject to a 48 percent tax each year on its net income (about DH60 million or U.S.$7.6 million in 1984; and a total of DH4.4 billion between 1987 and 1994). Other miscellaneous taxes during the 1987–94 period paid by ONPT to the government were expected to reach DH2.2 billion, and import duties on equipment (approximately 44 percent of foreign exchange cost) would provide another DH2.8 billion. The total financial resources transferred from ONPT to the government between 1987 and 1994 were estimated to reach DH12.8 billion (approximately U.S.$1.4 billion).

The Tunisian PTT requested funding in a manner similar to that of the Moroccan (until 1984) and Algerian entities, that is, submitting the investment and operating budgets to the central government for review and enactment into financial law. But beginning with the VIIth National Plan (1987–91) the Tunisian PTT was granted the right to retain earned revenues for direct reinvestment in the sector while transferring a portion of earnings to the public treasury. Between 1987 and 1990, PTT transferred TD3 million (approximately U.S.$3.3 million). The amount of investment financed by NICG increased steadily throughout the 1970s and 1980s: 33 percent of total investment in the mid-1970s; 48 percent during the Vth Plan (1977–81); 67 percent during the VIth Plan (1982–86); and 83 percent (TD314 million) during the VIIth National Plan (1987–91).

1.3.2.1 Telecommunications Investment

Telecommunications investment as a percentage of total gross fixed capital formation (GFCF) or total national investment during the 1980s was generally less than 1 percent in developing countries and about 2 percent in the advanced, industrialized countries. In Algeria, telecommunications investment averaged 1.44 percent of GFCF during 1980–89. Telecommunications investments from 1989 through 1991 totaled DA6.5 million. In Morocco, telecommunications investment averaged 1.6 percent of GFCF during 1981–85 and rose to 3.1 percent for 1986–88. In Tunisia, investment rose from an average of 2 percent of GFCF (1982–86) and 2.5 percent (1987–91) to 3.3 percent of GFCF in the VIIIth Plan (1992–96), when telecommunications investment doubled that of the previous plan period (TD960 million or about U.S.$1 billion in current prices, as opposed to TD340 million or U.S.$382 million during the previous plan).

Developing countries invested on average about 0.3 percent of GDP in the telecommunications sector in the early 1980s, compared with 0.6 percent for industrialized countries (Wellenius 1987). Algeria invested on average a higher percentage of its GDP in the telecommunications sector during the period

1972–86 than did either Tunisia or Morocco. But during the 1980s, Algeria's investment was actually lower than Tunisia's or Morocco's as a percentage of GDP. In Algeria, the average total annual investment in telecommunications (including land and buildings) during the period 1974–82 was 0.58 percent of GDP, compared with Tunisia at 0.45 percent and Morocco at 0.36 percent for this same period. Algeria invested heavily in the mid- and late 1970s (for the expansion of its domestic telecommunications infrastructure with satellite communications), but for the period 1983–88, Algeria's telecommunications investment dropped to an average of 0.32 percent of GDP, while Tunisia and Morocco's telecommunications investment increased during this period to 0.44 percent and 0.5 percent, respectively. For the period 1988–94, Morocco's telecommunications investment reached 1 percent of GDP.

Telecommunications investment per capita was highest in Algeria, the richest of the three countries, during the period 1972–86. Investment per capita in Algeria in 1972 was close to that of Tunisia (U.S.$1.40 in Algeria; U.S.$1.19 in Tunisia). But by the early 1980s (1980–84), Algeria's per capita telecommunications investment was U.S.$6.33 compared with Tunisia's U.S.$4.79 and Morocco's U.S.$1.98 per person. By 1986, per capita investment in Algeria had risen to U.S.$9.21, in Tunisia it was U.S.$6.50, and Morocco U.S.$4.32. In 1990, Morocco took the lead in per capita investment with U.S.$11.90, followed by Tunisia with U.S.$8.90. Algeria's telecommunications investment per capita was lowest of the three states at U.S.$6.80 (all current U.S. dollars).

1.3.2.2 Rates and Tariffs

The total estimated yearly cost of telephone service to subscribers in the 1980s was higher in Algeria (U.S.$316.20 in 1982) than in Tunisia (U.S.$246.40 in 1982) or Morocco (U.S.$270.47 in 1986). Algeria and Tunisia offered off-peak discount pricing. Morocco had no off-peak pricing. All three states used revenues from urban areas and nonresidential subscribers to subsidize rural and residential services.

In Algeria, telecommunications tariffs are proposed by P&T and require final approval by the Ministry of Finance. Since the 1980s, P&T has been required by law to cover all operating expenses, including depreciation and debt service, from operating revenues (unlike the policy of the 1970s). This has required regular increases in tariffs to cover the cost of providing service. Tariffs were increased in 1965, 1975, 1983 (by 50 percent), and once during the 1985–89 national plan period. In 1992, the cost of a local call was DA0.90 every six minutes. An international call (to France) was DA0.90 every 1.9 seconds.

In Morocco, prior to 1986, the telecommunications entity obtained permission to increase charges from the Council of Ministers. After 1986, ONPT had the right to increase tariffs gradually based on the financing requirements of its investment program with no prior approval by the government. Morocco increased tariffs in 1979, 1982 (by 12 percent), 1985 (by 12 percent), and 1987 (by 20 percent). The base charge for a local call rose incrementally from DH0.50 in 1985 to DH0.80 in 1992. Local calls were not metered.

The Tunisian PTT proposes tariff increases and obtains final approval from the minister of Finance. According to a PTT source, tariffs for telephone, telex, and data services tend to more than cover the cost of providing service. Tariffs were modified in 1965, 1975, 1986, 1989, and 1991. The cost of a local call in 1991 was TD0.070 to 0.100 (U.S.$0.11), depending on whether the call was made from a subscriber phone or a public phone, respectively. An operator-assisted call to France was TD2.940 (or U.S.$3.30) for the first three minutes and TD0.630 (or U.S.$0.69) for each additional minute. Such a call to the United States was TD4,200 ($4.60) for the first three minutes and TD2,100 ($2.30) for each additional minute. In 1997, Tunisia Telecom reduced its international rates by 25 percent to Europe and 45 percent to other countries. This reduction was mainly due to traffic diversion by callback users.

1.3.2.3 Collection of Payments

The authority to collect payments from all users, including government subscribers, was stronger in Algeria than Morocco or Tunisia throughout the 1970s and 1980s. The Algerian P&T had the authority to discontinue service to any customer who was delinquent in payment of bills by more than one month. The 1978 decree states that government agencies can be forced to pay through direct deduction from their treasury accounts, even when budget allocations for telecommunications usage are exhausted. In 1982, the payment deficit was equivalent to only one and a half months of total receipts. The Algerian P&T accounts receivable was described by the World Bank in 1984 as excellent. Nonetheless, some flexibility was shown toward payment delays by various government users. The principal delinquent payers of arrears in 1985 were the Ministry of Defense, the Ministry of the Interior, and the Ministry of Radio and Television (RTA). While the Ministries of Defense and Interior paid, however belatedly, RTA, with arrears of DA89 million accruing since its initial use of P&T-controlled satellite channels, did not pay until payment of arrears by RTA to P&T became a condition for a World Bank loan.

In Morocco, the payment deficit of government users in 1986 was fifteen months of billing or 75 percent of total accounts receivable. While delinquent residential or business subscribers were disconnected after the sixth week, government users were not disconnected for unpaid bills. In 1984, total arrears of government users represented a loss to ONPT of DH1.2 billion (U.S.$43 million). Budget allocations for telephone service by public agencies typically fell far short of actual usage. Under pressure from the World Bank, the Ministry of Finance increased allocations from DH90 million to 160 million per annum to partially cover an estimated annual government telecommunications bill of DH220 million in 1987 and beyond. During negotiations with the World Bank, the government granted that future telecommunications bills would be paid within three months of billing. Government arrears dropped from DH800 million 1988 to zero by 1992.

In Tunisia, there was a chronic unrecoverable deficit of payments. The central government estimated telecommunications usage prior to each national plan and paid a fixed sum to the PTT in advance. But these estimates were typically lower than the actual usage by government agencies. The PTT was not entitled to collect its due and simply forfeited these revenues, which could amount to several hundred thousand dinars during a national plan period. Residential users are usually disconnected after the third month.

1.3.3 Network Development

As noted earlier, while nationwide telephone density in Algeria was slightly lower than that of Tunisia, the quality of service was higher and the distribution of service between urban and rural areas was more equitable than in either Tunisia or Morocco. Nonetheless, both Morocco and Tunisia began to undertake ambitious projects for rural expansion in the early 1990s, and they are slowly surpassing Algeria.

1.3.3.1 Provision of Service

The number of main telephones or direct exchange lines (DELs)[4] per 100 inhabitants (DEL density) in Algeria was slightly lower than in Tunisia from the early 1970s to the early 1990s. In the mid-1980s, DEL density in Algeria was 2.5 percent (1986), in Tunisia it was 2.9 percent, and in Morocco 1 percent. The average DEL density for all less developed countries in 1986 was 3 percent; for the continent of Africa it was 1 percent. In 1994, DEL density for Africa was 1.6 percent. Algeria's density was not only below the average for less developed countries, it compared poorly with developing countries with similar economic activity, such as Jordan (7 percent), Iraq (5.9 percent), Syria (5 percent), and Hungary (8 percent). The Algerian P&T's dependence on local equipment suppliers, who were unable to meet network expansion requirements, slowed growth in the 1980s. Nonetheless, the average annual growth rate in DEL density (1972–86) was highest in Algeria (12 percent), followed by Tunisia (8 percent) and Morocco (4 percent).

In the ten years between 1976 and 1986, Morocco's DEL density rose only from 0.64 percent to 1 percent. This is two to three times lower than developing countries with similar per capita income. Algeria generally met a higher proportion of total demand for telephone service during this period (76 percent in 1986) than either Morocco or Tunisia (59 percent and 46 percent in 1986, respectively). The tables turned in the late 1980s, with substantial investment increases in Morocco and Tunisia. In 1990, the DEL density in Algeria was 3.3 percent, compared with Tunisia's 3.8 percent and Morocco's 1.6 percent. The average DEL density among Arab states in 1990 was 7 percent. In 1992, DEL density had risen in Algeria to 3.5 percent, in Tunisia to 4.1 percent, and in Morocco to 2 percent. The DEL density growth rates in the early 1990s jumped to an unusual, but successful, 22 percent in Morocco and to a more typical 12 percent in Tunisia. Tunisia achieved an average growth rate of 14.2 percent during the 1992–96 plan period. By 1994, DEL density in Tunisia was 5.5 percent, with 7.3 percent projected for 1996.

A growth rate of 15 percent is generally considered the peak at which a country can successfully manage increases in subscriber connections. In Algeria, P&T was unable to keep up with subscriber demand in the early 1990s; only 54 percent of total demand was satisfied in 1992. Tunisia, on the other hand, met 65 percent of demand and Morocco met 74 percent in 1992.

During the 1970s and 1980s, the waiting time for telephone connections was shorter in Algeria than in Morocco or Tunisia. Between 1972 and 1980, waiting time in Algeria averaged around 31.7 months (or about two and half years). Although this rose to 58 months (about 4.8 years) in the early 1980s, it was still lower than Morocco or Tunisia at that time. The waiting time for a subscriber connection in Morocco in 1984 was six years. It dropped dramatically after the late 1980s: from eighty months in 1987 (seven years) to thirty-nine months in 1989 to eighteen months in 1991. Algeria's waiting time for a telephone in the 1990s increased steadily as investment slowed.

1.3.3.2 Quality of Service

The quality of telephone service, measured by most indicators, was highest in Algeria during the 1970s and 1980s. But by the 1990s, Algeria was surpassed by Tunisia, as shown by several indicators. In Algeria, during the 1970s and 80s, exchange fill was below recommended limits; completed call rates were high; the number of telephone faults (breakdowns) per year per subscriber and the percentage of breakdowns repaired within forty-eight hours, although not exemplary, were better than Tunisia or Morocco. In the area of network modernization and digitization, however, Algeria lagged behind Tunisia and Morocco in the 1980s but was catching up in the 1990s. The recommended level of exchange fill for less developed countries is 85 percent. In general, an exchange fill below 85 percent of network capacity allows for peak-hour traffic and transfers and mitigates against circuit congestion.

In Algeria, exchange fill was below 85 percent throughout the 1970s and 1980s; in 1977, exchange fill was only 54 percent. It rose slowly to 62 percent in 1982 and reached only 72 percent in 1986. Unlike most less developed countries, Algeria had an excess exchange capacity in the network, that is, it actually underutilized switching capacity. Oddly, Algeria still suffered some congestion due to poor network architecture planning, specifically, efficient routing of traffic between cities. Nonetheless, call completion rates were good; during peak hours for local calls in Algeria in the 1980s the call completion rate was 95 percent (nationwide average).

In Morocco, exchange fill nationwide in the early 1980s was below the recommended limit, ranging from 77 percent to 84 percent. After 1984, however, exchange fill nationwide crept above the 85 percent limit, climbing steadily from 86 percent to 88 percent in the years 1984–86. In the commercial districts of Casablanca and Rabat, exchange fill was as high as 96 percent (1986). Call completion rates in these two principal cities were below 60 percent and 50 percent, respectively.

In Tunisia, exchange fill was below 85 percent outside the principal cities but was 88 percent in Tunis as early as 1969 and even higher in some other urban centers in the early 1970s (Sfax, 96 percent; Sousse, 93 percent; Nabeul, 92 percent; Moknine Ksar Hellal, 87 percent). Exchange fill in Tunis was still high at 86 percent in 1978. The nationwide average call completion rate during peak hours was 37 percent in 1984. But in 1982, PTT began major investments in new central switches, and annual gross investments for telephone switching equipment increased 97 percent between 1982 and 1986. The absolute growth in switching capacity in just those four years was almost 73 percent. This increased connection capacity brought exchange fill in Tunis down to 77 percent by 1986 and increased the nationwide average call completion rate to 65 percent by 1988. Nonetheless, chronic congestion of downtown sections of the city persisted into the 1990s. Congestion was alleviated with the installation in 1993 of thirty-nine new digital switches for national service.

An acceptable level for telephone breakdowns (faults) is about 0.5 per subscriber per year. None of the three states reached this level in the 1980s or 1990s. In Algeria (1983) the fault rate was 1.2 (nationwide average), in Algiers 0.9. In Morocco, telephone faults dropped from 1.3 nationwide in 1987 to 0.95 in 1990. In Tunisia, faults averaged 1.4 in 1987 nationwide; in Tunis they were 1.1; in the suburbs of Tunis they were 1.9. By 1993, faults dropped to 0.99 per subscriber per year in Tunisia. The percentage of faults repaired within forty-eight hours was 74 percent in Algeria (1983), 60 percent in Morocco (1990), and 50 percent in Tunisia (1993).

The proportion of digitization in the network has traditionally been higher in Tunisia than in Morocco or Algeria. Digitization began in earnest in Tunisia in 1982. By 1993, 75 percent of the network was digitized. Morocco rose from 5 percent network digitization in 1983 to 70 percent in 1992 (with 77 percent digital switching equipment). In Algeria, digitization got off to a slower start (not until 1984) due to the country's commitment to building a local telecommunications industry as early as 1972 when analog technology was state of the art. In 1996, 90 percent of the Tunisian network was digitilized and 98.3 percent of the Moroccan network was digitilized.

1.3.3.3 Distribution of Service

Equity between urban and rural distribution of telephone services was greatest in Algeria during the 1970s and 1980s. Telephone density in the urban centers of Algeria in 1980 was 4.5 percent while the rural density was 0.9 percent. Two years later (1982), urban density was only slightly higher at 4.7 percent, while rural density grew more rapidly to 1.4 percent. In Morocco, telephone service was highly concentrated in urban areas: 80 percent of the DELs in service were located in nine large cities and provincial centers with 42 percent of the population. In 1986, telephone density in the three principal cities of Rabat, Tangier, and Casablanca—with 19 percent of the total population—was 3 percent. In the six secondary cities (Agadir, Fes, Kenitra, Marrackech, Meknes, and Tetouan)—23 percent of the population—telephone density was 0.54 percent. In the rest of the country—58 percent of the population—telephone density was 0.24 percent.

Nonetheless, some form of telecommunications service was available in most provinces and rural areas of Morocco, with only 4 percent of rural communities without access to a telephone. In 1991, multiaccess radio links were part of a major expansion of the transmission network to serve rural areas in southern Morocco. By 1995, ONPT served a total of 10,000 subscribers using TDMA (time-division multiple access) radio systems. Service was to be extended to an additional 7,000 subscribers in rural areas using remote concentrators. By 1995,

ONPT had installed one hundred of these systems. The trunk microwave network was extended into the thinly populated, mountainous interior of the country with the installation in 1991–92 of a microwave transmission system by Alcatel.

In Tunisia, the capital city, Tunis, and its suburbs (with approximately 8 to 10 percent of the total population) had a very high telephone density relative to the rest of the country, including other cities. In 1977, with a national average DEL density of 1.34 percent, DEL density in Tunis was 6.6 percent. The next highest density was 1.2 percent in the secondary city of Sfax, and density for the rest of the country was just under 0.7 percent. Between 1972 and 1985, 62 percent of the DELs were in Tunis. In 1988, with a national DEL density of 3.5 percent, density in Tunis was 12.7 percent and in the rest of the country 2.3 percent. For the VIIIth Plan (1992–96), Tunis was allotted 100,000 of the 400,000 new telephone installations; the rest of the country was allocated the remaining 300,000. But in 1993, Tunis still had three times the DEL density (14 percent) as the national average (3 percent) and thirteen times the lowest density in the interior. The total subscribers in 1997 was 608,000.

Tunisia began a rural communications project during the VIIth National Plan (1987–91), scheduled for completion by the end of the VIIIth Plan (1996). Tunisia awarded contracts to two companies, one French, the other Canadian. A microwave base station in each provincial capital (at a cost of U.S.$120,000) connects to the PTT switch, linking 640 subscribers per repeater station (at a cost of U.S.$15–20,000). In 1992, 700 out of 1,055 planned rural sites were operational.

1.3.4 Technology Transfer

1.3.4.1 Technical Expertise

One of the most important institutional changes in the 1976 reorganization of the Algerian P&T was the abolition of the civil service statute. After 1976, P&T offered competitive salaries and promotion opportunities in order to attract and keep highly qualified expertise and leadership in the organization. As a result, since the mid-1970s, Algeria has had a higher proportion of technical to total staff than either Morocco (until recently) or Tunisia. On average, between 1977 and 1986, the Algerian P&T maintained a level of 66 percent technical staff. In Morocco, the proportion of technical staff between 1977 and 1982 (just before the civil service statute was abolished) was 41 percent. After competitive salaries were introduced, the proportion rose to an average of 59 percent between 1983 and 1986. In Tunisia, where the civil service statute was still in effect in 1995, the average proportion of technical staff between 1972 and 1985 was 33 percent. Between 1985 and 1990, the average proportion of technical staff had fallen to about 27 percent of total staff. In 1992, technical staff was down to 21 percent of total staff in Tunisia.

The World Bank recommends a ratio of about 20 staff members per 1,000 DELs for adequate staff productivity in developing countries, noting that a larger staff size increases the wage bill without significantly increasing the output of the organization, especially when that staff is clerical rather than technical or managerial.

Algeria increased its staff levels during the 1973–77 expansion of the telecommunications network, including domestic satellite service. By 1977, there were 42 staff members per 1,000 DELs, a ratio that fell within the range of many other developing countries but was high compared with countries having similar staff wage levels. Nonetheless, almost 60 percent of that staff provided technical expertise. By 1986, staff ratios were down to 30 per 1,000 DELs, with over 65 percent providing technical expertise, and by 1992 ratios were 20 per 1,000 DELs.

In Morocco, there were 29 staff members per 1,000 DELs in 1986 with 59 percent technical staff. By 1994 they were down to 18 per 1,000 DELs. In Tunisia, there were 44 staff members per 1,000 DELs in 1974, and only 12 percent of total staff were technical experts. By 1992, the staff ratio was a more productive 17 per 1,000 DELs, but the proportion of technical staff was still very low, at 20 percent of total staff. The target for 1996 is 14 staff per 1,000 DELs, with a higher proportion of technical staff due to competitive salary structure made possible by conversion to a public enterprise. While Tunisia's labor force is well educated and literacy is high, technical training is inadequate. The low level of technical expertise in the PTT is a major stumbling block for modernization of the network, but efforts are under way to increase the number of graduates in technical training.

1.3.4.2 International Services

International services for all three states are provided by undersea cables linking them directly with Europe and with each other and by international and regional satellite systems (Intelsat and/or Intersputnik and Arabsat, respectively). All three states are members of Intelsat, as well as Medarabtel and Panaftel, the regional terrestrial networks of the Mediterranean and African countries, respectively.

Algeria is a founding member of Intelsat and introduced international satellite communications services via the Intelsat network in 1975, with the installation of a standard A earth station at Lakhdaria (outside Algiers) operating with an Intelsat IV satellite located over the Atlantic Ocean for communications with countries to its west. In 1976, Algeria installed another standard A earth station operating with an Intelsat IV satellite over the Indian Ocean for communications with countries to its east. In 1979, Algeria purchased a station operating with the Intersputnik satellite system of the former Soviet Union.

Algeria was the first developing country in the world to use Intelsat for domestic communications. In 1973, in an innovative move, P&T's then director general for telecommunications equipment, Abdelkader Bairi, proposed the design and implementation of a domestic satellite network using fifteen small (eleven-meter diameter), less expensive earth stations (U.S.$500,000 per station, or about U.S.$6 million for the network of fifteen stations in 1974 dollars) and a transponder on an Intelsat satellite (U.S.$10 million per year from 1975 to 1979 on a preemptible basis). In 1974, Algeria contracted with an American company, GTE, to adapt the existing standard B earth stations to make them suitable to the harsh desert climate of southern Algeria (these modifications meant that the Algerian domestic stations no longer conformed to Intelsat standards). The first stations of the nonstandard domestic network were placed in operation with Intelsat II in the Indian Ocean in 1975. In 1983, Algeria awarded a contract to another American company, Scientific

Atlanta, for two more of the same type of stations. In 1988, a Canadian company, Northern Telecom, won the contract to expand the domestic network by another nineteen stations, which would provide access to a subsequent generation (Intelsat V) of satellites. The network facilitates communication between the major urban areas and population centers of Algeria's northern coast and the natural gas fields, mining areas, and industrial production complexes scattered across vast rural areas of sparsely populated desert in the south.

Morocco (with a standard A station near Rabat) and Tunisia (with a standard A station in Dakhila, near Tunis) also use and belong to the Intelsat system for international service. Morocco introduced domestic satellite service in 1983 with the lease of an Intelsat transponder. Neither Morocco nor Tunisia was a member or user of the Soviet satellite system, Intersputnik. In 1992, almost 60 percent of Tunisia's PTT revenues came from international traffic, and circuits were often blocked due to insufficient switching capacity. Morocco increased international transmission capacity by adding another standard A earth station, installed by GTE Spacenet in the late 1990s.

All three states also belong to the Arab satellite network, Arabsat. Dakhila, outside Tunis, is home to the second technical control center of the system. Algeria, Tunisia, and Morocco were initially cool to the idea of the regional satellite (circa 1976), and consequently their investment shares in 1978 were very small: 0.5 percent for Morocco, 0.6 percent for Tunisia, and 0.9 percent for Algeria. Each Maghrebi state purchased its Arabsat earth station in 1985 when the first-generation satellite (Arabsat IA and B) was launched. But for economic reasons Algeria did not transfer its domestic communications traffic from the Intelsat satellite to Arabsat. Algeria qualified under the Intelsat Agreement Article IIIb (II) of the Intergovernmental Articles of Agreement as a country that could earn return on investment using an Intelsat transponder for domestic communications because of the geographical barriers typical of international communications (high mountains, large bodies of water, vast distances). Under Article IIIb, if the implementation of wideband terrestrial facilities is impossible because of natural barriers of an exceptional nature, the provision of domestic telecommunications services via an Intelsat satellite transponder would be considered on the same basis as international public telecommunications services. In other words, the country earns return on investment share "according to use" for domestic as well as international traffic on the Intelsat satellite. This "return on investment share according to use" clause did not exist with the Arabsat network until two years after launch, that is, until 1987. Nonetheless, Algeria continued to use the Intelsat satellite for domestic communications.

Adding further capacity, the three countries agreed in 1997 to participate in AFRICA ONE, the 35,000-km submarine fiber-optic ring that will surround the African continent.

1.3.5 Equipment Manufacture and Installation

All three Maghrebi states have some domestic equipment manufacturing and installation capability (usually through public enterprises) for certain telecommunications equipment, typically telephone sets and transmission cables. Algeria has manufactured switching equipment since the 1970s. Neither Tunisia nor Morocco manufactured switching equipment during the 1970s and 1980s, but in the 1990s both countries christened joint venture partnerships for the local manufacture of some switching equipment accessories.

The Algerian government placed great emphasis on the development of domestic production capability in telephone and telex technology because it is a large enough domestic market to support a local industry. Transmission cable has been manufactured domestically in Algeria since 1928 (nationalized in 1968). The Société Nationale de Fabrication et de Montage du Matériel Electrique et Electronique (Sonelec) was established in 1974 and took over production of cable equipment. In 1978, it began to produce analog switching exchanges at a newly built manufacturing complex in Tlemcen. Sonelec was restructured into two state corporations in January 1984: Entreprise Nationale de Télécommunications (ENTC) for the manufacture of telephone handsets and analog (electromechanical) switching equipment in Tlemcen, and Entreprise Nationale pour la Fabrication du Câble (Enicab), located in Algiers, to produce small-capacity pair cables and small-diameter coaxial cable and to lay and joint cable for P&T. These two electronics manufacturing corporations operated under the supervision of the Ministry of Heavy Industry.

While this local manufacturing industry helped Algeria build its domestic telecommunications infrastructure during the 1970s and 1980s, it also delayed the country's transition from analog to digital technology. When Algeria began local production of telecommunications equipment in the early 1970s, analog technology was prevalent worldwide. Algeria established a manufacturing industry and related training programs around analog switching and transmission equipment. These sunk costs in manufacturing, which meant sunk costs in manpower training as well, made it very costly to retool for the new digital technologies. Moreover, the production levels of the state enterprises (ENTC and Enicab) were dropping steadily (to about 80 percent capacity in the early 1980s) through bureaucratic inefficiencies, as P&T demands for switching and transmission equipment spiraled. By the early 1980s, domestic production was clearly insufficient in the long run to meet network requirements. At the same time, digital technology had become more widespread, with such clear advantages as lower costs and greater service options. Therefore, starting in 1984, P&T began importing and installing electronic digital switching and transmission systems into the analog network.

To revitalize local industry and master the new technology, Algeria established a joint venture company called Sitel with Ericsson of Sweden in 1987 for the local manufacture of digital switching equipment. Algeria holds the majority shares in Sitel through the Algerian bank, BEA, and through ENTC and Sonatite (Société Nationale des Travaux d'Infrastructure pour les Télécommunications). Until production began in 1990, Ericsson sent about 90,000 local lines and 57,000 transit lines to Algeria, along with an international exchange and transmission equipment. Sitel began assembling ready-made kits in 1990 at the Tlemcen plant operated by ENTC, with vertical integration expected to increase over time. Initial production levels were to reach 50,000 lines per year, increasing to 200,000 lines annually by 1994. The production of the Pentaconta analog switches (by ENTC) was to be phased out slowly.

The installation of cable and switching equipment and PABXs (private automatic branch exchanges) is the responsibility of Sonatite, a public enterprise under the supervision of P&T. Sonatite is also in charge of the country's private branch telephony, specifically installation, connection, and maintenance of PABXs and other subscriber terminal equipment.

In Morocco, the supply and installation of telephone and telex equipment for government offices and private businesses is privatized and sometimes in competition with ONPT. There is a small telecommunications manufacturing, assembly, and construction industry (urban cable manufacture comprises less than 10 percent of total purchases) consisting of one government-owned company, Société Nationale des Télécommunications (SNT), and four small private companies. Société Nationale des Télécommunications participates in the construction of ONPT's local and urban networks and has equity participation in the four private firms, which are local subsidiaries of foreign suppliers. Three of these assemble or produce minor switching, radio, teleprinter, and power equipment components. The fourth factory produces cables and fulfills a substantial part of ONPT's need for cables and accessories. In addition, SNT installs and maintains private telecommunications equipment, such as PABXs. The Moroccan government has encouraged local private-sector entrepreneurs in the installation of cable and subscriber plants (including PABXs, telex, modems, terminal equipment, mobile telephone sets, and facsimile machines). Northern Telecom and its sister company, Bell Canada International (BCI), have established a partnership with ONPT, but this arrangement will likely result primarily in transfer of know-how through a training center and does not include local production of switching equipment. Bell Canada International has stated that it intends to establish a plant in the future for the production of spare parts. Northern Telecom and Alcatel CIT share the largest shares of ONPT's digital switching orders. Alcatel produces components for its E10-B digital switch in Morocco. Ericsson of Sweden is the third supplier.

In Tunisia, Ericsson has supplied switching equipment since 1959. Its AXE digital switch is imported from Sweden, but AXE-10 cables and connectors will eventually be manufactured locally entirely for re-export, and AXE-10 software will be updated by Tunisian engineers. In 1990, Tunisia entered into agreements with both German and French digital switch manufacturers. The Siemens Group (consisting of Siemens AG and Belgian subsidiary ATEA) established the Centre Tunisien des Télécommunications (CTT) to oversee delivery, installation, and cutover of Siemens EWSD digital switches. Initially, only interconnecting cables and accessories were produced locally, but Siemens agreed that after 1992 about 20 percent of the value of the EWSD system, including frames and nonproprietary line cards, would be produced in Tunisia. The PTT may seek to enter into a partnership with the Siemens Group at any time, but its share may be limited to 40 percent; la Société Tunisienne d'Entreprises des Télécommunications (Sotetel), a public enterprise created in 1981 under the Ministry of Communications and 51 percent owned by PTT, may enter into partnership with the approval of PTT and Siemens. Tunisia also established a joint venture company with Alcatel of France, which has been exporting its E10-B digital switch to Tunisia since the mid-1980s. Alcatel Tunisie consists of Alcatel; SAT (a French microwave radio supplier); Youcef Letaief, a local construction magnate; and Tunisien Télécom Electrique (a local producer of key telephone systems). Northern Telecom was the next entrant into the Tunisian digital switching market.

Sotetel installs the majority of cable for PTT and carries out research, installation, and maintenance of public switching, local and long distance transmission, mobile telephone, telex, telematic services, and PABXs for PTT. In addition to the 51 percent PTT share, other shareholders of Sotetel are public authorities such as transport, maritime, electric, construction, and government-owned financing agencies. A private bank, BIAT, also holds a small percentage of shares. At least three other companies install cable and provide civil engineering services: ECR, Somatra, and Sarrt.

1.3.6 New Services and Markets

Algeria has liberalized its telecommunications market for certain services and equipment. While there is no import tax on computers, the total number of computers imported is limited. The price for an IBM microcomputer, around U.S.$3,000, is therefore high. Nonetheless, computers are widely used by government officials, private entrepreneurs, and professionals. The facsimile machine is also very popular. The private sector is permitted to provide fax as well as desktop publishing services (similar to the Kinko's chain in the United States). Algeria's P&T introduced packet-switched data services in 1991 and cellular telephony in 1990 for the northern coastal area, serving the cities of Algiers, Oran, Constantine, and Annaba, with a capacity for 5,000 subscribers. The cost of a mobile phone was about DA60,000 (or U.S.$1,500 to $3,000 in 1992) at the official exchange rate, with a call charge fifteen times that of fixed telephone service. There is nevertheless strong demand for this service, especially from the private sector.

The telecommunications market in Tunisia has been liberalized substantially since 1987. During the 1992–96 plan period, public telephone service was opened up to the private sector; the number of public telephones managed by private companies rose rapidly from 1,097 in 1992 to 4,730 in 1994. Facsimile and paging services were also privatized. The import tax in 1992 for computers was 17 percent and the value-added (sales) tax was the lowest possible at 6 percent. In 1992, an IBM-compatible computer (386 processor at 25 MHz with 50 MB hard disk) cost about U.S.$2,000, including tax. A Macintosh Classic II (80 MB hard disk) cost U.S.$3,200, including tax. Regulation of the import of computers has been eased so that a simple import certificate is all that is required of the importing agent. Previously, the agent was required to seek preapproval through the Agency for Promotion of Industry. In 1992, there were about 100 importers of computer and other telecommunications equipment in Tunisia. Formerly, computer purchases were routed for review through the National Center for Informatics in order to assure "intelligent" choices in the public sector. However, this official procedure caused tremendous bottlenecks in the ordering process and was eliminated.

Growth in the import of microcomputers has been exponential, from a few score in 1980 to over 17,000 in 1992. Just between 1988 and 1991, growth in the number of computers rose 400 percent. In 1986, approximately 87 percent of the banking and insurance industry was computerized; trade, transport, and hotel management were only about 9 percent computerized; and 7.5 percent of the manufacturing sector computerized. In 1991, PTT introduced "Tunipac," a packet-switched data transmission network with the X.25 transmission standard (DPS 2500 with Alcatel CIT of France), which can accommodate up to 1,545 ports. Tunipac grew to 1,800 subscribers by 1992, when most of PTT's data service clientele were transferred from the public telephone (circuit-switched) network to the packet-switched network, thereby providing more reliable and efficient service. Import duty on modems was 25 percent plus 17 percent value-added tax, bringing the total cost of a typical modem to about U.S.$1,000 in 1994.

Although the facsimile machine does not enjoy the same low import taxes as computers nor the accessibility of fax in Algeria, it has tremendous popularity in Tunisia. Customs duty on fax machines in 1992 was 27 percent plus a value-added tax of 17 percent, bringing the retail price to about U.S.$800–1,200, depending on the features of the machine. Until 1991, only post offices offered fax service; however, the private sector thereafter was permitted to provide facsimile service.

Cellular telephone service in Tunisia, introduced in 1985 and managed by Sotetel, consists of five base stations situated at Bourknine, Bouficha, Sousse, El Jem, and Sfax, covering two-thirds of Tunisia and ranging from Tunis to Sfax with a radius of fifty to sixty kilometers around each of the base stations. The cost of cellular phone subscriber equipment has come down steadily over the past five years but was still high. Calling charges were three times that of local fixed telephone service. In 1997, the rate per minute was reduced by 50 percent to about 20 cents a minute. Handset prices fell to an average price of $1,100. Digital service (GSM) was introduced in 1998.

In Morocco, mobile telephony subscribers doubled between 1990 and 1992 to 1,800 subscribers, located in the urban centers of Rabat, Casablanca, Marrakech, Fes, Meknes, and Kenitra. Like Tunisia and Algeria, Morocco expects to adopt the GSM (groupes de systemes mobiles) European transmission standard for future mobile telephony. In 1991, Morocco introduced a packet-switched data transmission network called Maghripac, with transmission speeds of from 300 to 48,000 bits per second. The small satellite antenna market in Morocco was legalized only at the end 1991. The cost of a 1.8-meter dish was high at about U.S.$2,500 including license fee in 1991, but subsequent deregulation has dropped costs dramatically. In 1995, a 90 cm dish cost only about $200 and could often be seen in the countryside.

Small satellite receiving dishes have also seen big sales increases in Tunisia, especially after the Gulf War (when CNN was very popular viewing) and the deregulation of the satellite dish industry and ownership laws in 1991. A 1.8-meter fixed dish cost about U.S.$850, including tax but excluding the license (U.S.$80), and the support and transport costs (U.S.$45). The fixed dish could receive a maximum of twenty channels on one satellite. A mobile 1.8-meter dish, which could receive up to sixty channels from six satellites, cost about U.S.$2,300, plus license and support and transport fees.

All three countries offer some form of access to the Internet. Algeria, Morocco, and Tunisia have dial-up access to the Internet. Unlike Morocco or Algeria, Tunisia is an Internet host site registered with the top-level domain name system (Internet's Domain Name System). The Institut Régional des Sciences Informatiques et Technologiques (IRSIT) located in Tunis, Tunisia, has offered full Internet Protocol (IP) Internet connectivity since 1991 over the X.25 packet-switched network Tunipac to EUnet in Europe, based in Amsterdam. Since 1993, the connection to Amsterdam is completed via a 19.2 kilobits per second (Kbps) leased line. Bitnet is accessible through EARN (European Academic Research Network, the European bitnet network). There is one bitnet node in Tunisia, operated by IRSIT. Also, IRSIT offers access via Unix-to-Unix Copy Protocol (UUCP) and dial-up IP to users with no IP connectivity. In 1994, there were sixteen full IP sites in Tunisia and about 1,000 electronic mail users (UUCP and dial-up IP). Of the sixteen IP sites, four were commercial, four academic, and eight were nonacademic sites.

In 1993, IRSIT was awarded a contract from the secretary of state for Scientific Research and Technology in the Prime Minister's Office to build a national research network for Tunisia (Réseau National de Recherche de Tunisie, or RNRT). Phase one of the project was completed and provided full Internet connectivity to eight applied research centers around the country (six in Tunis, one in Sfax, and one in Mednine in southern Tunisia). Phase two and three of the project consist of interconnecting about fifty additional centers. Subsequently, other ISPs were licensed. Two private firms, Planet Tunisie and GNET, were given licenses to offer Internet access to consumers. They were permitted access to the Internet backbone through ATI (Agene Tunisienne d'Internet), but cannot seek international access to the Internet with international carriers. ATI is a government agency controlled by the Ministry of Interior and the Ministry of Communications.

In Morocco, IP connectivity began by dial-up access via EUnet based in Paris. The center in charge of Internet access, registration, and user support is the Ecole Mohammadia d'Ingenieurs, an engineering school in Rabat. Access in 1995 was through a UUCP arrangement that accommodated only electronic mail, since communications via dial to Europe were very expensive. The Moroccan PTT, in 1995, provided full IP connectivity and took over the work of the Ecole Mohammadia for user registration and support.

Algeria offers Internet access through the Center for Information Science and Research (CERIST) in Algiers, a governmental academic organization in charge of IP connectivity. This organization was established in 1985 to promote the exchange of scientific and technical information, new technologies, and networking at the national level and to form a link with outside researchers. Since 1993, CERIST has maintained a connection to the Internet via EUnet and its gateway in Amsterdam. In 1994, CERIST initiated a second IP connection to the Internet using a 9600 baud leased line, paid for by the Algerian government, to a gateway host in Pisa, Italy. CERIST hoped to establish itself as an Internet gateway host. It worked on the establishment of a national academic research network with government funding.

1.4 The Future

All three countries can be expected to become more and more open to foreign investment. Although France is the major trading partner of all three states for historical and geographical reasons, Morocco and Tunisia are placing special emphasis on attracting American investment and American goods. Tunisia's foreign investment codes were all liberalized in the wake of the adoption of the structural adjustment program in 1986–87. The law encouraging foreign investment in the services sector is aimed at data processing, among other services. In 1991, Morocco created a Ministry of Foreign Investment to provide foreign investors with information on prospects and regulations relevant to their interests. Telecommunications equipment, including computers and software, are among the best prospects for foreign suppliers. The prime contractors for the big growth areas of the next ten years, such as installations of fiber-optic connections to France, Spain, Portugal, and the United Kingdom, earth stations, modernization of switching equipment, and a large number of new lines, are likely to be European, Japanese, and Canadian.

American products dominate the computer and computer software market, although a large share is purchased from European subsidiaries of U.S. companies. Both Morocco and Tunisia have large tourism sectors that will continue to enjoy priority in national economic planning. The tourism industry is a major user of telecommunications equipment and services. Moreover, switching markets in North Africa are expected to grow significantly over the next ten years. For the entire North African region (which includes Libya and Egypt along with Algeria, Morocco, and Tunisia), the switching market is expected to reach over one million lines per year by the end of the decade. Clearly, local production capability in Algeria, Morocco, and Tunisia is geared for such regional growth.

Tunisia committed itself to liberalization in the WTO Basic Telecommunications Services Agreement reached in 1997 for telex and packet-switched data in 1999; mobile telephone, paging, teleconferencing, and frame relay in 2000; and local voice telephony in 2003. The government considered the licensing of private local telephone franchises. It also scheduled to sell, in 1998, 78 percent of Sotetel. Morocco, similarly, prepared to privatize its state-owned national Telecom company during 1998.

Algeria will eventually restructure its telecommunications entity, probably into two separate national offices, one for post and one for telecommunications, following the French model, as Tunisia did.

Notes

I am very grateful for support from a Fulbright Research Fellowship and for special assistance from Mohamed Hedie Lahouel, Jeanne Jeffers Mrad, M. Tijani Azzabi, Mondher Makni, Hedi Hamza, A. J. Maaloui, Salem Missaoui, Abdelkader Bairi, Kareem Boussaid, Choukhman Larbi, and Fatima Kaidi. I am responsible for any errors that remain.

1. U.S. Embassy, Economic Section, memo, Rabat, Morocco, 1990.
2. Presidential Ordinance No. 76–168, October 24, 1976.

3. Decree No. 2-84-20, January 10, 1984.
4. A direct exchange line (DEL) connects a telephone set to the public-switched network.

Bibliography

Algeria. 1972–89. *National Plans.* Algiers.

Algeria, Ministère des Postes et Télécommunications. 1984, 1986. *Annuaire Statistique,* 1974–1984, ed. 2, and 1986, ed. 4. Algiers: Direction de la Planification.

Anderson, Lisa. 1987. *The State and Social Transformation in Tunisia and Libya.* Princeton, N.J.: Princeton University Press.

"Battle for North Africa Switching Market." 1990. *Télécommunications Development Report* 5, no. 2 (February): 1–6.

Hermassi, Elbaki. 1972. *Leadership and National Development in North Africa.* Berkeley, Calif.: University of California Press.

"Informatique: Quels créneaux pour l'avenir?" 1986. *Enterprise Arabo-Africaine,* no. 22 (April).

"L'Informatique en Tunisie." 1991. *Communications et Informatique* 9, no. 38 (September): 16–62.

Intelsat. 1984. "Feasibility Study on the Algerian Domestic Satellite Communications Network." Attachment no. 1 to BG.58.79E. Washington, D.C.

"L'Informatique en Tunisie. 1992. *Communications et Informatique,* no. 44 (October). Special Issue.

"L'Informatique en Tunisie." 1993. *Communications et Informatique,* no. 50 (October). Special Issue.

International Telecommunication Union. 1992. *Yearbook of Telecommunications Common Carrier Statistics.* Geneva: ITU Publications.

Journal Officiel de la République Tunisienne. 1991. No. 22 (March 29).

Makni, Mondher, Khaled Sellami, and Karima Bounemra. 1994. "A National Research and Technology Network in Tunisia: A Must for Innovation and Technology Transfer." Paper presented at IDATE Conference, June.

Ministère des Communications. 1972–91. *Rapports Annuels.* Tunis: Republique Tunisienne.

______. 1991. "Synthèse du Rapport du Comité des Communications et Informatique." Internal memorandum, Tunis.

Ministère des Transports et des Communications. 1992. *Statistiques des Transports et des Communications, 1972–90.* Tunis: Rèpublique Tunisienne.

Morocco. 1972–91. *National Plans.* Rabat.

Morocco, Ministry of Planning. 1981. "Plan to Stimulate the Economy: 1981–85." Rabat.

Morocco, Office National des Poste et Télécommunications. 1988. "Realisations Plan et Stratégie a Moyen Terme." Rabat.

______. 1990a. "Rapport d'Activité Anneé 1990." Rabat.

______. 1990b. "Special [Report]." March 3.

______. 1992a. "La Marche des Télécommunications Marocaines." Internal memorandum, Rabat.

______. 1992b. "Special [Report]." March 3.

"Les Télécommunications en Chiffres" (1994). *Communications et Informatique* No. 56 (November), pp. 13–21.

"Telecommunications Sector Fares Better than Most." 1989. *Middle East Executive Reports* (May): 12–15.

Tunisia. 1972–96. *National Plans.* Tunis: République Tunisienne.

______. 1992–96. *VIIIeme Plan de Développement Volume II: Contenu Sectoriel.* Tunis: République Tunisienne.

Wellenius, Bjorn. 1987. "Telecommunications Policies for LDCs." Paper presented at Telecommunications Sector Restructuring and Management Conference, Kuala Lumpur, November 17–19.

The World Bank. 1994. *World Development Report.* New York: Oxford University Press.

2
Egypt

GEHAN RACHTY

The changes experienced in the field of telecommunications since the mid-1980s have had a palpable and positive effect on the social and economic development of modern Egypt. In the 1990s, Egypt's telecommunications policy reflects its attempt to keep pace with the global revolution in communication while satisfying the specific telecommunications needs of its population. This chapter discusses Egypt's early telecommunications development in the colonial and postcolonial eras; the role and structure of the government telecommunications agency in shaping and implementing Egypt's present-day telecommunications policy; and the significant projects and challenges of Egypt's telecommunications modernization efforts in the 1990s and beyond.

2.1 History of Telecommunications in Egypt

Perhaps because of its geographic position between Europe and the Far East, Egypt has historically maintained a substantial lead over most of the Near and Middle Eastern countries in keeping pace with the rest of the world in the development of its "wired" and "wireless" communications. Telegraph traffic between Europe and the Far East in the nineteenth century, for example, although not directly involving Egypt, had to pass through the Eastern Telegraph system by way of cables in the Red Sea. As a result, the history of telecommunications in Egypt began long before the foreign intervention in the Egyptian administration in 1876 (intended to secure payment of the debts in which Khedive Ismail had drowned the country) or the British occupation of 1882. The benefits to Egypt of this early progress in telecommunications have been enormous.

Despite the absence of reliable documentation and statistics on the years before 1889, it has been established that telegraphic communication between Cairo, Alexandria, and Suez existed as early as 1856. Egypt was ruled at that time by Mohammed Ali's second successor, Said Pasha, who, like his successors, had inherited Ali's goal of modernizing Egypt along Western lines. In this period, two Englishmen obtained permission from the Sublime Porte to lay a cable from Istanbul to Alexandria. The cable expanded on the link that was already present, which ran from Alexandria to Suez through Cairo, and ensured that all parts of the country had communications connections to the outside world. From Suez, this link was later extended to India (the exact date is unknown), though not without many technical and financial setbacks. Egypt was thus linked first to Europe and Istanbul and then to Aden and India via the Red Sea.

By 1870, there were sixty-six internal telegraph lines in Egypt, sixteen of which were in Sudan. The telephone was introduced eleven years later when the American company Edison Bell constructed the first telephone line between Cairo and Alexandria. The period that followed was one of rapid cable expansion, and England's entry into Egypt in 1882 facilitated the development of Egyptian telecommunication. Although such development had begun many years before, it was significantly furthered by the British administration's self-interested desire to link Egypt to the United Kingdom and the world.

Among the major developments of this period was the introduction by a telegraph company of the LeKlanche battery in 1883, which replaced the Minotto battery then in use. This new "Duplex System" was then used in the telegraph lines linking Egypt with Syria. In 1884, moreover, both Reuters and the French news agency Havass were allowed to send their telegrams from Alexandria to Cairo free of charge.

After taking over the telegraph lines south of upper Egypt in 1885, the British military returned these lines to the Telegraph and Telephone Authority in 1887. In the years that followed, more telegraph lines were established, which led to an increase in telegraph services and the establishment of service regulations.

The first telegraph and telephone lines in Egypt were owned by the state's Railway Authority. Trunk lines were also government owned and leased to the telephone company in return for a 70 percent share of the company's income. In 1918, the government bought all telephone and telegraph lines, save for a few small offices, and a separate independent department for telegraph and telephone was later established. Investment in telephone service rose from £E780,000 (Egyptian pounds) in 1920 to £E2 million in 1930. This was accompanied by an increase in profit in the same period from £E220,000 to £E700,000.

Automatic exchanges were installed in Cairo in 1926 and in Alexandria in 1935. This benefited not only the two cities but also the smaller governorates where the amortized telephone equipment from Cairo and Alexandria were transferred. In 1932, an underground telephone cable linking Cairo to Alexandria was introduced. The project improved communication between the two cities, reducing delays in telephone communication, for example, to five minutes at the most. Although the advent of the telephone created competition for the telegraph, for domestic business transactions the telegraph was preferred over the telephone because it offered the benefit of written documentation. Domestic telegrams were handled by telegraph offices owned by the government and operated either by railway workers or by the Delta, Fayoum, or Bahreya railway stations licensed by the Egyptian government. The telegraph was also used externally between Egypt and such Middle Eastern countries as Saudi Arabia, Yemen, Ethiopia, and Sudan. It was used additionally for communication between Egypt and Palestine, eastern

Jordan, and Syria before these countries made direct communication with Europe through the Haifa-Cyprus cable owned by the Eastern Telegraph Company, or through the Beirut Wireless Station that belonged to Radio France. International telegrams were handled by Marconi, Eastern Telegraph Company, or Radio France.

These early telephone and telegraph facilities considerably encouraged the expansion of commerce and industry, as well as government business and private intercourse. The state, moreover, was able to make an easy profit of many thousands of pounds in royalties annually from telegraph cable fees. A glance at any cable map will show that the main cord of the Eastern Telegraph system passed through Egypt; hence, the larger portion of Egypt's annual royalty income was derived from telegraph traffic between Europe and the Far East.

Because radio services differ fundamentally from cable services with regard to transit traffic, the emergence of radio presented serious competition for cables. Radio traffic usually travels from the country of origin to the country of destination with perhaps short extensions at either end of the route. The countries over which radio signals pass naturally receive no payment, and only the terminal countries are able to demand royalties. Because cables, on the other hand, often physically pass through several countries, each state is able to charge the operating company for permission to run cable operations within its borders. As a result, while cable messages from London to India and China generated royalties for Egypt, radio services began passing free of charge.

In Egypt, wireless came to be owned by three organizations. The first was the Egyptian government, whose stations were run by the Telegraph and Telephone Authority. In addition to communicating with other parts of the country, these stations—located in Alexandria, al Kaser, al Tour, and al Arish—exchanged messages with commercial ships in the Red Sea and the Mediterranean and with airplanes entering and leaving Egypt in conjunction with stations of the Coast Guard and the Docks and Phares Authority. The second organization with radio stations in Egypt was the Suez Canal Company, which guided ships passing through the Suez Canal, and the third was the Marconi Company, which sent commercial messages to Europe, paying the Egyptian government a specified sum for every word sent.

When the service between Egypt and London was fully developed in 1932, London had links with India, Canada, Japan, South Africa, South America, Australia, and New Zealand. With the excellent trunk facilities that existed between London and most countries in Europe, it was thus possible for a subscriber in Cairo to speak to over 90 percent of the telephone users of the world.

Although the original radiotelephone service between England and America was a long wave, disabling weather conditions made this type of service impossible for Egypt. Consequently, Egypt's new radiotelephone service in this period was a short wave that consisted of two waves whose use was dictated by the time of day and season of the year. So that they would not interfere with each other, two radio transmitters were installed in Abou Zaabal and two receivers at Maadi. The receivers and transmitters were then linked by telephone lines to terminal equipment situated in a room near a trunk exchange. Here, they were combined so that they could be connected to the main telephone system.

In addition to the Cairo-London channel, direct radio links between Cairo and Paris, Cairo and Berlin, and Cairo and Rome were later opened. With two complete transmitting and receiving installations, it was possible to keep one circuit open on a constant basis to London during the hours of service, while the other could be used to arrange any calls required over the other circuit. The cost was 150 piasters for one minute, and calls were not supposed to exceed three minutes. Prices were not fixed, however, and were subject to change at any time.

2.1.1 The Postcolonial Period

The British occupation of the Suez Canal Zone ended two years after the overthrow of the monarchy in 1952 and the subsequent military takeover by the Free Officers led by Colonel Gamal Abdel Nasser. During President Nasser's regime, Egypt's relationship with the West was not harmonious. Despite the socialist tendency of this new regime, however, there were impressive industrial gains.

The present-day electronics industry began during the Nasser regime as a group of nine private shareholding companies providing service and maintenance for electrical appliances. There were also eight cable manufacturing companies, and later three electronics assembling companies were created: Phillips, established in 1956, which assembled radio receivers on a large scale; the Arabic Transistor Radio Company, which assembled roughly the same products as Phillips; and the El Nasr Television Company, established in 1961, which assembled television sets.

To provide the electrical and electronic components required by these three assembling companies, Satron Company was created in 1969. Two research centers were also established around this time to design and develop new equipment and to solve electronics-related problems.

Other noteworthy telecommunications achievements of the postcolonial period included the establishment of the Wired and Wireless Telecommunication Authority (1957) and the introduction of the first coaxial cable linking governorates (1961). This was followed by the operation of the first automatic (crossbar) exchange in 1962 and the introduction of the first telex exchange in 1963.

When Anwar Sadat became president after Nasser's death in 1970, he liberalized the Egyptian economy with an "Open Door" policy to encourage the influx of Western and Arab capital and the introduction of many foreign aid programs. The telecommunications sector expanded further during this period with the opening in 1972 of the first marine cable linking Egypt and Italy (with a capacity of 480 circuits) and the introduction in 1979 of an international exchange with an initial capacity of 160 circuits. Two mobile earth stations for communication via satellites were also constructed during Sadat's presidency—one at Kobba in 1974 and the other at Maadi in 1978. With the installation of a microwave network between governorates in 1975, wireless telephones in automobiles became available to users in Cairo.

2.2 The Present

2.2.1 Institutional Structure of Telecommunications Operations and Regulations

The Arab Republic of Egypt National Telecommunications Organization (ARENTO) was established by National Assembly Law No. 153 in 1982 as an autonomous public utility organization under the direct supervision of the Ministry of Transport, Communications, and Civil Aviation (which replaced the Telecommunication Authority established in 1957). The legislation defined ARENTO's regulatory authority and role and allowed it to enter into joint ventures with other parties for the promotion of telecommunications. In 1997, a new law provided for the partial privatization of ARENTO, and for its renaming as Telecom Egypt.

With branches all over the country, ARENTO (Telecom Egypt) is Egypt's central telecommunications organization, and exclusively responsible for establishing and operating all cable and wireless communications networks on a national level. Its offices administer and maintain all establishments and equipment needed for telephone services, and its responsibilities include connecting local networks with international ones, constructing cable and wireless networks throughout Egypt, and supplying cable and wireless telephone services. Additionally, with the approval of the minister of Communication and Transportation, the company can establish joint stock corporations either alone or with other partners. As soon as such corporations are established, shares can be bought and sold, with employees receiving first priority in the purchase of shares.

2.2.2 ARENTO's Network Modernization Plans

2.2.2.1 Universal Service Goals

As formulated in Egypt's third National Five-Year Plan (1993–97), ARENTO in the mid-1990s set forth an ambitious program of network expansion and modernization. Because obsolete exchanges had a negative impact on the traffic flow and consequently reduced the operator's revenue, ARENTO ranked the replacement of this equipment as its highest priority. To this end, in the mid-1990s ARENTO began installing new electronic digital exchanges in order to increase the number of subscriber lines and hence augment the country's telephone density, which remains modest relative to international standards (see Table 2.1).

Within the framework of the Five-Year Plan, ARENTO's stated goal was to increase telephone density from less than 5 percent of the population to 7 percent, in addition to reducing the waiting time for provision of service to a maximum of four years. The number of telephone subscriber lines installed by mid-1997 was expected to be over 4.5 million.

The goal of providing universal service outlined in Egypt's Five-Year Plan also included expanding services to both rural and remote areas by (a) replacing manual exchanges with semiautomatic exchanges and (b) adding remotes to

Table 2.1. Telephone Lines

Date	Number of Lines	Density per 100 Inhabitants
1981–82	650,000	1.4
1982–83	700,000	1.5
1983–84	875,000	1.8
1984–85	1,150,000	2.4
1985–86	1,350,000	2.7
1986–87	1,600,000	3.1
1987–88	1,700,000	3.2
1988–89	1,800,000	3.5
1989–90	2,000,000	3.9
1990–91	2,350,000	4.3
1991–92	2,500,000	4.4
1992–93	3,000,000	5.0
1993–94	3,550,000	6.0

digital exchanges and connecting them to the national long distance network (see Table 2.2).

By the mid-1990s, telecommunications service had been introduced to the remote areas along the Cairo-Alexandria desert road and provision had been made for emergency telephone service using solar energy.

2.2.2.2 International Network and Value-Added Services

Modernization of the international network and enhancement of value-added services were also included in ARENTO's plans for the mid-1990s and beyond. In hopes of meeting the increasing demand for international circuits and other media that represent alternatives to satellite communications, ARENTO decided to complete the implementation of the Transoceanic Fiber Optic Cable (SEA-ME-WE 2) linking Southeast Asia, the Middle East, and Western Europe. The capacity of the Cairo International Exchange was also increased to 3,680 international circuits (versus 160 in operation in 1980). In 1990, a second international gateway in Alexandria with a capacity of 1,000 circuits was put in service, bringing the total number of Egypt's international circuits to 4,680.

These projects were completed because ARENTO recognized that international traffic represented the main source of its revenues. The marked growth in interna-

Table 2.2. Growth of Telephone Service in Rural Areas

Date	Number of Subscribers (in Thousands)
1989	20
1990	50
1991	60
1992	100
1993	240
1994	310

Table 2.3. International Circuits and Traffic

Date	Circuits	Traffic
1981	800	30,000,000
1982	850	50,000,000
1983	1,700	90,000,000
1984	2,000	110,000,000
1985	2,000	135,000,000
1986	4,600	160,000,000
1987	4,500	175,000,000
1988	4,700	200,000,000
1989	5,000	230,000,000
1990	6,000	275,000,000
1991	6,000	290,000,000
1992	7,750	320,000,000
1993	7,750	400,000,000

tional traffic after 1981 shown in Table 2.3 reflects the installation of the international circuit extensions just described, as well as the enhanced efficiency of the switching and transmission facilities of both local and international networks.

In 1990, an Arabsat earth station with a capacity of 348 circuits and two television channels was placed into service, providing Egypt with better communication facilities and broadcast exchange with other Arab countries. In the same period, ARENTO sought to extend the mobile telephone network and to introduce portable mobile sets with nationwide coverage.

By the mid-1990s, mobile telephone service coverage had been extended to all cities and roads in the delta region and to the Suez Canal Zone, with a capacity of 56,000 mobile sets for Cairo alone (versus only 400 in 1981). Facsimile service was also increased from 347 sets in 1987 to over 8,000 in 1993, while paging service—with an initial capacity of 7,000 sets—was introduced in the mid-1990s for greater Cairo. In 1994, the capacity was 16,000 subscribers, mainly business and professional users. However, since the growth experienced was slower than forecasted, ARENTO decided to liberalize the Pager Equipment Market. It is hoped that the private sector's participation will spur uptake of this service. In 1994, there were 9,500 subscribers to ARENTO's paging service.

2.2.3 Development of the Telecommunications Sector

By the mid-1990s, there were indications that Egypt's telecommunications sector was progressing rapidly toward a more liberal environment and a more market-oriented approach. Originally ARENTO had claimed that the fulfillment of the present demand for basic telephone service was the focal point of its development policy. Competing with this goal, however, were the increasing demands made primarily by the business community and public agencies for value-added services such as paging services, mobile telephone services, and data network services. As these calls for changes in the national telecommunications policy developed in the mid-1990s, the Egyptian government was faced with the challenge of providing services to specific, vital segments of the population. It was this very attempt to offer value-added services that prompted the government to liberalize the terminal market by inviting leading private-sector companies to participate actively in the marketing of these services and in the sale of the terminals to the public end users. As with the creation of the national paging system (see Table 2.4), selected companies were first evaluated by ARENTO, the terminals were type approved, and the companies were then granted the license to operate.

In the 1990s, the service requirements of Egypt's national security agencies put high demands on network performance and on the lead time required for the provision of the value-added services. To respond to these demands, ARENTO established the information network EGYNET (see Table 2.4), which boosted interagency communication considerably.

The project took two years to finish and cost £E18 million. The network's capacity started at 2,300 users. By 1994, EGYNET served over 11,000 users. The International Telecommunication Union approved the X.25 protocol for EGYNET rather than the X.75 protocol used in the international networks because of its ease of connection and its widespread use around the world. In addition to one converting station in Alexandria, several stations for converting were established in Cairo, each connected to the other by microwave, coaxial cable, and optical fibers, which can transfer data at a speed of 4.6 kilobits per second. The EGYNET network also contains two concentrators in Cairo to concentrate the messages from data of 32 lines and arrange them in a single line reaching the nearest converting unit.

With EGYNET, users in Egypt can send or receive data through specially leased lines at 4.6 kilobits per second or through general telephone lines at 1.2 kilobits per second. Data can also be transferred throughout Egypt and abroad through telex lines via the network. The network is already connected with TRT American gateway, which gives EGYNET users direct access to all data networks in the United States, Germany, and Spain. The network is also connected to the TRANSPAC French network and the universal transit station ITL in Paris. In 1994 ARENTO was reviewing plans to invite private-sector participation in EGYNET through

Table 2.4. Subscribers

Date	Mobile Telephones	Paging Service	EGYNET
1985	—	—	—
1986	600	—	—
1987	2,100	—	—
1988	2,700	—	—
1989	3,300	400	120
1990	4,400	3,000	320
1991	5,500	4,400	600
1992	5,800	5,000	840
1993	7,000	8,000	1,040
1994	7,700	9,600	11,160

"qualified service providers." If these plans are approved, investors will be allowed to invest in an electronic mail service on a nationwide basis.

In the midst of these developments in Egypt's telecommunications, there was a growing recognition in the mid-1990s that a new legal framework was needed to address liberalization and deregulation. Toward this end, an ongoing process for restructuring ARENTO was begun, with emphasis placed on the organization continuing to develop administratively in order to efficiently meet the increasing and diversified demand for services. In the mid-1990s, there were also reports that ARENTO was involved in preparing a long-term national policy for the telecommunications sector that would attempt to secure a balance between the interests of the private investors and the public.

2.2.4 Egypt's Telecommunications and Electronic Equipment Market

Since 1961, Egypt's public telecommunications agency has been involved in the establishment of a national public telecommunications industry for crossbar exchange equipment. By the mid-1990s, ARENTO was manufacturing, through its subsidiary companies, digital public exchanges, PABX equipment, and network material.

Nevertheless, ARENTO continued to rely on foreign suppliers for the implementation of its numerous projects, all of which were deemed essential for the development of Egypt's national telecommunications network. The organization bases its purchasing policy and selection criteria for world suppliers and manufacturers of telecommunications equipment on a technical evaluation of the supplier's system and equipment in addition to the associated commercial terms. Project financing, based on governmental credit facilities or their equivalent, also plays a central role in ARENTO's selection process. In such cases where the financing offer requires that the equipment be supplied out by the home suppliers of the country offering financing, ARENTO tendered the projects to the selected national suppliers and awarded the projects on the basis of competition. While financing offers have been made by numerous countries, including France, Germany, and the United States, the Egyptian government has required ARENTO to reimburse the loan's principal and interest at the prevailing commercial interest rates.

2.2.5 Financial Health and Independence of Telecommunications

Although ARENTO's funds were public, its budget was made according to internal regulations and, with the exception of five-year development programs, it did not adhere to the laws and regulations stipulated in the governmental budget. The organization had its own independent bank accounts to which its profits are channeled, and its fiscal year starts and ended with the government's.

The Egyptian government allocated investments for ARENTO as a source of funding for the latter's projects. The funds came from gifts, loans, and self-finance. Although ARENTO's service charges were below the actual cost of providing service, income from local and international services was ARENTO's main source for covering its expenses. ARENTO (and subsequently Telecom Egypt) considered the following actions as a means to generate revenue:

- Raising installation fees.
- Introducing time limits for local calls.
- Charging business users significantly higher subscription fees.
- Introducing a priority fee for subscribers requesting immediate service.

2.2.6 *Quality of Service*

Due to an extreme deterioration of service during the 70s and 80s, in the 1990s, ARENTO developed a scheme for assessing the indicators of quality for various telecommunications offerings. Among the criteria were service availability and call processing for local and international calls. These indicators presumably reflect detected service impairments, the impact of fault repair measures, and network performance as a whole. Examples of these indicators—usually compiled on a zone basis all over the national network—are:

- Outstanding faults per month.
- Average fault reported per day (over a month).
- Average fault repair per day (over a month).
- Outstanding faults at the end of the month.
- Percentage of outstanding faults out of the number of lines in operation.
- Percentage of unsuccessful local, national, and international calls out of total calls made.

These same indicators are used to assess the quality of service of the telex, the junction network, long distance lines, and the like.

Although ARENTO is not bound by any legal obligations to compensate subscribers for any impairment of service quality or service interruptions, it has invested in network operation and maintenance through the creation of a number of fault repair centers in large metropolitan parts of Cairo and Alexandria in order to process fault reporting, repair, and follow-up.

Under the auspices of the national five-year plan (1993–97), ARENTO had the following service goals:

- To increase telephone density to 7 percent.
- To reduce waiting time for provision of service to a maximum of four years (in 1992, the average waiting time was 6.1 years).
- To improve the quality and level of service in rural areas.

2.3 Future Outlook

During the 1980s and 1990s, technological developments in Egypt's telecommunications led to substantial advancements in the utilization of space stations, optical fiber cables, and electronic computers, as well as in the transition to digital systems. In 1994, 59 percent of the system was digital. The next decade, however, will also bring with it some equally important developments. New services will be available, the telecommunications system will be completely transformed into digital technology, and, as a result, integrated service networks can be constructed, allowing information to be exchanged at much higher speeds than were achieved previously.

The national telecommunications organization attempted to anticipate future telecommunications trends: the underlying assumption of all ARENTO's studies on the development plans of the 1990s has been the transition to digital systems; the manufacture of electronic digital exchange equipment in Egypt began at the end of 1990; and ARENTO's program for the 1990s includes the creation of optical fiber networks between governorates and the extension of new telecommunications services. For example, in 1992, ARENTO introduced fiber optics to the Cairo metropolitan areas and soon awarded a U.S.$1 million contract to Northern Telecom for a 140 Mbits/s optical fiber link connecting the Assait area of Upper Egypt.

In the meantime, ARENTO was faced with the challenge of maintaining the telecommunications system's efficiency and performance during the 1990s. It had to sustain the current initiative to digitize the network while fulfilling the existing demand and reducing backlog—a task that will prove especially difficult in light of Egypt's rapid population growth.

Also, ARENTO had to continually raise the efficiency and technical competence of the telecommunications workforce (engineers, technicians, and laborers) through training on the latest technology in the field. The goal was complete reliance on trained local labor for the installation, operation, and maintenance of telecommunications equipment, as well as for the manufacture of electronic digital exchanges.

Providing basic services remained the priority of ARENTO's development goals. It continued in the mid-1990s to seek to strengthen the role played by the private sector in expanding services to a wider portion of the public. Because private enterprises can mobilize large amounts of resources in a competitive environment, they can have a positive impact on the development of services while simultaneously creating a new source of revenue. In hopes of tapping this potential, the government, beginning in the mid-1990s, sought to award licenses to selected operators who provided value-added services through the network infrastructure, such as mobile telephones and EGYNET.

Telecom Egypt set ambitious development goals given the backlog in demand. Egypt had a total number of telephone subscribers of six phones per 100 habitants. If Telecom Egypt is to achieve its four-year plan, five million telephone lines will have to be installed by the year 2002, or one million per year. The initial major contractor for this upgrade project was Siemens of Germany.

In 1997, Egypt awarded the first mobile license to MobiNil, a consortium of France Telecom, Motorola, and several Egyptian partners. A second cellular license was awarded to a consortium led by Vodaphone of the UK. A group led by France Telecom won a license for pay phones. Next, the government took on the privatization of ARENTO. Yet the state kept 80 percent of the firm, renamed Telecom Egypt, selling only a minority share to private investors. But it also loosened restrictions on other operators for the future.

Increasing demand for special and advanced telecommunications services, as well as the continuing trend toward technological development, has made changes in Egypt's existing telecommunications and related regulatory legislation imperative. A process of review must include the building of popular support in favor of deregulation and the associated privatization schemes. Only through deregulation, privatization, and continuing technological modernization can Egypt hope to keep up with the global revolution in communications technology and adapt from it what is suitable for her needs.

Bibliography

Saba, Belkais El Said. 1994. Public Telecommunications in Egypt: Planning and Development. Paper presented at Africom 1994 Conference.

TELEC240.308, ARENTO. 1994. "Study of the Development of Telecommunications in Egypt."

3
Ethiopia: Past, Present, and Future

ABII TSIGIE AND GIRMA FEYISSA

The history of telecommunications in Ethiopia—from its introduction to the present day—spans one hundred years. Since very little of this history has been documented, this chapter will provide some basic facts and figures as well as a brief analytical account of the development of the country's telecommunications sector. The chapter is divided into three main parts: (1) an introduction to Ethiopia giving brief accounts of its geography, economy, resources, and population size and distribution; (2) a history of the development of the telecommunications system, including the introduction of telecommunications services into the country and the milestones of the early years (1894–1941), the sector's postwar rehabilitation and expansion work, the evolution of the system's organizational structure, and the accomplishments and problems of subsequent package development programs; and, finally (3) a discussion of the future of telecommunications services in Ethiopia in light of the objectives of the economic policy of Ethiopia's new transitional government. We will conclude the chapter with some general observations and recommendations for the future development of the sector.

3.1 An Introduction to Ethiopia

3.1.1 The Country

Ethiopia is located in what is generally known as the Horn of Africa, at the intersection of the busy sea routes and crossroads that connect the African continent to the Middle East and India. With an area of 1,112,032 square kilometers, Ethiopia is one of the seven largest political entities in Africa and is bounded by Sudan to the west and northwest, Eritrea to the northeast, Djibouti and Somalia on the eastern coast, and Kenya to the south. The country's topography ranges between hilly uplands and low-lying valleys, and the climate is divided into a dry season (November through February) and two rainy seasons (a moderate rainy season between March and May and a heavy rainy season between June and October).

3.1.2 The Economy

Agriculture is the mainstay of the economy, accounting for about 48 percent of the country's GDP in 1994. The industrial sector, which includes mining and quarrying, manufacturing, small-scale industries and handcrafts, electricity, and water, as well as construction, maintained a share of only 11 percent of GDP in 1994, showing a drop of 3 percent from the level it had reached a decade earlier. Services, consisting of trade, hotels, transport and communications, banking, insurance, real estate, public administration and defense, education, and health and other related services, have increased their share of GDP from 34 percent in 1984 to 41 percent in 1994.

Throughout the 1980s, Ethiopia's socioeconomic development was seriously affected by repeated droughts and civil strife. The droughts of 1984–85, in particular, resulted in large-scale famine affecting millions of people. The internal conflict, which reached its peak in 1990–91, was also a major cause for the country's poor economic performance of that period. By 1991, industries were on the average operating at 25 percent of capacity.

3.1.3 Resources

The most promising element in Ethiopia's physical resource base is its potential for agricultural production. Only 14.5 percent of the country's 80 million hectares of arable land is under identifiable crop production. Ethiopia has some 75 million cattle, camels, sheep, and goats—the largest livestock population on the African continent. In fact, livestock contributes about 40 percent of the total value of Ethiopia's agricultural output, which is the equivalent of the combined value of its wheat cereals, oilseeds, and other field crops. In the mid-1990s, there was a substantial unrealized potential for increasing the value of the country's livestock output.

In the mid-1990s, Ethiopia's energy sources were just beginning to be explored for development. Hydroelectric energy from newly commissioned projects contributed about 85 percent of the country's electric power in the mid-1990s, and a natural gas resource was discovered in the Ogaden region. Previously underexploited minerals, especially primary gold, are also known to exist, along with potential resources of soda ash, tantalum, marble, potash, and base metals.

3.1.4 Population

The first census conducted in Ethiopia in 1984 showed the population to be 42.2 million, including Eritrea, which became a separate state in May 1991. Assuming a 3.1 percent growth rate and other factors remaining constant, the population was estimated to have reached 52.5 million at the end of 1991. Forty-six percent of the country's population are under fifteen years of age, and in 1984 the country's total fertility rate was about 7.5 per woman with a death rate estimated at around 15.2 per thousand population. According to a 1984 Central Statistics Office report, life expectancy at birth in Ethiopia was about 51.9 years.

Eighty-six percent of Ethiopia's population was engaged in rural-based economic activity in the mid-1990s, mainly in agriculture in the highland areas. About 3 percent of the rural population lived in the lowlands and was engaged primarily in raising cattle. The urban population, which is concentrated in a few towns, was growing at an annual average of 5.5 percent in the mid-1990s, and the total urban population was about 8.4 million. Roughly 26.1 percent of Ethiopia's urban population lives in Addis Ababa.

3.2 Ethiopia's Telecommunications: The Past

3.2.1 Early Years (1894–1941)

Because Ethiopia had no colonial history—except for a brief five-year occupation period between 1935 and 1941—the history of the country's telecommunications sector can be roughly divided into the "early years" and a "postwar period." Telecommunications service was introduced in Ethiopia in 1894 during the rule of Emperor Menelik II. The first major telephone line construction spanned a total distance of about 477 kilometers and connected Harar, a major trade center in the eastern region, with Addis Ababa, the capital city. The line, which took only two years to construct, also interconnected small towns situated along the route. Immediately after the telephone line, a telegraph line was installed following the construction of the first and only railway line in the country—the Ethio-Djibouti railway. Within two years, an 880-kilometer north-south telephone line connecting Asmara, the capital of Eritrea, to Addis Ababa was constructed and made operational in 1904. The "verbal repeater" system was used to facilitate long distance calls, making use of the several intermediate stations opened at the small towns and villages along the route.

Several routes branching out from Addis Ababa to connect provincial administrative centers and major towns were being extended in advance of the construction of the road network. Pack animals were used to transport material and equipment. By 1930, a route distance of 7,000 kilometers was completed and over 170 towns were being served by the telephone network. The development of Ethiopia's long distance telephone network, particularly in reaching the country's strategic areas and border towns, was a remarkable feat given the rugged terrain and the absence of modern transport systems. International communications services, however, took longer to develop. Until the end of 1930, Asmara and Djibouti, both under colonial rule at the time, were the only two locations with international connections.

Administratively, Ethiopia's communications system was run by an office in the Imperial Palace, where it was accorded the direct attention and supervision of the emperor aided by the assistance of foreign experts (who in 1907 were replaced by Ethiopians). In the early years of the nineteenth century, a group of French experts undertook a project to study and restructure Ethiopia's telecommunications and postal administrations. This took more than two years (1909–11) and became the cornerstone for the establishment of the country's Ministry of Posts, Telegraph, and Telephone. As the century progressed, demand for telephone service grew at a rapid pace, and new stations extending in different directions were added in various parts of the country.

Meanwhile, hostilities among the competing colonial powers—Britain, France, and Italy—were growing in the Horn of Africa, putting Ethiopia's sovereignty and independence in a precarious position. The emperor was forced to move fast and took action to safeguard the country's sovereignty. A member both of the League of Nations and since 1932 the International Telecommunications Union (ITU), Ethiopia consequently took steps to free itself from a dependence on the foreign administration of its international traffic. By 1934, Ethiopia had established direct radiotelephone links with Cairo, Djibouti, Aden, and London and soon after established a radio communications training center for Ethiopians in order to replace the expatriates handling the nation's traffic (which included confidential state affairs).

When war inevitably broke out and the fascist powers invaded Ethiopia, telecommunications facilities were targeted for destruction to deprive the resistance forces of access to information. Most of Ethiopia's telecommunications installations and facilities were destroyed and both local and international communications disrupted. Realizing, however, that telecommunications links were vital to their operations, the Italian forces soon made efforts to restore what they had destroyed. In their turn, Ethiopian resistance forces put these rehabilitated lines out of service and disrupted restoration efforts. Despite their earlier efforts, toward the end of the period of occupation the Italians had begun to restore Ethiopia's telecommunications system, installing automated telephone exchanges in Addis Ababa and Asmara (with a capacity of 1,500 and 1,200 lines, respectively). By 1941, as the Italian forces finally fled the country, they destroyed the telephone service in approximately one hundred Ethiopian towns, which they themselves had restored. As a result, by the end of the war only a handful of Ethiopia's stations were functioning, and then very poorly.

3.2.2 Rehabilitation and Expansion of the Network (1941–1973)

3.2.2.1 The Early Postwar Period

The history of the initial period of postwar telecommunications in Ethiopia is essentially a history of network rehabilitation and restoration. The Ministry of Posts, Telegraph, and Telephone was reestablished, and the enormous task of reconstruction began shortly thereafter. Some of the country's distant stations were provided with temporary service by radiotelephone, and the radio transmitting station at the southwest end of Addis Ababa was rehabilitated and placed back in service. But the task of rehabilitating the damaged infrastructure in all sectors of Ethiopia's economy was huge and almost insurmountable. Since funds and skilled manpower were unavailable in sufficient numbers, international aid agencies had to be approached.

In 1950, the first International Bank of Reconstruction and Development

(IBRD) mission (which consisted of the organization's founder, Eric Beecroft, and other bank officials) came to Ethiopia to conduct an investigation of the possibility of reforming the ministry and creating an organization entrusted with the sole responsibility of restoring and extending Ethiopia's telecommunications services. (The technical part of their study was actually made by a group of experts of the International Telephone and Telegraph Corporation, ITTC.) The study proposed a short-term (three-year) U.S.$2.2 million investment program to be carried out as an initial rehabilitation project. It also proposed the establishment of a semiautonomous telecommunications body charged with the maintenance and development of the country's telecommunications network.

Establishment and Evolution of IBTE. In 1952, the proposals made by the ITTC's technical experts were accepted, and the establishment of the quasigovernmental Telecommunications Board was approved by the Ethiopian government. Subsequently, the Imperial Board of Telecommunications of Ethiopia (IBTE) was established as a chartered organization by Proclamation No. 131 in October 1952. The organization was to be independent of the Ministry of Posts, Telegraph, and Telephone and had the following main objectives:

- To rehabilitate, extend, and maintain Ethiopia's telecommunications facilities and to engage on a for-profit basis in the civilian telecommunications business.
- To act as an agent for the Ethiopian government in all matters relating to telecommunications services in and outside the country.
- To establish appropriate training procedures and a training institute for present and future board personnel.

In addition to these specific objectives, the IBTE's overall purpose was to provide and efficiently maintain satisfactory telecommunications services for the general public. In order to achieve these objectives and meet the ever-growing demand for services, the IBTE has periodically undertaken structural reforms and modifications to its organization.

Reorganization Efforts. The Imperial Board of Telecommunications of Ethiopia (IBTE) began operations as an autonomous body on January 1, 1953, with an organizational structure appropriate to its functions at the time. This first organizational arrangement, which took effect in March 1953, remained in place for about a decade. Under this structure, decision making was highly centralized—partly because of the shortage of highly skilled manpower and partly due to the need for both closer follow-up and stronger control of activities.

The major change to the IBTE took place in 1971. A new hierarchical structure extending from the general manager to departments, divisions, regions, branches, areas, subareas, sections, and units or offices was established under the name Ethiopian Telecommunication Authority (ETA).

Since the organizational structure was created in 1971, a considerable number of changes have taken place in the field of telecommunications. The number of subscribers, for example, has almost doubled, and in the 1990s demand continued to grow quickly. The types of services offered and the area of coverage provided has also expanded, and rapid advances in telecommunications technology have presented new challenges and responsibilities. The ETA will have to reorganize itself to face these challenges and live up to the expectations of the coming years. To this end, a thorough study was carried out by the Authority in the early 1990s and submitted to the board of directors for consideration. After a long delay, the proposed structure was rejected and a new group was formed in 1994 to carry out a new study and come up with an alternative proposal.

3.2.3 Recent Telecommunications Development

It has been nearly one hundred years since the telephone was introduced in Ethiopia and about four decades since a systematic approach to the development of the country's telecommunications was adopted. In that time, many observers, including the World Bank (see, for example, its *World Development Report,* 1983), have rated Ethiopia's achievements in the telecommunications sector in relation to other developing countries as good. Indeed, in terms of telephone density, quality of service, labor productivity, and affordability, the Ethiopian telecommunications service compares favorably with many African countries.

In the mid-1990s, however, Ethiopia's economic development, as measured by indicators such as GDP, was still one of the lowest in Africa. The structure of the country's economic production in the mid-1990s has changed very little since the early 1980s, with agriculture's share of GDP at 48 percent, industry's (consisting of manufacturing, electricity, water, construction, and mining) at only 11 percent, and services at a respectable 41 percent. By comparison, according to the *World Development Report* of 1989, the GDP share of agriculture and industry in sub-Saharan Africa was 31 percent and 26 percent, respectively, in 1987.

As the figures in Table 3.1 clearly indicate, Ethiopia's industrialization is still in its infancy. The agricultural sector is dominated by subsistence production, for example, and the value-added element of the services sector comes mainly from wholesale and retail trade, as well as from public administration and defense.

In 1991, telecommunications contributed about 1 percent of Ethiopia's GDP—not an unexpected figure for an economy characterized by subsistence agriculture. However, the vital role telecommunications services play in economic development can only be determined by accounting for both its direct and indirect contributions, such as the rise in productivity and efficiency of other economic sectors as a result of the increased availability of reliable telecommunications services.

Ethiopia's pattern of investment in telecommunications closely follows the movement of the country's gross domestic investment. Because Ethiopia has been the scene of a series of protracted internal conflicts and recurrent droughts over the last two decades, it is not surprising that its rate of gross fixed capital formation (GFCF) as a percentage of GDP has been four percentage points lower than the rate for sub-Saharan Africa (roughly 11 percent versus 15 percent, respectively). During the same period, Ethiopia's rate of investment in telecommunications has fluctuated at around 0.4 percent of GDP.

Table 3.1. GDP, GFCF, and Telecommunications Investment from 1983 to 1993 (in millions of Birr)

	1983	1985	1987	1988	1989	1990	1991	1992*	1993*
GDP at market price	9,760	9,924	11,399	11,851	12,414	12,586	12,772	12,363*	13,884
GFCF at market price	1,240	1,540	1,797	1,873	1,660	1,534	1,420	1,088	2,221
Telcom investment	10	16	70	50	60	23	38	17	68
Telcom investment as % of GDP	0.18	0.16	0.61	0.42	0.48	0.18	0.30	0.14	0.49
Telcom investment as % of GFCF	0.8	1.0	3.9	2.7	3.6	1.5	2.7	1.6	3.1
Telcom contribution as % of GDP	0.61	0.76	0.88	0.92	0.73	0.85	0.91	9.92	0.98

Sources: ETA's Statistical Bulletins (1987–93); ONCCP Plan documents (1985–93); revised series of national accounts statistics of Ethiopia Advance Summary Report (a draft).

*Excludes Eritrea.

Note: The Birr is the Ethiopian currency. Before its devaluation in 1992, 1 U.S.$ = 2.07 Birr. The official exchange rate in 1995 was 1U.S.$ = 5.90 Birr.

Since its inception, the ETA's investment activities have been carried out through the development program approach. The achievements of this approach and the problems the ETA has encountered in the investment process are described briefly in the following sections.

3.2.3.1 The First Four Telecommunications Development Programs (1958–1973)

Since the establishment in 1953 of what was then known as the Imperial Board of Telecommunications of Ethiopia (currently known as the Ethiopian Telecommunication Authority, or ETA) as a semiautonomous organization, six development programs have been carried out in the country. During the first four development programs, implemented between 1958 and 1973, a total investment of U.S.$50.2 million was made. These development programs were financed partly from the IBTE's own funds (45.4 percent) and partly from external sources: 28.8 percent from an International Bank of Reconstruction and Development (IBRD) loan; 22.3 percent from a Swedish International Development Agency (SIDA) loan; and 3.5 percent from a USAID grant.

By the end of the Fourth Telecommunications Development Program, Ethiopia's telephone service had reached 343 cities and towns, and there were 47,263 telephone subscribers, a total of 63,689 telephones, and 271 telex subscribers.

3.2.3.2 The Fifth Telecommunications Development Program (1974–1979)

The Fifth Telecommunications Development Program (FTDP) was originally planned to be implemented during the 1974–79 five-year period. The actual implementation of the program, however, began in 1975 and lasted until 1984. During this period Ethiopia experienced major political and economic difficulties. Widespread internal conflict and foreign aggression had already exacerbated the country's shortage of such critical development inputs as construction materials and foreign exchange. On the international scene, the late 1970s and early 1980s was a period in which the price of both oil and manufactured goods was rising rapidly. The value of the Japanese yen was also rising, making Japanese equipment harder for the ETA to afford.

Although the total planned capital outlay for the Fifth Program was close to U.S.$62 million, the actual expenditure was over U.S.$72 million. Urban exchanges alone required 52 percent more than the planned outlay.

The achievements of the FTDP can be summarized as follows:

- Ten new automatic exchanges with a total line capacity of 32,200 were installed.
- The capacity of existing exchanges in eighteen towns (including the capital city) was raised by a total of 22,000 lines.
- Subscriber trunk dialing service was introduced into six towns.
- A total route length of 1,917 kilometers of microwave radio relay system, linking a number of regional capitals to Addis Ababa, was established.
- A satellite earth station and an automatic telex exchange were put into service to cater to international routes. By June 1985 Ethiopia had direct satellite links with a total of thirteen cities in Africa, Asia, Europe, and the

United States; and direct microwave links were established with Djibouti, Kenya, and Tanzania.

Financing the Fifth Development Program. The FTDP required about U.S.$38.6 million in foreign exchange and U.S.$21.8 million in local currency. The foreign exchange was mainly raised through external borrowing from the International Development Association (IDA), which extended a soft loan amounting to U.S.$34 million. The rest was provided by internal resources.

Ethiopia's previous development programs had been financed by the ETA's own funds, as well as by loans from the IBRD and, to a lesser extent, SIDA.

3.2.3.3 The Sixth Telecommunications Development Program (1984–1991)

The Sixth Telecommunications Development Program (SXTDP), which was originally planned for 1984–88, was extended to 1993 for a number of reasons, the most important being inadequate implementation capacity, shortage of construction materials and foreign exchange, and delays in the mobilization of funds from donors. The SXTDP differed significantly from the FTDP in the amount of capital expenditure planned, in the targets it set out to achieve, and in the level of technology employed.

The total investment of the SXTDP amounted to U.S.$150 million, 150 percent higher than that of the FTDP. Of the total expenditure of U.S.$150 million, imports of equipment and machinery (switching, network, transmission, and auxiliary equipment) and vehicles and supporting materials accounted for 65 percent.

The SXTDP set out to raise the country's telephone exchange capacity from 123,900 at the beginning of 1984 to 195,000 lines in 1988. The actual figure for the target year was 125,665—35.6 percent short of the goal. Similarly, subscriptions were expected to grow at an average annual rate of 12 percent, from 89,544 in June 1984 to 140,000 direct exchange lines (DELs) in June 1988. However, the actual number of subscribers in 1988 was 24.3 percent below target.

Given the overall weakness in Ethiopia's economic performance in the second half of the 1980s and the ETA's inability to attract adequate funds in time, the actual achievements of the SXDTP can be rated as satisfactory. Other achievements of the SXDTP included the following:

- Subscriber trunk dialing facilities were extended to twenty-seven towns—an increase of 69 percent on the figure at the beginning of the program.
- Rural telecommunications penetration was raised by opening an additional 150 public call offices.
- International direct dialing was introduced in some exchange areas.
- The second satellite earth station (standard A) working with the Indian Ocean satellite space segment was established in 1986.

Financing the Sixth Development Program. The SXTDP was financed partly by the ETA's own funds and partly by external credit. The ETA's own funds covered all of the local expenditures, amounting to U.S.$48.4 million. The foreign exchange component of the total investment—amounting to U.S.$102 million—

was acquired from four external sources. The main sources were the IDA, which contributed U.S.$40 million, the African Development Bank (ADB) and the Government of Italy, which each extended a loan of U.S.$25 million, and the Government of Sweden, which had a share of U.S.$12 million.

3.2.3.4 The Seventh Telecommunications Development Program (1992–1997)

In line with the ETA's long-standing practice of guiding its development activities via five-year development programs, the Seventh Telecommunications Development Program (STDP) received the government's approval for implementation in the early 1990s. Originally planned for implementation between 1990 and 1994, the STDP was rescheduled for implementation during the period 1992–93 to 1996–97. Implementation actually began in 1993. Its major objectives were the following:

- To extend basic telecommunications services to as many rural communities as possible.
- To provide telephone connections to as many urban customers as would help narrow the existing demand/supply gap.
- To improve the quality of service by upgrading as many switching offices as was economically viable from manual to automatic operation and maintaining the fault rate at a realistic minimum.
- To improve the international telecommunications service by increasing direct links to additional countries in keeping with Ethiopia's economic and cultural ties.

To accomplish these objectives, the STDP outlined a number of strategies that included the following steps:

- Expanding existing facilities and installing new ones.
- Reviewing the organizational structure with a view to adapting it to changing needs.
- Improving the quality of service through proper maintenance of plants and proper circuit dimensioning.
- Establishing a centralized maintenance workshop to effectively undertake the maintenance of the new generation of equipment and plants.
- Expanding and upgrading the Training Institute so that more extensive and higher level training could be offered.

Under the STDP the total number of telephone stations in Ethiopia, including public call offices and manual and automatic exchanges, was projected to increase from 522 in 1992 to 672 in 1997—an increase of 150 new stations (see Table 3.2). The number of stations with automatic exchanges alone was projected to rise from 35 in June 1992 to 58 in 1997. Total exchange capacity was to grow at an average annual rate of 15 percent to reach 340,070 lines in 1997. Most of the projected increase (158,592 lines) would be from automatic exchanges.

STDP: International Telephone Service. Using trend analysis, the STDP also assumed that the number of telephone subscribers in Ethiopia would grow at the rate of 12 percent per year. However, because the supply of telecommunications

Table 3.2. Major Targets of the Seventh Telecommunications Development Program

Item	1992–93	1996–97	Growth Rate (%)
Telephone subscription	140,959	248,512	12
Automatic	122,628	233,080	14
Manual	18,331	15,432	–3
Telephone	175,168	309,018	12
Exchange capacity	165,598	340,070	15
Automatic	141,108	299,700	16
Manual	24,490	40,370	10
Telephone stations	522	672	5
Automatic	35	58	11
Coin boxes	1,130	2,273	15
Telex subscription	1,006	1,605	10

Source: ETA, *Seventh Telecommunications Development Program* (1992–93 to 1996–97), vol. 2.

facilities has historically been highly constrained by a number of factors, notably the shortage of foreign exchange, projecting growth rates on the basis of historical data is far from an accurate way of reflecting the dynamism of the market. The size of Ethiopia's telephone subscriber waiting list is extremely high (standing in the mid-1990s at 92 percent of total connected lines and growing at 14 percent per year), and at the rate subscriptions were forecast to increase, about seven years would be required just to clear the backlog. It would be a commendable achievement indeed if the STDP succeeded in maintaining the current gap between expressed demand and main line connections.

In the mid-1990s, two standard A earth stations working with the Intelsat system—a semi- and fully automatic gateway digital telephone exchange and a fully automatic electronic telex exchange—provided international telephone and telex services in Ethiopia. Both satellite and microwave circuits were used for international links, the latter mode being limited to links with neighboring countries. In 1991, Ethiopia had direct satellite links with seventeen countries. Of these, only one (Ivory Coast) was in Africa, seven were European Economic Community member states, three were Middle Eastern countries, and six represented the rest of the world. Over 75 percent of Ethiopia's satellite circuits in 1991 were with France, Italy, Saudi Arabia, the United Kingdom, and the United States.

During the STDP, most of the investment allocated for international service was expected to go toward increasing the number of circuits on existing links. The number of direct telephone circuits was expected to increase from 249 in 1992 to 349 by the end of the program. Direct telecommunications links with Sudan and Somalia, based on microwave radio relay systems, were also planned. In addition, Ethiopia accounted for two missing links on the Panaftel network that Ethiopia and its two neighbors were expected to complete as part of the Panaftel Project.

STDP: Telex Subscription. With the expansion of microwave and UHF links, additional Ethiopian towns were expected to be provided with telex service under the STDP. However, most of the growth in telex connections would result from the addition of subscribers in cities and towns already having telex service. The STDP envisaged a 10 percent annual growth in telex subscriptions.

Ethiopia's international telex service was also expected to show significant growth as a result of the diversification of international links under the STDP. Under the plan, the satellite link that carried the largest volume of international telephone and telex traffic was to be increasingly augmented by microwave and submarine cable links. Since 1991, Ethiopia has acquired circuits on the submarine cable system that links Southeast Asia, the Middle East, and Western Europe (popularly known as SEA-ME-WE). The average annual rate of growth of international telex traffic during the STDP was forecast at 9 percent.

The STDP also included other services such as facsimile and data communication, which although already introduced as public services had not been well developed through the mid-1990s. In addition, mobile telephony, which is not yet part of the services provided by the ETA, was being considered for introduction in the near future. Facsimile service, which is officially provided by the ETA and unofficially by a few private operators, was expected to be upgraded during the STDP from the current speed standard of one A4-size page every three minutes to one A4-size page per minute. Data communication, which is currently limited to low- and medium-speed transmission on point-to-point leased circuits, was expected to be expanded to include switched data services under the STDP. In the early 1990s, Ethiopia's telex exchange was improved to accommodate data services at 4,800 or 9,600 bits per second.

Financing the Seventh Telecommunications Development Program. In the early 1990s, the ETA estimated that implementation of the STDP would require a total investment of about U.S.$250 million, of which U.S.$170 million would be in foreign exchange. According to the financing plan laid out in the program document, the foreign exchange component of the investment was expected to come from external lending agencies. Since preceding development programs had been financed through external loans, the ETA was optimistic about the possibility of employing a similar mode of financing for the STDP.

The remaining U.S.$80 million, which will be required in local currency, was expected to be drawn from internal resources—for the most part from the ETA's own funds. As laid out in the program's financing plan, the ETA is capable of generating adequate funds from its net earnings and depreciation funds to cover the capital expenditures required in local currency. Components of capital expenditure in local currency include local purchases, civil works, labor, and other expenses that do not call for importation of goods and services.

The STDP was expected to attract adequate funds from external sources. However, despite the ETA's efforts to secure loans with favorable terms and conditions, in 1994 it managed to obtain only U.S.$64 million from the African Development Bank. There were indications, however, that the other financing institutions involved in past programs, such as IDA, might also once again assist in financing the STDP.

3.3 Current Status of Telecommunications in Ethiopia

3.3.1 Services

Major telecommunications services offered by the ETA include telephone, telegraph, telex, and facsimile. Ethiopia's telephone service is by far the most important telecommunications sector service in terms of revenue generation and coverage (see Table 3.3). In 1991, local, long distance, and international telephone services had a combined share of 90 percent of the country's total telecommunications gross revenue. Telegraph and telex together accounted for a mere 7 percent of the 1991 revenue. Other services including facsimile had a negligible share of revenue in the early 1990s.

Ethiopia's local, long distance, and international telephone traffic grew at an average annual rate of 10 percent, 9 percent, and 25 percent, respectively, after 1980 (see Table 3.4). Local and long distance telephone services, although more mature than the international services, have more potential for growth, as the huge unsatisfied demand for these services indicates. With the installation of the first satellite earth station in 1979, the importance of the international service grew. Today, it is the fastest growing service, with two earth stations, microwave links, and a submarine cable used for international services.

In the 1990s, the telegraph service, which in the remote past was the only mode of message communication, was no longer a growing industry. While inland telegraph traffic increased at an average annual rate of 10 percent between 1980 and 1987 (followed by a downward slide of 4.6 percent per year between 1987 and 1991), outgoing international telegraph traffic decreased, falling at an average annual rate of 13.2 percent between 1980 and 1991. Ethiopia's telegraph service is expected to become even less significant as the availability of improved telephone, telex, and facsimile services grows.

Telex traffic used to be one of Ethiopia's fastest growing services until internal conflict along the main trunk route linking the capital and the second largest city, Asmara, disrupted services, including all economic and social activities. Between 1980 and 1986 inland and international (outgoing) telex traffic grew at an average annual rate of 18.9 percent and 11.3 percent, respectively. Although both services experienced cyclical patterns in the following years, 1991 was the worst year, with inland and international telex traffic diving by 29.5 percent and 27.5 percent, respectively, from the previous year, mainly as a result of the civil war that engulfed half the country. In May 1991, after years of bloody battles, the former government was forced out and a new interim government formed.

Facsimile service is in its infancy in the Ethiopian telecommunications network. Introduced as a public service in 1988, it had attracted over 700 subscribers by 1993, and this figure was increasing at a rapid rate. Although it is too early for a clear growth pattern to have emerged, facsimile service is expected to be an area of rapid growth.

Another recent addition to services provided by the ETA is data communication. With a handful of subscribers served with leased circuits and generating limited data

Table 3.3. Revenue and Expenses by Types of Services (in millions of Birr)

Item	1985	1987	1989	1990	1991	1992	1993	Annual Growth Rate (%)
Revenue	105	119	127	152	187	192	215	10
Urban telephone	44	52	53	62	71	72	83	8.3
Interurban and international telephone	43	48	52	70	97	10	155	14.5
Telegraph and telex	14	15	19	17	14	10	12	0
Others	3.5	3.5	2.9	2.6	5.2	3	3.7	6.8
Expenses*	84	97	111	130	146	148	204	9.7
Urban telephone	15.4	19.4	24	30.3	30.5	28	34.6	12.1
Interurban and international telephone	11.5	12	15.4	16.9	16.6	16.7	19.1	6.3
Telegraph and telex	5.5	5.6	5.4	6.2	6.8	6.4	6.9	3.6
Others	51	60	67	77	92	97	144	10.3
Net profit	21	22	16	22	41	44	50	12

Sources: ETA, *Statistical Bulletins* (1989–93); ETA, *40th Annual Report* (1991–92).

*Includes profit tax.

The 1992 and 1993 figures do not include those of Eritrea and hence are not included in the growth rate calculations.

Table 3.4. Telephone, Telegraph, and Telex Traffic

	Average Annual Traffic							
Item	1953	1980	1985	1989	1990	1991	1992*	1993*
Telephone traffic								
Local (mn)	10	194	295	312	362	396	424	47,110
Long distance (TH)	190	3,428	4,318	4,644	4,364	4,427	3,000	37,009
International (mn min.)	—	0.8	2.1	4.7	4.3	8.9	10.1	10.925
Telegraph traffic								
Inland (TH)	125	158	252	269	291	257	140	1,502
International (outgoing mn)	77	47	17	16	14	13	7	5–13.2
Telex traffic								
Inland (mn)	—	329	730	1,097	1,114	860	340	3,409
International (TH min.)	563	1,057	1,121	1,020	800	670	570	3

Sources: ETA, ***VIIth Development Program***, vol. 2; ETA, ***Annual Statistical Bulletin*** (1992–93).

Key: mn = million calls; min = minutes; TH = thousand.

* Figures do not include those of Eriterea; as a result, some of the figures show sharp falls.

traffic, public data networks are just beginning to be important. Since its introduction in 1987, only low- and medium-speed data transmission service based on point-to-point leased circuits has been provided in Ethiopia. Switched data network service is expected to be provided in the near future as the demand for connections to such service has increased rapidly—20 percent per year in the mid-1990s.

3.3.2 Network

The Ethiopian telecommunications network consists of an integrated system of cables, manual and automatic exchanges, VHF/UHF and microwave radio relay systems, satellite earth stations, and customer premises equipment.

3.3.2.1 Telephone Density

The total number of DELs in Ethiopia in 1993 was 132,000, resulting in a density of 0.25 DELs per 100 inhabitants (see Table 3.5). In the same year the total number of telephone apparatuses was 160,000, equivalent to a density of 0.31 telephones per 100 inhabitants—a very low penetration rate even by African standards. In 1987, the average density of DELs for Africa was 0.76 per 100 inhabitants.

Of Ethiopia's total 1993 DELs, 89,752, or 68 percent, were in Addis Ababa, and 3,785, or 2.9 percent, were located in Diredawa, the second largest city. These two cities together accounted for only 4.5 percent of the country's population and 36 percent of the urban population.

In 1993, there were only 475 public telephone stations in Ethiopia—or one telephone station for every 2,341 square kilometers of land surface and over 90,000 rural inhabitants. Because of the nonuniform distribution of these stations, however, there are areas of Ethiopia where one telephone station must cover an area greater than 7,800 square kilometers.

The growth of exchange capacity in Ethiopia is far below the growth of demand for telephone services. In 1993, there were 141,000 registered waiting subscribers—the equivalent of 107 percent of the total connected lines. At the planned connection rate of about 21,000 DELs per year, it would require about seven years just to clear the backlog of registered telephone demand.

3.3.2.2 Switching Capacity

Exchanges used in Ethiopia's telecommunications network can be categorized into three types: manual, electromechanical, and electronic (digital). In 1993, of the country's total exchange capacity of 169,000 lines, 16 percent were manual, 46 percent were electromechanical, and 38 percent were digital. The first digital exchanges were installed during the Sixth Telecommunications Development Program.

3.3.2.3 Transmission Facilities

The Ethiopian telecommunications network uses a combination of traditional and modern transmission media. Most of the 506 cities and towns with telephone services are interconnected with open wire lines, as are all links carrying light traffic between small- to medium-sized towns. The backbone of Ethiopia's long

Table 3.5. Telecommunications Trends: Selected Indicators of Growth

Item	1953	1980	1985	1989	1990	1991	1992	1993	Rate (%)
Telephone subscription*	4	64	96	116	125	133	127	132	10
Apparatus*	5	86	119	146	156	164	154	160	10
Exchange capacity*	—	80	124	162	171	176	164	169	7
Telex subscribers	—	416	670	920	971	1,003	880	912	8
Tel. stations	65	398	476	494	506	512	466	475	6
Waiting subscribers†	—	—	—	91	109	123	22	141	16
Fax subscribers	—	—	—	111	233	247	506	745	49
Staff/1,000 DEL	—	71	56	50	48	45	43	40	—
Fault/DEL	—	—	—	1.5	1.4	1.3	1.3	1.3	—
DEL/100 population	—	0.17	0.22	0.23	0.25	0.26	0.25	0.25	—

Source: ETA, *Annual Statistical-Bulletins* (1990–93).

*In thousands.

†The 1992 and 1993 figures do not include those of Eritrea, hence the sharp drop in 1992.

distance transmission system is the 960-channel microwave radio relay system. Secondary routes, far from open wire lines, are served by VHF and UHF systems.

3.3.2.4 Quality of Service

Faulty equipment, inadequate capacity (resulting in congestion in automatic exchanges and canceling of booked calls in manually switched public offices), and operational inefficiency affect the quality of Ethiopia's telecommunications service. With respect to other African countries, the ETA's network had 1.53 faults per DEL in 1986 according to the ITU, which was lower than the fault rate in nine other African countries. However, this was not necessarily an acceptable fault rate even for a developing country: in 1986, more than ten sub-Saharan African countries had fault rates per DEL of 1.0 or less. In 1993, the ETA reported a fault rate of 1.3 per DEL.

3.3.2.5 Financial Performance

The 1980s was a period of financial difficulty for many state-owned enterprises (SOEs) in Ethiopia. Some SOEs in the manufacturing sector, for example, reported losses for consecutive years, and others had rates of return below the prevailing interest rate.

The ETA is one of the few SOEs that has registered a reasonable rate of return for consecutive years since 1980. Table 3.6 shows the ETA's financial position since 1975. Its gross revenue grew from Birr 35 million in 1975 to Birr 187 million in 1991, an average growth rate of 11 percent per year. The increase in revenue derived mainly from telephone services, which accounted for 90 percent of gross revenue in 1991. Expenses (excluding interest) increased from Birr 27 million to Birr 130 million in the same period and grew at an average annual rate of 10 percent.

The ETA's average rate of return between 1975 and 1980 stayed at around 13 percent, a couple of percentage points higher than the prevailing interest rate. With the introduction of a new tariff in 1980, the rate of return started to improve, maintaining since 1983 an average annual rate of 20 percent. Table 3.6 also shows two financial ratios that illustrate the ETA's ability to meet its currently maturing debts and to potentially borrow additional funds on a long-term basis.

The liquidity ratio, which had generally stayed over 2 since the mid-1970s, rose sharply beginning in 1985 as a result of the buildup in inventories of investment items at the beginning of the Sixth Telecommunications Development Program.

Since 1987, the ETA's asset coverage ratio has declined rapidly after reaching highs of 4.2 and 3.4 in 1985 and 1986, respectively. This decline was not an indication of a long-term financial difficulty, however, but a natural by-product of Ethiopia's development programs: the rise and fall of the asset coverage ratio following the sharp rise of one of the components of the ETA's assets and the phasing out of a long-term debt are cyclical patterns observed around the beginning and end of the development programs.

The ETA has steadily improved its bill collection performance. Collection of sales as a percentage of the collectible amount rose from 72 in 1985 to 80 in 1988. The rather high outstanding balance was mainly caused by bills not settled by government organizations who complained of budgetary constraints.

The ETA is legally expected to contribute to government revenue in a number

Table 3.6. ETA's Financial Performance Indicators (in millions of Birr)

Description	1975	1980	1985	1989	1990	1991	1992	1993	Annual Growth Rate (%)
Revenue	35	48	105	127	152	187	192	215	11
Expense* (without interest)	27	37	75	95	91	130	132	150	10
Net profit	8	11	30	32	61	57	60	65	13
Avg. net fixed asset	72	81	133	166	204	229	222	225	7
ROR on asset (3/4 × 100)	11	13	23	19	30	25	27	29	—
Current asset	39	72	137	174	219	211	200	190	11
Current liability	12	41	56	54	56	66	217	236	11
Current ratio (5/6)	3.2	1.8	2.4	3.2	3.9	3.2	0.92	0	
Total asset	122	213	282	462	484	567	586	477	10
Long-term debt	29	62	43	186	186	193	161	133	13
Net-asset [7 – (6 + 8)]	81	110	183	222	242	163	190	239	8
Asset coverage ratio (9/8)	2.8	1.8	4.2	1.2	1.3	1.4	1.2	1.8	—

Sources: ETA, *Annual Reports* (1990 and 1993) and *VIIth Development Program.*

*Expense includes profit tax.

The 1992 and 1993 revenue and expense figures do not include those of Eritrea.

of ways. These include customs duty and municipal tax on imported goods, income tax, sales tax, capital charge, and residual surplus. Of these, the ETA has dutifully settled only customs duties, income tax, and sales tax. Customs duties amounted to 24 percent until 1992 when it was lowered to 17 percent and income tax to 50 percent of gross profit. Because of the controversy surrounding the government decree (Proclamation No. 163 of 1979 on Public Financial Operations) on payment of capital charge and residual surplus, the ETA's customs and income tax obligations have not yet been fully met. According to this proclamation, the ETA (as well as all public enterprises and financial agencies) is expected to pay the government an annual capital charge amounting to 5 percent of the state capital plus the general reserve fund. According to the Proclamation, residual surplus, which amounted to 90 percent of the ETA's profit after income tax, was also payable to the government annually.

Since 1992, a new public enterprises proclamation has been in place. Among other things, this proclamation repealed the article on the payment of capital charge and replaced the one regarding payment of residual surplus with payment of a state dividend, the amount of which will be decided by the owner (the Government).

While the tug of war was going on between the government trying to enforce the proclamation and the various public agencies trying to resist payment (of residual surplus, in particular), the ETA consistently drew on these funds for its investment requirements, based on government-approved plans. All in all, by the mid-1990s, the ETA had paid to the Ethiopian government Birr 426 million in the form of income tax and capital charge since the mid-1970s—nearly Birr 19 million annually.

3.3.2.6 Tariff

A major tariff revision for telecommunications services in Ethiopia was made in 1980. This revision brought a new tariff into effect with the following specific changes:

- Urban call charges were raised by 50 percent.
- Interurban call charges were raised by 15 to 20 percent (20 percent being applied to the lower side of the range).
- Subscription and rental rates remained unchanged.
- International calls were to be based on international and bilateral agreements.

In 1991, when the government introduced a new sales tax policy, the ETA raised all its charges by 12 percent. The extra revenue from this new tariff, however, actually went to the government. In 1994 (fourteen years since the last tariff revision), the government approved a new tariff structure. The tariff revision focused on telephone subscription installation charges and rentals, domestic telephone and telex calls, and international telephone and telex calls.

The 1994 tariff revision raised charges as follows:

- Urban call charges increased by 43 percent.
- International call charges increased by 75 percent.
- Subscription charges increased by 165 percent.

Table 3.7. Training by Types of Studies

Training Category	1954	1970	1980	1985	1988	1989	1990	1991
Technician	163	82	95	65	283	325	256	222
Traffic	78	113	102	232	430	208	272	8
Administration	—	59	16	43	291	447	409	615
Total	241	254	254	440	1,004	980	937	845

Source: ETA, *Statistical Bulletins.*

Operator-assisted interurban call charges, however, remained unchanged.

3.3.2.7 Staff Training

At the time the Imperial Board of Telecommunications of Ethiopia (IBTE) was established it had 642 employees, of whom 96 were expatriates, including the general manager and the high-level trained manpower engaged in administrative, financial, and technical activities. By 1974, however, after years of effort by the ETA and the government, all of the 5,620 employees on the IBTE's payroll were Ethiopian nationals.

In addition to the higher institutions of learning Ethiopia has developed over the years, the Telecommunication Training Institute has played an important role in the development of the country's telecommunications human resources. The Training Institute, whose activities are broadly divided into preservice and in-service training programs, trained 5,620 staff persons between 1954 and 1991, of which 3,262 were in the technical, 1,930 in the traffic, and 528 in the administrative and financial areas (see Table 3.7). Foreign nationals sponsored by their respective employers also attend the regular technician courses offered by the Training Institute.

At the national level, Ethiopia's technological backwardness is partly manifested in a general shortage of trained manpower. The telecommunications subsector may be one of the exceptions, however, in that it has developed adequate institutional capacity to produce the trained manpower required to install, operate, and maintain the system.

3.4 The Future of Telecommunications in Ethiopia

3.4.1 The Policy Environment: Nationalization and Centralization

Although telecommunications in Ethiopia has been state owned since its inception, state ownership of important manufacturing industries, banks, insurance companies, and many other firms became a priority item following the military's seizure of state power in 1974. Between January 1 and February 3, 1975, alone, all Ethiopian banks, thirteen insurance companies, and seventy-two of the country's largest industrial enterprises were nationalized by government decrees. Since then, only the government has been involved in major investments in the agricultural, industrial, and service sectors. The private sector dominated only small-scale peasant agriculture, small-scale and cottage industries, retail trade, and road transport operations.

State control of economic activities intensified during the following decade and culminated in the launching of a Ten-Year Perspective Plan (TYPP) for the period 1984–85 to 1993–94. The TYPP, which had "Expanding and Strengthening Socialist Production Relations" as one of its major objectives, was to serve as the leading economic policy document in the years that followed. According to this document, all major investments in all sectors were to be the responsibility of the state, and an increasing share of the country's wholesale and retail trade was to be handled by the state trading organizations.

The total investment envisaged in the TYPP was about U.S.$15 billion, of which agriculture's share was 23 percent. The share of the manufacturing industries and transport and communications was 14 percent each, and telecommunications claimed its fair share of 1.2 percent of total planned investment. The lion's share of the TYPP investment went to the state sector, which accounted for over 90 percent of the total investment.

The TYPP, among other things, highlighted the government's intention to lay the foundation for Ethiopia's electrical and electronics industry. Among the 216 industrial projects planned for implementation during the plan period, five were related to the manufacture of electrical and electronic goods. Radio and television and electric motor and electric bulb factories, among others, were to be established during the plan period.

The other important feature of the TYPP was its strong advocacy for the building of a science and technology capability in Ethiopia. About U.S.$0.6 billion—3.7 percent of total national investment—was allocated in the TYPP for the development of this capability. Policy and institutional measures were also laid down toward the realization of this objective.

In hindsight, it is now clear that the TYPP was too optimistic. During the first seven years of the TYPP's existence (1984–90), only U.S.$4.5 billion worth of investment out of a total projected gross investment of U.S.$10.5 billion was actually carried out—an implementation rate of only 43 percent. Almost all projects suffered shortages of foreign exchange and other capital inputs.

In those first seven years, the TYPP experienced very weak economic performance due essentially to recurrent drought, war, and inappropriate policy. However, although the poor economic performance during this period was reflected in all sectors of the economy, including telecommunications, the investment performance of the telecommunications sector was far better than the rest of the economy, with 82 percent of the projected investment of the TYPP actually realized.

3.4.2 Economic Reform

The first major economic reform after the TYPP was the new Economic Reform Program announced by the then ruling party in March 1991. The major elements of this economic reform were the following:

- The promotion of a mixed economy.
- The creation of an appropriate atmosphere for the market mechanism to guide economic decisions.
- Active encouragement of the private sector through a series of new incentives.
- Reorientation of public-sector management toward competitiveness and profitability.
- Formation of cooperatives on a strictly voluntary basis.
- Greater decentralization of economic decision making.

The most distinct feature of the new economic policy was its intention to move to a market-oriented economy, expanding the role of the private sector and streamlining state-owned enterprises to make them profitable. The implications of these objectives on agriculture, industry, and the service sector were far-reaching. For example, large-scale private commercial farming, which had been totally nonexistent prior to the new economic policy, was expected to attract a large number of investors. Typical incentives offered to investors included exemptions from customs duties and income tax for up to five years.

The new investment code opened up a large number of activities for private entrepreneurs. With the exception of the defense industry, postal and telecommunications services, air, rail and large-scale shipping transport, as well as radio and television broadcasting services, all other sectors were, in principle, now open for private investment.[1]

Before the Economic Reform Program could be fully implemented, however, the government was overthrown by the Ethiopian Peoples Revolutionary Democratic Front (EPRDF) in May 1991, which shortly afterward formed the Transitional Government of Ethiopia (TGE). The new economic policy announced by the TGE in December 1991 represented Ethiopia's move toward institutional and policy reform at the national level.

The new economic policy comprises the following general provisions:

1. The state sector will be limited to economic activities that are instrumental in the overall economic and social development of the country and to those areas that, for various reasons, do not attract private capital. More specifically, the state develops the country's economic and social infrastructure, human resources, and research institutions; safeguards the well-being of society through price stabilization mechanisms; and creates an enabling environment for the people, in general, and the private sector, in particular, for their wider participation in development.
2. The private sector will be encouraged to engage in diverse economic activities without limitations on the amount of capital to be employed.
3. All laws and policies, including investment, tax, and labor laws, as well as monetary, credit, and interest policies, will be revised to the extent that they facilitate the implementation of the new economic policy.

In addition to these general provisions, the new policy stated the following with respect to communications: "Since telecommunications and postal service are essential public services, these services will remain under state ownership. However, the possibility of private sector participation will be explored and appropriate policies and regulations will be issued to that effect." According to the new economic policy, the role of Ethiopia's private sector in the telecommunications sector was to be defined later. But it can be surmised that Ethiopia's basic telecommunications services are unlikely to be open in the future for private-sector competition.

3.4.3 Liberalization of Telecommunications Services

Since the beginning of 1991, the ETA has relaxed some aspects of its long-held monopoly position. For the first time in its history, the ETA issued a policy that allows subscribers to operate their own facsimile equipment on the ETA's network. This policy further stipulated that the private sector could import facsimile machines that meet the ETA's specifications and distribute them to subscribers.

This move was the beginning of a new liberal policy toward private-sector participation in the provision of Ethiopia's telecommunications services. Although the facsimile terminal equipment was the only telecommunications facility open to private ownership and distribution in the mid-1990s, telephone equipment, teleprinters, data modems, PABXs, and the like were also possible candidates for the liberalization policy.[2]

In 1990, the ETA was encouraged by the government's announcement of a new economic reform program that provided for increased participation by the private sector in many areas of the economy. Historically, the growth of Ethiopia's telecommunications services has been seriously hampered by the shortage of facilities, from customer terminal equipment to exchange capacity. The emerging trend of encouraging private participation in the provision of terminal facilities will therefore undoubtedly ease some of the burden carried by the ETA.

In order to increase the significance of the liberalization policy, the ETA will first have to allow wider private participation to cover at least the telephone apparatus and teleprinters. Second, private participation will have to be extended to cover the provision of certain telecommunications services on value-added networks (VANs). The areas of data communication and information may also be included in the list of services for private-sector participation.

Internet services, for example, is one area in which the government is considering opening to private participation in an effort to advance Internet connectivity. As one step in this direction, Ethiopia hosted the first meeting of the African Technical Advisory Committee (ATAC) of the African Information Society Initiative (AISI) in 1997. The committee comprised experts whose mandate is to advise the implementation of new policies and new initiatives such as Internet development.

3.5 Conclusion

Telecommunications in Ethiopia is almost as old as the technology of electrical communication itself. Ethiopia adopted telephone technology fairly quickly, and in the early days of its telecommunications development it was not very far behind the rest of the world. The first long distance telephone line in the world was installed between Boston and New York in 1885; it was only nine years later that Ethiopia's long distance telephone line between Addis Ababa and Harar (spanning almost 480 kilometers) became operational.

The development of Ethiopia's telecommunications should be assessed primarily in relation to the overall economic and social development of the country. The political environment, national development policies, and socioeconomic order have influenced the growth of the telecommunications sector. Despite the resource and policy constraints that the telecommunications sector has endured in the past, there has been progress. Ethiopia has a relatively efficient telecommunications services sector, affordable to several thousand citizens, and a telecommunications administration run by an adequately trained, all-Ethiopian staff.

Still, the telecommunications sector is characterized by very low telephone accessibility and penetration rates—even by African standards. The number of registered waiting subscribers reached 87 percent of the total actual number of connected subscribers in 1990, and in early 1995, the figure was still climbing. While the rural service suffers from a shortage of physical facilities, which is characteristic of the network as a whole, the low financial returns in rural areas is an additional obstacle to extending service throughout the country.

From the analysis of this chapter, it is clear that the most critical issue facing Ethiopia's telecommunications today is the huge gap between the demand for the most basic telecommunications service and the existing capacity. Other problems of increasing importance include quality of service and the demand for new services. The provision of efficient, adequate, and reliable telecommunications services requires the availability of adequate resources—financial, human, and material. But the most serious resource shortage is finance, particularly foreign exchange. This constraint stems from Ethiopia's inability to expand its export base and, in reference to telecommunications, from the absence of a domestic capability for manufacturing telecommunications equipment.

Overcoming the constraints described in this chapter is a prerequisite for modernizing Ethiopia's telecommunications network and extending services to the rural areas. This will be a protracted process, calling for a short-term measure for removing immediate hurdles and a long-term plan that takes into account the need for a sustainable development of local telecommunications manufacturing capabilities.

3.5.1 Future Demand for Telecommunications Services

Even before the Ethiopian government's economic reform of the early 1990s, the growth in demand for both traditional and new telecommunications services was extraordinary. But the increasing number of waiting subscribers is only one indicator of the magnitude of the demand for that basic communications facility—the telephone. The opening up of Ethiopia's economy through increased private-sector participation in all areas, and in the service sector in particular, will, undoubtedly, bring about more dramatic changes in the magnitude and structure of demand for telecommunications services.

In addition to basic telephone service, significant increases in the demand for data communication and information services, as well as for telecommunications services for entertainment (such as cable TV), are expected in the short to medium term. Moreover, the ETA has been approached by some customers requesting new services such as packet-switched data communication, faster facsimile, and mobile telephone. Demands from the public are not limited to the provision of new services by the ETA itself, however, but also include requests for relaxation of the ETA's monopoly right in such areas as importation of customer premises equipment.

In summary, the vitality of Ethiopia's telecommunications sector will depend on the effectiveness and clarity of the government's economic policy in general and its telecommunications policy in particular, as well as on the responses of the private sector and all other parties concerned with the development of the country's telecommunications.

3.5.2 Future Telecommunications Policy

3.5.2.1 The Role of the Private Sector in Services and Manufacturing

In 1996, the government's policy maintained that telecommunications is a public utility that should remain under state ownership, with the provision that some aspects of the service may be opened to private participation in the future. Wholesale privatization or some similar measure may not serve the country's long-term interests, however. On the one hand, there is the need for an adequate, reliable, and modern telecommunications network and, on the other, the ever-nagging aspiration to develop the capability to manufacture telecommunications equipment locally. The policy that should evolve must ensure, for the short term, the influx of private capital needed to compensate for the shortfall in public investment for the improvement and expansion of telecommunications services and, for the long term, the development of a telecommunications equipment manufacturing capability.

The role of the private sector in the provision of Ethiopia's telecommunications services must be clearly defined. Private-sector participation may be limited in the beginning to the terminal equipment market and value-added networks and then gradually move to areas currently known as basic telecommunications services. Such a move might be seen as an infringement of the ETA's long-cherished monopoly right and as a loss of an important revenue source for the government. But at the same time it should come as a relief to the ETA to be spared the massive burden of being the sole satisfier of customer demand—a responsibility the ETA alone can never hope to fulfill in the near future. With respect to government revenue, the potential tax revenue from the private sector should be more than able to offset the loss of revenue that will result when the private sector takes over some of the ETA's line of activity.

The new government economic policy of the early 1990s provided for wider private-sector participation in the manufacturing sector. Only heavy engineering and metal industries, as well as plants producing basic drugs, fertilizers, and chemicals, are singled out for state ownership. From this, it appears that electronics industries in general and telecommunications equipment manufacturing in particular are now open to the private sector. Even so, the new economic policy is only a broad guideline that should be translated into concrete laws and regulations to be meaningful for the private sector.

To develop Ethiopia's electronics industry, which should start almost from scratch, a policy of active promotion should be employed. Modern telecommunications equipment manufacturing is knowledge-intensive, and as such it requires the close collaboration of foreign investors who can bring in both technology and capital. This will require an appropriate incentive mechanism for attracting foreign investors.

3.5.2.1 Restructuring the ETA

Regardless of whether the private sector's entry into the telecommunications sector is instituted or not, the ETA will remain the most important carrier in Ethiopia's telecommunications industry. Serious thought should therefore be given to the periodical review of the organizational structure of the ETA with a view to meeting the challenges of the future.

If the private sector is to have an increasingly significant role both as a provider of telecommunications services and as a supplier of equipment, the role of the ETA must be redefined. The ETA's current regulatory function combined with its role as a provider of telecommunications services can only be acceptable in a monopoly situation. Given the government's intended liberalization policy, a new structure that separates regulatory powers from commercial roles will therefore be necessary.

Once the roles of the ETA vis-à-vis the telecommunications industry as a whole are defined, the other crucial aspect to be considered is management of the ETA itself. During the first two decades of its existence, the ETA enjoyed a certain degree of autonomy: its board of directors determined its annual capital expenditures, appointed the general manager, and so forth. However, following the nationalization drive of the previous government, the management autonomy of the ETA, along with all other SOEs, had been drastically curtailed. It took a new enterprise law issued in 1992 to pave the way for hundreds of SOEs to be run on commercial principles.

Future restructuring of the ETA would restore the level of management autonomy that prevailed in its early years. Finally, the tariff issue is another important element related to the question of the ETA's autonomy: setting tariffs may remain the responsibility of the government, but an efficient tariff revision mechanism should be established nonetheless.

Notes

1. However, the code stated, "Investments in the provision of electric light and power, processing of tobacco, banking and insurance, and the supply of potable water activities shall require the prior authorization of the Council of Ministers."

2. In 1995, further liberalization of the CPE (customer premises equipment) market was decided upon.

Bibliography

Alemayehu, Kibret. 1985. *Price Policy and Revenue Effectiveness of Tariff Rates in the Ethiopian Telecommunication Authority.* Addis Ababa.

Asfaw, Fikru. 1990. Management of Telecommunications in Ethiopia. Presented at the African Telecommunication Development Conference, Harare, Zimbabwe.

Eshete, Aleme. 1979. *The Construction of Italo-Ethiopia Telephone and Telegraph Line 1898–1974.* Addis Ababa.

International Telecommunication Union. 1990. African Telecommunication Statistics. Presented at the African Telecommunication Development Conference, Harare, Zimbabwe.

Ethiopian Science and Technology Commission. 1988. Working papers presented to the First National Conference on Science and Technology Policy of Ethiopia. Addis Ababa, June 20–25.

______. 1990. *Policy Consideration on Licensing of Private Customer-Premises-Equipment for Accessing ETA's Networks.* Rev. 2, March.

Ethiopian Telecommunication Authority. 1982, 1989. *Telecommunications Development Programs.*

Lee, William. 1973. *From Bush Telegraphy to Microwave System: Telecommunication in Ethiopia 1897–1973.* Addis Ababa.

Lemma, Taye. 1987. *Price of Telecommunication Service in Ethiopia.* Addis Ababa.

Ministry of Post, Telegraph, and Telephone. 1969. *Telephone Service in the Eritrea Province 1953–1963: Asmara II Pilgraphics.*

Niswonger, C. Rollin et al. 1970. *Accounting Principles.* 12th ed. Cincinnati, OH: South-Western Publishing Co.

Office of the National Committee for Central Planning of Ethiopia. 1990. National workshop on Population Policy Development. Addis Ababa, June 12–13.

______. *Ten Year Perspective Plan (1984/85–1993/94).* Addis Ababa.

Pankhurst, Richard. 1975. "Transport and Communication in Ethiopia 1835–1938." *The Journal of Transport History,* vol. 2.

Seyed, Mohammed. 1986. Unpublished notes on Financial Analysis for Project. Addis Ababa.

Transitional Government of Ethiopia. 1990. *Economic Policy of the Transitional Government of Ethiopia.* Addis Ababa: Transitional Government of Ethiopia.

World Bank. 1990. *Ethiopia's Economy in the 1980's and Framework for Accelerated Growth, No 8062-ET,* March.

______. *World Development Report, 1983–1989.*

4

Kenya: Facing the Challenges of an Open Economy

MICHAEL TYLER, JANICE HUGHES, AND HELENA RENFREW

This chapter provides an overview of telecommunications in Kenya and discusses the major policy issues facing this sector in the 1990s in the wider context of the challenges of development strategy as a whole. It also considers aspects of Kenya's telecommunications experience that may be of special interest to other countries pursuing or considering market-oriented strategies for a national economy similar to Kenya's. Finally, it reviews research conducted in Kenya over the last decade on the economic role of telecommunications and the benefits of investment in telecommunications infrastructure.

The challenges faced by the makers of telecommunications policy in Kenya are exceptionally demanding. To meet economic needs, it will be necessary to expand the network, enhance service quality and features, and upgrade operational efficiency and productivity. Kenya has a rapidly expanding economy, but it also has one of the world's highest population growth rates—by the year 2000 its population is expected to reach 38 million. Kenya will also need to invigorate agriculture and enhance the lives of those in its rural areas to stem the tide of migration into the towns. Five million new jobs will be needed in the urban areas if the country is to avoid massive unemployment and social unrest.

Kenya's government has responded to these challenges with a market-oriented economic policy that emphasizes openness to the world economy and export-led growth. This policy necessitates a more universal and reliable telecommunications network than would be needed had Kenya attempted a predominantly inward-looking, centrally directed economic strategy similar to those attempted by some other African countries.

As in other countries that rely to a high degree on exports for both job creation and foreign exchange, economic policy in Kenya must ensure that the export sector is fully competitive in the global marketplace. As this chapter will show, the mere availability of a commodity for export (or of a tourist attraction to draw in visitors) is less and less a sufficient condition for economic success. Quality, productivity, effective marketing and distribution in global markets, superior customer service, and speedy and appropriate responses to changing market conditions are all essential. An efficient and reliable telecommunications infrastructure is essential to achieve these goals.

Moreover, successful export economies need the participation of global corporate leaders to set the pace for quality, technology, productivity, and innovation by implementing global "best practices." Their direct investment, though useful, is not as indispensable as their broader role as innovators, pace setters, and conduits for the transfer of technology and best practices. In Kenya, these global companies directly and indirectly support hundreds of smaller companies and tens of thousands of employees. The operating methods of such global companies require extensive use of both voice and data telecommunications, domestically as well as internationally. Experience shows that global companies will focus their management efforts and their investments where adequate telecommunications (as well as other preconditions for productive, effective operations) permit them to remain globally competitive.

The history of economic development from the 1970s to the 1990s, especially the spectacular success of export-led growth in certain newly industrialized countries in Asia such as South Korea and Thailand, and the equally spectacular failure of many national economies in Africa and Asia under nationalization and central planning, makes the Kenyan case of wide interest and significance for those concerned with development strategy. The economic role of the telecommunications sector in Kenya has been the subject of significant economic and business research. Based on that research and on a series of field interviews, we have drawn several conclusions:

- Expanding the scope and enhancing the quality of the telecommunications services offered to rural and urban businesses yields economic benefits far in excess of the costs incurred.
- Despite major expansion of the public network during the 1980s and early 1990s, there are still unserved or underserved user requirements of major economic significance.
- There are large direct and indirect benefits in foreign exchange earnings to be derived from improving telecommunications services; these benefits are particularly valuable to a country like Kenya with an economy strongly linked to international trade.
- The substantial net in-payments of hard currency accruing to Kenya from telecommunications carriers in other countries through the international settlements process could be used as collateral for the financing of major investments in telecommunications. This approach could help sustain the high rate of telecommunications sector investment that is clearly required—a rate that might otherwise be difficult to sustain because of the financial state of the Kenya Post and Telecommunications Corporation (KP&TC).

This chapter reviews the efforts that have been made in Kenya to understand and meet the telecommunications needs of economic development. It draws conclusions about the challenges that must be overcome if the telecommunications sector is to play its essential role in supporting and enabling continued economic growth—especially the continued growth of exports. It also offers some ideas regarding the future of the telecommunications sector in Kenya.

4.1 Telecommunications in Kenya: Historical Evolution

4.1.1 Development of the Public Telecommunications Network

Kenya's earliest telecommunications connections to the outside world were the submarine cables linking Zanzibar, Mombasa, and Dar es Salaam laid by the Eastern & South African Telegraph Company in 1888. Internally, the construction of a telegraph network began with a 200-mile coastal line linking the port city of Mombasa with Lamu. Extension into the interior of the country began in 1896 in conjunction with the building of the railway system, forming a dual backbone for Kenya's communications infrastructure. The extension of the telegraph line even overtook railway construction, reaching Nairobi in 1898 and Kampala and Entebbe in Uganda in 1900. Telephone service soon followed. In 1908, the public telephone network began service in Nairobi, the capital, and in Mombasa. In Nairobi that year, eighteen telephone subscribers were connected.

The subsequent history of Kenya's network was one of gradual but sustained expansion. By 1980, there were 73,932 direct exchange lines (DELs) in use in the public telephone network; just over 84 percent were connected to automatic switching equipment and 75 percent had direct long distance dialing (STD or subscriber trunk dialing) capability. There were 1,228 telex lines in use and 50 leased data transmission circuits in use. The network of 1980 represented a solid foundation for future expansion even though it had significant shortcomings: 33 percent of long distance call attempts failed due to congestion, and at any given time 15 percent of exchange lines were not in working order (KP&TC Annual Reports; Tyler and Jonscher 1982).

In the 1980s, growth of Kenya's network occurred on a larger scale. The Kenya Post and Telecommunications Corporation undertook three telecommunications development programs: the First Program ran from 1979 to 1983; the Second Program began in 1984 and was completed in 1988; and World Bank funding for the Third Program was negotiated in 1985–86, with disbursements beginning in 1987 and completion achieved in 1992.

The First Program called for the addition of 58,800 exchange lines of capacity, a 60 percent increase over the system capacity available at the end of 1979. It also called for the provision of public telephones in 200 previously unserved locations, urban and rural. External funding was provided by the World Bank and bilateral development assistance programs, notably those of Japan and the Netherlands. Although the ambitious targets were by no means fully met, substantial growth was achieved (e.g., the number of working DELs rose from 69,996 at the end of 1979 to 95,000 at the end of 1983).

The Second Program stressed the expansion of service in Kenya's rural areas, with the emphasis on "District Focus"—installation of new digital switches in nine locations to ensure that all forty-one "District Headquarters" locations in Kenya had automatic telephone service. This goal was achieved in 1988.

The Third Program largely continued the approach established by the first two but included two significant innovations: extensive replacement of small manual exchanges in rural areas with digital switching equipment and the introduction of optical fiber transmission for the links (known as "junctions") connecting nearby exchanges.

As Table 4.1 shows, the three programs succeeded in achieving rapid growth of the network, especially since 1983. The network doubled from just under 96,000 working exchange lines at the end of 1983 to nearly 214,000 in 1993, a compound annual growth rate of almost 8 percent.

The available data do not indicate improvement in service quality. In 1980, congestion of the long distance network was a major problem: 33 percent of all call attempts failed due to congestion (Tyler and Jonscher 1982). Based on interviews with KP&TC management, the overall call completion rate for long distance calls in 1991 was only 48.1 percent, suggesting (though not definitely confirming) that the degree of congestion remained similar to that of 1980. Congestion of the long distance network and other service quality problems continued to be a major concern for many users (see Section 4.4).

Another problem area that became evident during the Third Program concerned the financial management and financial condition of Kenya's public telecommunications organization and its management's relations with international (and certain bilateral) lending and development assistance agencies. The issues involved are discussed in Section 4.5.

4.1.2 Institutional Structure

The historical evolution of the institutional structure of Kenya's telecommunications has been shaped by political developments in East Africa as a whole. During the 1920s and 1930s, the British colonial administrations in Kenya and Uganda, and the British-administered League of Nations administration of Tanganyika, became more and more closely linked. By 1933, the postal and telegraph services of the three countries had been fully amalgamated with a single postmaster general responsible for all three postal and telecommunications services. In varying forms, the joint operation of posts and telecommunications for the three countries continued until 1977, through independence and other major political changes, such as the union of Tanganyika and Zanzibar to form Tanzania.

In the 1960s and early 1970s, it was widely believed that the advantages of a large-scale common infrastructure and economic union would be reconciled with national sovereignty through an East African Community (EAC) broadly analogous to the European Community. By the late 1970s, however, the desire to maintain the East African Posts and Telecommunications Corporation (EAP&TC) as a going concern was in direct contradiction with the political realities of the three participating countries, which had divergent political orientations and development

Table 4.1. Expansion of the Public Telecommunications Network in Kenya

	1978	1980	1983	1985	1987/88[e]	Jan. 31, 1991	1992[f]
DELs in use	65,344	73,932	95,749	118,361	151,964	184,583	207,328
Percentage of DELs in use served by automatic switching	86.2%	84.5%	86.7%	88.5%	91.4%	92%	92%
Public telephones	1,490[a]	—	734[b]	2,189[b]	3,630[b]	5,631[b]	5,631[b]
Telex lines in use	1,017	1,228	1,750	2,188	2,536	2,357	2,031
Data modems[c]	—	50[d]	169	216	307	—	—

Source: All data are from KP&TC Annual Reports, except as otherwise indicated.

[a]Includes rented coin phones (coin boxes) on private premises as well as KP&TC pay phones.

[b]Does *not* include rented coin phones on private premises.

[c]Modems provided by KP&TC and used on leased lines for data transmission.

[d]See Tyler and Jonscher 1982, Appendix B.

[e]KP&TC changed from calendar year accounting and reporting to a financial year ending in June.

[f]1992 data reported by the ITU.

strategies. In 1977, the EAC collasped and a separate Kenya Post & Telecommunications Corporations was established.

Today, little remains of the former collaboration within the framework of the EAP&TC. Tanzania and Uganda have their own independent access to the Intelsat global satellite system, although they still make some use of Kenya's earth stations. The long distance dialing arrangements (in which calls from Kenya to Tanzania and Uganda are routed directly rather than through the international gateway) and certain collaborative training arrangements are the only major parts of the old three-country collaborative facilities that remain.

The Kenya Post and Telecommunications Corporation has proved to be very durable; indeed, its first managing director, Kipng'Eno Arap Ng'eny, was still running the corporation in 1991. It is a wholly government-owned enterprise (referred to in Kenya as a "parastatal") but is intended to be run on a commercial basis. Its board is appointed by the government, and the minister of Transport and Communications exercises broad policy-making powers. In practice, however, management appears to enjoy substantial autonomy.

Kenya's international telecommunications services have a somewhat distinct history. In the colonial era, these, like similar services in other British colonies, were operated by the Cable & Wireless Company. In 1964, control of these services passed to the newly formed East African External Telecommunications Company Limited (EXTELCOMS), jointly owned by the government of Kenya and Cable & Wireless. This company continued as a joint venture until 1974 when KP&TC purchased the 40 percent share owned by Cable & Wireless and renamed the entity KENEXTEL. In 1982, KENEXTEL was merged with KP&TC, which is now responsible for both national and international telecommunications.

Notwithstanding the collapse of the EAC, other forms of international cooperation have thrived. Kenya is an active member of the International Telecommunication Union (ITU). In 1981, Kenya promptly completed its national component of the ITU's Panaftel program, which involved the interconnection of African countries' national networks by means of new or extended microwave transmission routes. Kenya's Panaftel links were financed collaboratively by the World Bank (a loan covered the foreign exchange cost) and KP&TC, which covered the local currency cost. In 1994, KP&TC became a member of the Regional African Satellite Communications System Corporation (RASCOM), which has the goal of launching a dedicated African satellite system. Also, KP&TC is active in regional training programs, and Kenyan telecommunications specialists have played a major role in the international exchange of ideas over telecommunications planning and policy (Okundi, Ogwayo, and Kibombo 1977; Okundi and Evans 1975).

In 1968, Kenya became a member of the Intelsat global satellite communications consortium, with EXTELCOMS (and subsequently KENEXTEL and ultimately KP&TC) responsible for operating earth stations to access Intelsat's satellites. Kenya's first major earth station came into operation at Longonot northwest of Nairobi in 1970. There are now two such stations at Longonot, each accessing Intelsat satellites in the Atlantic Ocean and Indian Ocean, with a third earth station in Nairobi and a fourth in Kericho.

4.1.3 Personnel

An essential component of telecommunications evolution in Kenya was the country's successful effort to "localize" its personnel, which greatly reduced its need for expatriate specialists that traditionally were recruited mainly from what was then the British Post Office (BPO).[1] In 1961, of the total EAP&TC staff numbering just over 7,000, 7.6 percent were European, 21.9 percent were Asian, and 70.5 percent were African. Training programs made it possible to reduce the recruitment of expatriates to the small number needed to fill posts requiring very specialized qualifications. By 1975, just before EAP&TC split and KP&TC was formed, less than 1 percent of EAP&TC's employees were expatriates.

4.2 The Economic Context and Its Implications

The Kenya Post and Telecommunications Corporation operates in a rapidly changing economic environment. In the early 1990s, Kenya was experiencing rapid export-led growth and urbanization and a correspondingly high growth in demand for infrastructural services such as telecommunications:

- Real GDP has been growing at nearly 5 percent annually in recent years, considerably faster than the recent growth of the global economy, and faster than the advanced Organization for Economic Cooperation and Development (OECD) countries as a group.
- Kenya has one of the highest population growth rates in the world, averaging 3.5 percent a year from 1985 to 1992. Of the total population of 28 million in 1994, 20 percent live in the urban areas and just over 50 percent are under fourteen years of age. Consumer tastes and spending patterns are changing rapidly.
- Kenya's volume of exports grew at a remarkably high rate in the late 1980s and early 1990s (export volume increased 22 percent between 1989 and 1990 alone), with companies in value-added agriculture (fruits, vegetables, and flowers) and tourism playing a major role.[2] Meeting the telecommunications requirements of these exporting sectors is exceptionally challenging and crucial to the success of Kenya's national economic strategy (see sections 4.4 and 4.5).

Superimposed on this promising but challenging situation—a rapidly growing economy supporting one of the world's fastest growing populations—have been some special difficulties arising from world market conditions:

- The falling world price of coffee, which dropped 56 percent between 1986 and 1990.
- Problems with the marketing of tea exports.
- The increase in prices of oil imports as a result of the 1990–91 Gulf crisis. In December 1990, the price per barrel was U.S.$25.42, 35 percent above the level of a year earlier.

- The continued depreciation of the value of the Kenyan shilling (KSh) relative to the U.S. dollar, dropping from a value of U.S.$0.048 in 1989 to U.S.$0.017 in 1993.

Agriculture continues to play a very major role in Kenya's economy. In serving the agricultural economy, KP&TC has to build its network and operate its services across a vast and difficult terrain. While there are districts with high population densities (such as Kisii with over 400 people per square kilometer), much of Kenya is hilly or semidesert with large areas of low population density.

Notwithstanding Kenya's high economic growth rate, per capita incomes are still very low. World Bank data, for example, show Kenya at U.S.$320 per capita, 42 percent below the level in Zimbabwe and 58 percent below Senegal (International Telecommunications Union 1993). Very few rural households can afford a telephone, yet rural telephone service is important for Kenya's economic development. Numerous rural enterprises need telephone service; thus, both public and private telephones play a vital economic role. The government is committed to stimulating small- and medium-sized businesses outside the major urban centers, and the provision of telecommunications infrastructure is essential to the efficient operation of such businesses.

In 1987, on the tenth anniversary of the founding of KP&TC, the then minister for Transport and Communications, Arthur Magugu, stated: "It is the policy of this government to provide the necessary infrastructure that will stimulate economic activity especially in rural areas." The managing director of KP&TC, Kipng'Eno Arap Ng'eny, has observed that a "direct consequence of the economic and social imbalance between urban and rural has been the unidirectional flow of population from the rural to urban centers. The lure of the city has resulted in a serious 'brain drain' from the rural areas, hampering any prospects for future development of the areas." Large amounts of resources have been invested with the aim of improving telecommunications in the rural areas.[3]

4.3 Telecommunications in Kenya: The Present

4.3.1 Regulatory and Policy Environment

As in most countries prior to the procompetitive changes in industry structure and regulation that became widespread in the 1980s, there was very little separation in Kenya of the regulatory and operational functions. Under the 1977 telecommunications act that formed KP&TC, the minister is empowered to give directions of a general nature to the KP&TC board. Although the minister would be expected to intervene on any controversial internal or international issues, in practice the corporation's management appears to enjoy substantial autonomy both in broad regulatory policy issues and in operational matters.

Nevertheless, KP&TC is subject to certain financial controls. External borrowing must be approved by the government since the treasury guarantees loans to the corporation. Any increase in tariffs greater than 10 percent and all salaries of the management and staff of KP&TC have to be approved by the government. As a result, KP&TC salary levels are not competitive with those paid in the private sector.

Kenya's telecommunications policy and telecommunications organizational structure has, until recently, followed the "PTT monopoly" approach traditionally employed in European countries and most of Africa. Telecommunications policy in Kenya appears to be just beginning to be influenced by the wave of change toward increased competition that swept the United States, Japan, Europe, and other advanced industrial areas of the world during the late 1980s and 1990s. There are significant early signs of a shift toward a more open and flexible industry structure: specifically, the customer premises equipment (CPE) market opening to competition; the emergence of a variety of special-purpose "private networks," or closed-use groups, operated by business corporations and by nonprofit and intergovernmental organizations; and the arrival in Kenya of various international value-added network services (VANs).

Prior to 1991, all terminal equipment in Kenya except small PBXs (with less than thirty extensions) had to be bought or rented from KP&TC, which also had a monopoly on CPE installation and maintenance. Since 1991, independent suppliers may provide CPE and independent contractors may install and maintain CPE and inside wiring, provided that KP&TC approves the type of equipment used, approves and licenses the installation and maintenance contractors, and conducts a postinstallation inspection of privately installed inside wiring.[4] The private marketplace was quick to respond to this relaxation, with several CPE companies advertising in the Kenyan press soon after the mid-1991 announcement, including GEC Plessey Telecommunications, Aztech Electronics (selling Sharp brand equipment), Kenya Microcomputers (selling Sanyo), and Samura Communications (selling Nitsuko).

In interviews, KP&TC's top management has stated that its motivation in initiating this move toward liberalization was a desire to improve efficiency by introducing competition and to have the private sector share the increasing financial burden of supplying terminal equipment, thus freeing KP&TC to concentrate its resources on major projects. One user interviewed gave strong evidence to suggest that KP&TC's resources were indeed overstretched in the early 1990s: KP&TC was unable to supply a new PBX on a timely basis, forcing the company to buy its PBX directly from a foreign manufacturer—a purchase that cost the company KSh200,000 (approximately U.S.$7,400) instead of the KSh10,000 (U.S.$370) per quarter rental that it would have paid to KP&TC.[5]

Although the relaxation of the CPE monopoly was welcomed and was expected to benefit users and the national economy as a whole, this 1991 deregulation was undertaken to solve specific, immediate problems rather than to reflect a general procompetitive policy trend. Neither major expansion of competition nor privatization appeared to be under serious consideration in the early 1990s. Senior managers at KP&TC have argued that such policies would be incompatible with the corporation's nonprofit objectives, especially the priority given to rural telecommunications investment. One senior manager interviewed during our fieldwork argued that it would be "inconceivable for the Corporation to be privatized until the country's network expansion program has been completed." Based on the experience of other countries, these are not convincing reasons for rejecting the option of privatization and competition.

4.3.2 The Inland Telephone Network

Kenya's inland network in 1993 was still, by world standards, very small: 184,583 working exchange lines in use. The size of the network had more than doubled since 1983,[6] with major efforts undertaken to upgrade it significantly. The benefits resulting from the most recent modernization (e.g., the installation of digital switches and establishment of a digital microwave transmission backbone between Nairobi, Mombasa, and northeast Kenya), in terms of gains in service quality and capacity, have been significant. In 1990, however, Kenya's long distance network was still severely congested: only 48.1 percent of call attempts on the long distance network were being completed successfully. Of all domestic call attempts made in 1990, the call completion rate was 53.7 percent—about the same rate achieved in 1986–87 (53.6 percent), but worse than the 57.2 percent achieved in 1987–88.[7]

It is not altogether surprising that congestion was, at the time of our fieldwork, a major problem in Kenya in view of the high rate of traffic growth: the increase between fiscal year 1986–87 and 1987–88 (the most recent year for which data on comparable measurements of traffic are available) was 21 percent, and the increase in the number of direct-dialed calls (though not available to the authors) was certainly much higher.[8] Matching traffic growth on this scale with an adequate expansion of capacity is a major logistical as well as financial challenge.

Kenya's extensive rural network has continued to grow rapidly. In many rural districts, telephone service continued to be based on manual exchanges for a number of years. From a social point of view, in a society faced with the problem of absorbing a rapidly growing population of the "educated unemployed," this slow modernization is not necessarily undesirable, even though manual exchanges utilize society's investment in transmission capacity less effectively than do well-planned and well-maintained automatic switches. In practice, despite substantial investments in new automatic switches for rural areas, the number of manual exchanges in Kenya rose from 269 in 1983 to 338 in 1991. More recently, a program has been under way to replace these exchanges with low-cost, small-capacity digital switches.[9]

In 1993, there were only 0.81 DELs per 100 inhabitants in Kenya, indicating that network expansion still has a long way to go. This level of telephone line penetration is lower than that of many Asian and Latin American countries but in the middle of the range in African countries. Telephone penetration is higher in Kenya than in neighboring Tanzania and Uganda, but it is 65 percent lower than the penetration level in Botswana and 38 percent lower than that of Zimbabwe. In 1993, there were 426,000 telephone sets connected to the public network in Kenya, yielding a density of about 1.58 telephones per 100 inhabitants.

In 1992, 61 percent of Kenya's 126,539 exchange lines were business lines. Since 1983, the number of business exchange lines has grown at an annual rate of 10 percent. In contrast, residential lines have grown 7 percent annually between

1983 and 1992, reflecting the increasing telecommunications demand by the self-employed and home-based small businesses, the high demand for purely residential service, and the efforts being made to clear a long waiting list. In 1991, the waiting list for residential lines in Kenya was reported to be twice as high as for business lines: 59,000 compared with 26,000.

A country's waiting list for telephone service remains an important measure of how effectively the country's telecommunications operator meets the demand for service. Problems in interpreting waiting list data, however, do exist. For example, when the wait for service is long, some people will put themselves on the waiting list on a speculative basis, even when they are not sure they will be able or willing to pay for service when the time comes. On the other hand, others may simply give up and not bother to join the list at all.[10] Nevertheless, a long waiting list is a strong indication that there is a large amount of unmet demand. And unmet demand, especially for business, means economic loss: applications of telecommunications services whose benefits exceed their costs have to be forgone.

A comparison of a country's waiting list with the scale of the existing telecommunications service provides a good indication of the size of the problem. The problem in Kenya is substantial and greater than in many other developing countries—though not on the crushing scale experienced by Egypt or Nigeria, where the size of the waiting list is comparable to the size of the *entire* existing telephone system.

4.3.3 Pay Phones: An Important Special Case

In developing countries, where the majority of people do not have access to a telephone at home or at work, public pay telephones (pay phones) play, or should play, an extremely important role in supporting the objectives of economic efficiency and distributional equity. Kenya has been an admirable exception to some countries' tendencies to ignore the social importance of pay phones because they are "unglamorous." Pay phones are of great value to poor households and rural communities in developing countries.

Pay phones are also especially important to small businesses that cannot yet afford (or, because of the waiting list, cannot yet obtain) their own telephone line. Economic policy throughout the world is belatedly recognizing the importance of such small enterprises in generating employment, income, and savings; in developing entrepreneurial attitudes, behavior, and skills; and thus in laying the foundations for future growth. Kenya is fortunate in having a strong tradition of such grass roots entrepreneurship.

Long queues and malfunctioning telephones impose frustrations and wasted time on both residential and business users of pay phones. Economic research in several countries has shown that many residential and business customers place high economic value on pay phone communications, are willing to pay for the service at a level at which the telephone operator can make a profit, but are prevented from doing so by their country's low level of pay phone investment.[11] Thus, a well-implemented public telephone program can simultaneously achieve high utilization, high profitability, and large social benefits.

Although it has been somewhat controversial (see section 4.5), KP&TC's pay phone program is one of its main success stories of the 1980s and early 1990s. Before 1982, public pay phones were rare in Kenya. The existing service was operator-controlled, not automatic.[12] On January 1, 1993, 5,613 public pay phones were in operation—one of the highest country totals in Africa. Moreover, KP&TC appears to have been fairly successful in keeping its pay phones operating reliably by clearing faults promptly and emptying the coin boxes regularly.

The corporation has worked with Danida, the Danish government's bilateral development assistance organization, to upgrade and expand the pay phone service. Starting in 1982–83, Danida funded the purchase of about 3,500 modern coin pay phones from Denmark through a mixture of loans and grants. In 1988, installation of card phones began, adding a further dimension of convenience to the pay phone service but planting the seeds of future controversy.

4.3.4 The International Network

Kenya's international services are provided via submarine cable systems, the Intelsat satellite system (accessed via the standard A satellite earth stations at Longonot, Nairobi, and Kericho), and terrestrial microwave radio links to neighboring countries. Time-division multiple access (TDMA) equipment was installed at the Intelsat earth stations to help meet the rising demand for international capacity.

The Intelsat Business Service (IBS) was introduced to meet a wide range of business telecommunications requirements for voice, data, telex, facsimile, and video conferencing services. The service provides private nonswitched communications (i.e., the functional equivalent of international leased lines) mainly via small- and medium-sized earth stations located near the end user's premises or through a larger country's gateway terminals.

Kenya is connected to the SEA-ME-WE submarine cable system linking Africa to Southeast Asia, the Middle East, and Europe. As of June 1988, there were 896 international circuits in operation between Kenya and the outside world, 810 via satellite, 39 via the SEA-ME-WE cable link, and 47 via Panaftel microwave. A total of 2,500 international circuits, not counting microwave links to neighboring countries, were in operation in the early 1990s.[13]

4.3.5 Mobile Communications and Radio-Based Services

Six hundred radio call subscribers in Kenya are estimated to be using KP&TC's high-frequency (HF) radio communications service to remote stations. In the early 1990s, this service was expected to be phased out officially, but in practice it is still highly valuable to farmers and to the travel and tourism industry; thus, HF is likely to continue operating for many years. In addition, KP&TC provides a VHF radio service for mobile communications in the capital and surrounding districts. This service has a relatively small customer base.

A VHF Community Repeater Service was introduced in 1980, enabling mobile users in the Nairobi area to communicate via a shared repeater station owned and operated by the KP&TC.[14] There were plans to extend the service to the Mombasa area.

A paging service was introduced within the Nairobi area in 1980. A full cellular service covering eight regions in southern Kenya was introduced in 1992 in collaboration with the NEC Corporation.

Mobile services, and radio-based telecommunications in general, could play a far more important role in Kenya and all of rural Africa than anyone expected in the 1980s. Although the initial infrastructure costs are high, the subsequent advantages from reduced maintenance costs and enhanced reliability are significant, not least because it eliminates thefts of copper overhead plant, a real problem in rural areas. A cellular service in Kenya could easily attract 5,000 to 10,000 subscribers in its early years, with the numbers rising substantially as costs are reduced and the scale of operations increased. It could perhaps even lead to a low-cost personal communications network (PCN) service, such as the Japanese "Personal Handyphone." Such radio-based services could be widely used for fixed communication (i.e., "wireless local loop"), as well as for mobile applications, once advancing technology and increasing scale make them cost-competitive with wired technology.

4.3.6 Data Communications

The dynamism of Kenya's economy and its links to the global marketplace are reflected in the strong demand for data communications. Until recently, data transmission requirements were met mainly through the use of modems over leased analog lines. Poor line quality and congestion made the use of switched voice connections for data transmission problematic, especially for international links; digital leased lines were not available. Growth in the use of leased analog lines for data has been rapid: fifty leased lines were in use in 1980; by 1983, 169 modems were in use to terminate leased data lines; and by the end of KP&TC's 1987–88 fiscal year (June 1988), 307 modems were used in this manner.[15] The total in the early 1990s may have been about 400, but achieving an exact count of modems has become more difficult since 1991, when the CPE market was opened to competitive supply.

No longer holding a monopoly on supply, KP&TC cannot know precisely what proportion of its leased lines or public network exchange lines are connected to independently provided data communications equipment. As of 1992, KP&TC reported that there were 525 pieces of data termination equipment attached to the public telephone network and 4,500 on other networks (International Telecommunication Union 1994). From the leased-line point of view, this count of data-terminating equipment probably underestimates the true magnitude of data users in Kenya, since each termination point could represent several end-user terminals.

In 1991, a new alternative to the use of modems and leased lines became available: the Kenpac packet-switched data network.[16] Kenpac represented a major technical and operational advance for KP&TC, but it was controversial in some parts of the user community at the time of its introduction. Some users voiced concern about the quality of service that could be expected and expressed fears that Kenpac would lead to higher prices for leased lines if KP&TC tried to force a migration from leased lines to public-switched digital services, as some European PTTs had attempted to do.

In practice, marketplace realities and users' bargaining power resulted in this fear being unfounded: because of resistance from users, Kenpac charges were less than half those originally announced.

With respect to ensuring service quality, Kenpac's marketing manager emphasized the company's determination that faults would be dealt with quickly after the customer reported them. To achieve prompt service, Kenpac relied on a team of dedicated technicians organized and managed separately from KP&TC's telephone network staff. During the day, at least two of these technicians were made available to support every Kenpac node. Five professional engineers per day and two per night were scheduled to staff a help-desk facility where two network management computers are located.

4.3.7 Value-Added Services

In 1990, KP&TC created a Value-Added Services (VAS) department. The VAS manager reports to the general manager for Telecoms Services and in certain matters has direct access to the managing director (an indication of the importance placed on this area of business). Value-added services are very broadly defined within KP&TC: the department runs the public telephone service (both coin and card phones), public telex facilities, and Kenpac, as well as being responsible for such services as electronic mail, which are "value-added" or "enhanced services" as these terms are generally used in Europe or the United States.

4.3.7.1 VAS Voice Services

In the 1980s, KP&TC provided no value-added telephone services in the strict sense of the term: there was no voice mail. Automatic toll-free (800 number) service was not available. Increasing numbers of digital switches were installed in the early 1990s, which would in time permit the widespread provision of advanced services, as well as convenience features such as call waiting and call forwarding.

4.3.8 Other Services

A variety of other services complete KP&TC's product line. The most important of these is facsimile. Telex was historically important; although telex is still used extensively, it is in decline.

Telex subscribers increased from 1,750 in 1983 to over 2,031 in 1992. Although the lack of capacity has hampered connections in Mombasa, where the exchange at one stage was 96 percent full, pressures of demand have been relieved by the rapid growth of facsimile. In 1991, telex demand began to drop.

Data are not available on the total number of installed facsimile machines since users do not need to register them with KP&TC. They were thought to number several thousand in the early 1990s, since almost all export companies, for example, had a facsimile machine. In addition, KP&TC has set up "bureaufax," which are facsimile service centers open to the public, in the major towns throughout Kenya. In 1992, KP&TC reported 203 telefax stations and four bureaufax stations. Of the pages transmitted by KP&TC facilities that year, 73 percent were transmitted to international destinations.

4.3.9 *Financial Performance of KP&TC*

Although KP&TC has achieved success in expanding the network and making its benefits widely available through the public pay phone program, two interrelated areas of major difficulty emerged in the early 1990s: the financial condition of KP&TC deteriorated substantially, and it became clear that KP&TC was drastically overstaffed. Although it was not possible to analyze these issues in great depth during the preparation of this chapter, clear danger signs were visible.

At the outset of the Third Telecommunications Program in 1986, the World Bank, a major funding source for the program, had few concerns about KP&TC's financial health. The corporation's *Annual Report and Accounts* for 1987–88 showed a surplus of about KSh0.5 billion (about U.S.$18.5 million at the 1991 exchange rate) on KSh2.9 billion (U.S.$107 million) of revenues. Corresponding figures for the telecommunications business considered on its own showed a KSh0.7 billion (U.S.$25.9 million) surplus on KSh2.6 billion (U.S.$96 million) of revenues. The 1987–88 annual report noted that KP&TC's return on equity had risen from 23.6 percent the previous year to 28.5 percent in 1987–88 but that this was "mainly attributed to the increased financial leverage." A large amount of borrowing was indeed taking place. Unfortunately, much of it involved high costs, which eventually would have an adverse effect on future years' financial results.

By 1989, the financial situation had deteriorated substantially: the net surplus, even before adjustment for foreign exchange losses, was less than KSh100 million (U.S.$3.7 million). The corporation's cash flow was negative in 1989. In the period 1987–89, revenue per unit of telephone traffic carried fell by 2 percent per year, while costs per unit rose at 11 percent annually, an increase partly explained by the growing financing charges associated with the large investment program.

The combined effect of several different factors seemed to be the cause of KP&TC's financial difficulties. One of these was a poor success rate in the collection of accounts receivable. The public sector was the worst offender, with unpaid bills amounting to about two years' outstanding billings from the government. The corporation offset this by not paying the government's telecommunications tax. In December 1990, KP&TC conducted a national crackdown on bill defaulters at the direction of President Moi. In the course of this effort over fifteen government ministries, including the Ministries of Water, Information, Public Works, and Health had their telephone service disconnected.

A second problem area was pricing policy. The government had not allowed domestic tariffs to be increased sufficiently to offset inflation and had not allowed international tariffs to be increased at all during 1989–90, despite the 40 percent fall in the Kenyan shilling against the U.S. dollar over this period.

A third source of financial difficulty appeared to have been a rapid escalation of the level of capital investment. Much of this increase appears to have been financed from high-cost sources of credit, and it is not clear whether strict economic tests were applied to assess the viability and rate of return of the additional investment projects. The original plans for the Third Telecommunications Program called for a total of U.S.$90 million in investments. The loan agreement for the World Bank funding for the Third Program was signed in May 1986 and loan disbursements began in 1987, with the intention that the program would be completed in 1991. By 1989, the total of actual KP&TC investments during the Third Program period was far above this figure and may have been in excess of U.S.$200 million.[17] The difference was apparently due mainly to previously unprogrammed rural telecommunications investments, many of them underwritten with costly financing provided by equipment vendors and other forms of high-cost bilateral credit. The annual level of KP&TC's capital spending rose from a level of KSh653 million (about U.S.$24 million at 1991's exchange rate) in 1983 to over KSh804 billion in 1987–88. In 1991, annual capital investment was running at an annual rate of over KSh1 billion (U.S.$37 million).

The ability of KP&TC to finance its capital program itself or to attract outside finance obviously depends on the profitability and economic efficiency of the corporation, which in turn reflects productivity, costs, pricing, and the way the capital investment program is financed. Unfortunately, the trends affecting KP&TC's profitability appeared to be quite unfavorable at the time of our interview fieldwork in 1991. To some extent, this was due to factors beyond KP&TC's control, notably government restrictions on pricing, which limited KP&TC's ability to adapt to the rapid depreciation of the international exchange rate for the Kenyan shilling (which fell 18 percent in 1989, 19 percent in 1990, 17 percent in 1991, and 14 percent in 1992 against the U.S. dollar). To some extent, it was also attributable to the decision made, on social and political grounds, to invest heavily in the rural infrastructure on a scale that was unlikely to provide favorable short-term financial returns.

The corporation's financial problem, however, seemed even larger than could be explained by these considerations alone. Overstaffing seems to have been a major factor in KP&TC's financial woes. There is a very high ratio of staff to the number of exchange lines and telephone sets KP&TC supports, although this low productivity has improved in recent years. In 1990, there were 120 staff members per 1,000 telephones, compared with the target of 62 agreed with the World Bank. Although KP&TC management argued during our program of interviews in Kenya in 1991 that the corporation had a social obligation to employ people to reduce unemployment, the very low productivity had a serious adverse effect on the corporation's financial condition and on the general business disciplines needed for economic efficiency. More recent statistics available indicate that KP&TC's workforce was 13,000 in 1992, or 30.5 staff members per 1,000 telephone sets. The corporation had also achieved staff levels of 62.5 workers per 1,000 working DELs (International Telecommunications Union 1994).

In its drive to modernize, KP&TC focused too much on additional hardware, thus overextending its investment resources, and did not sufficiently develop the "softer" elements of customer service—marketing, preventive maintenance, repair, and staff training—that would have enhanced revenue growth and reduced costs. The net result of these factors was a state of financial difficulty that was not possible to diagnose fully from the data available in KP&TC's *Annual Reports and Accounts.* Nevertheless, based on the available information, there was concern in the early 1990s that KP&TC might encounter difficulties servicing its foreign loans due to its large external debt burden, aggravated by the unfavorable trend of the exchange rate.

As KP&TC's position, viewed from a commercial point of view, declined in the late 1980s, its ability to raise capital on a purely commercial basis at a tolerable cost also became more limited. Its relationships with sources of funding on concessionary terms (both grants and loans) also appeared to deteriorate. The issues involved are discussed further in section 4.5.

4.4 Telecommunications Users and User Benefits

The discussion so far has primarily viewed telecommunications in Kenya from the point of view of the network operator, KP&TC, and public policy makers. All of this, however, is relevant only because of the need to produce an end product: service to the user. This section will address users' needs, how well these needs are being served, and how important the service of these needs is to the national and international economy. An in-depth discussion of these issues is possible due to the existence of extensive quantitative economic and business research on this subject carried out in Kenya between 1983 and 1991—possibly more than research on any other developing country.[18]

4.4.1 Research on Users and Telecommunications Benefits

In 1981, a study of the role of telecommunications in Kenya's economy was undertaken by the consulting and research firm CSP International Inc. under contract to the ITU. This work formed part of a wider research program on the role of telecommunications in economic development undertaken by the ITU in collaboration with the OECD.[19] Virtually all previous studies in this field used macroeconomic statistics and sought, with only limited success, to demonstrate a cause-and-effect relationship between the expansion of the public telecommunications network and economic growth, using data on broad economic trends: that is, a macroeconomic method. The 1981 study for the first time examined the relationship between the expansion of the public telecommunications network and economic growth at the level of the individual business enterprise: a microeconomic approach.

The CSP/ITU study analyzed the impact of telecommunications services (or the lack of such services or quality problems with such services) on the functioning of specific business activities such as purchasing, overcoming production stoppages, or operating vehicle fleets. It was based on nine in-depth case studies of enterprises in Kenya, ranging from a food processing/manufacturing firm to a large exporter of vegetables and a major transport company.

The study identified substantial costs and operational problems associated with limited availability of telephone exchange lines, severe network congestion, and unreliable service. The estimate of potential benefits to the case study companies (increased value added through expanded output and sales and/or decreased costs per unit of output) were large, ranging from 1.3 percent to 9.2 percent of each firm's total revenues. In no case was the potential benefit less than ten times the cost to the economy of expanding the capacity of the telephone system sufficiently to achieve the estimated benefits. Practical limitations of management and organizational behavior within the sampled organizations might prevent 100 percent the potential efficiencies theoretically possible through improved communications (e.g., better scheduling of trucks with return loads) from being fully achieved in practice. Nevertheless, the study demonstrated for the first time that the economic benefits to users, and hence to the national economy, that were achievable by expanding and upgrading the public telecommunications services could be estimated through hard data and a systematic methodology, and that the benefits are large compared with the costs.

In 1987, a further study, again carried out by CSP International for the ITU, was undertaken to determine the foreign exchange benefits that could be achieved by investment in the World Bank's Third Telecommunications Project for Kenya. This study included interviews with twenty export-oriented businesses. Five of these businesses were revisited in 1991 as part of a series of twenty interviews that encompassed five additional user companies; technical assistance, aid, and lending organizations; and KP&TC.

The 1987 study concentrated on Kenya's top telecommunications users within the agricultural, industrial, and tourism sectors. The interviews with additional companies in 1991 contributed further insights concerning the airline, newspaper/publishing, and banking sectors. All the companies interviewed were major players in their fields, making a significant contribution to the national economy. The twenty businesses interviewed in 1987 accounted for almost 20 percent of Kenya's total export earnings in that year; by 1991, they had expanded to contribute 27 percent of total export earnings.

It was evident from the 1987 study that the revenues of the KP&TC were constrained far below their potential level by inadequate capacity and service quality, reflected by the long waiting lists and by service congestion. Slow repairs to faulty facilities were another major cause of lost revenue. During the 1987 study, a technique was developed to estimate the foreign exchange earnings forgone by Kenya's economy as a result of the inability to meet demand for telecommunications services: this predicted shortfall combined estimates of direct effects (foreign exchange that KP&TC could have earned through the international settlements process if it had met the full demand for international calls[20]) and indirect effects (foreign exchange that telecommunications users would have earned through their normal business operations if these had not been constrained by inadequate telecommunications services). The annual foreign exchange "cost" to the economy of inadequate provision of telecommunications services in this sense was estimated at just over KSh186 million.

The 1991 interviews showed that a minimum threshold requirement for telecommunications services was being met, allowing companies in Kenya to operate in a modern manner and to do so more efficiently than in the majority of developing countries. However, beyond this threshold there are substantial gains that could be made from a higher investment in network infrastructure and service provision.

The national economy and the demand for telecommunications services in Kenya in the early 1990s were much larger than they were in 1987, and the gap between demand and supply was greater than ever. Consequently, when we applied the same methodology used in 1987 to estimate the foreign exchange earnings forgone as a result of underprovision of telecommunications services, we derived a substantially larger estimate (see section 4.4.6).

Such estimates cannot capture the full range of benefits to Kenya's economy from expanded and improved telecommunications. For example, improved communications would further enhance Kenya's attractiveness as a location for the "footloose" activities of multinational corporations in Africa.

The users that benefit the most from higher investments in telecommunications are exactly those who also contribute the most to Kenya's foreign exchange revenues. Kenya is in danger of being caught in a vicious circle involving frustrated telecommunications users, slower export growth, slower growth in KP&TC revenues, and hence greater difficulty on the part of KP&TC in sustaining an adequate level of investment in the expansion and upgrading of the public network.

The following sections assess the implications of inadequate availability and quality of telecommunications services on three key economic sectors. Difficulties encountered by end users are summarized in terms of lost managerial time, reduced sales, and higher inventory levels that are directly attributable to telecommunications problems.

4.4.2 Service Quality

In the earlier review of service quality data (specifically data on call completion rates, a measure of network congestion), we noted that these data did not show any clear improvement over time. Improvement targets set in the Third Telecommunications Development Program called for increases in call completion rates to 75 percent for local services, 60 percent for long distance, and 50 percent for international calls. In 1989, the overall call completion rate was only 54 percent. By the early 1990s, these targets had still not been met.

The general view among users is that service has improved substantially since 1987. For example, the users we interviewed focused more on the total disruptions of service that are periodically experienced than on congestion, a significant but lesser consideration. Between 1989 and 1991, our interviewees told us, the problems previously encountered in the rainy season, when water often entered the cable plant and service was sometimes completely interrupted for long periods, had been greatly reduced in most areas.

The continuing concerns expressed by users focused mainly on the availability of service and the degree of service reliability and congestion in rural areas; specific local problems in the Nairobi industrial area (where a large amount of industrial activity takes place) and the Jomo Kenyatta Airport area outside Nairobi; delays and unpredictability in the installation of new exchange lines and leased lines; and delays in repairing faults.

In the Nairobi industrial area, network congestion was still severe at the time of our interview fieldwork in Kenya in 1991. This has continued to be a concern for some companies located in this area.

Our interview respondents felt that the situation in the industrial area had not improved. The situation at the airport had its own specific problems: old exchange equipment; a shortage of "junction" circuits to Nairobi, with consequent congestion; and frequent outages of junction circuits, including those used as leased lines. Respondents attributed these outages to water ingress in the rainy season and to cable theft.

Most of the companies we interviewed expressed concern about the time delay in installing additional exchange lines and the unpredictability of the delays. Sometimes companies that planned for a long lead time were caught out by an unusually rapid installation. One of the banks interviewed requested a leased data line one year before it calculated that the software it was developing would be ready. In fact, the line was delivered after three weeks and was consequently lying unused. In interviews, users also expressed dissatisfaction with KP&TC's maintenance procedures and maintenance delays, although in the early 1990s there were indications that the situation had improved.

Overall, then, the telecommunications picture in Kenya is one of significant but uneven improvement in service quality, with the most extreme problems of service interruption being overcome in most locations (with important exceptions) and congestion, slow installation, and repair as continuing concerns. It is indicative of the significance of these problems that telecommunications difficulties figured prominently in the controversy in the early 1990s over an (unsuccessful) proposal to relocate the world headquarters of the United Nations Environment Program (UNEP) from Nairobi to Geneva.

In the following discussion, we draw on details gained from our interview program with individual businesses to assess the economic significance of problems resulting from the availability and quality of telecommunications services.

4.4.3 Agriculture

The export of high value-added cash crops is one of the most dynamic elements of Kenya's economy. While total exports grew (in volume terms) by 22 percent in 1990, horticultural exports grew by a record 40 percent in volume terms and nearly doubled in value terms from KSh79 million in 1989 to KSh140 million in 1990. This success story is as important from a social standpoint as it is from the standpoint of the achievement of macroeconomic goals. Export-oriented agriculture is an important direct source of rural employment and indirectly generates a large amount of economic activity in other rural occupations. Kenya's rural areas today employ over 75 percent of the labor force, and it is the rural economy that will have to create most of the several million new jobs that will be needed by young Kenyans in the mid- to late 1990s.

Kenya's horticultural export subsector, which grows and exports fruits, vegetables, plants, and flowers, has become a highly dynamic and internationally oriented business community, with a mixture of both large and small farmers. Kenya has several of Africa's largest flower farms and is the single largest supplier to the Dutch flower market. Kenya has also become a premium fruit and vegetable grower for top supermarket chains throughout Europe. Local businesses grow, process, clean, trim, and package the produce—highly labor-intensive activities. One farm ships 180 tons of flowers weekly to Europe, employing over 2,000 staff either directly on the farm itself or indirectly in the supporting transportation, freight, and baking activities that support the business.

The agricultural export businesses are highly dependent on communications to keep in touch with prices in the European market and to monitor the movement of their extremely perishable goods across Kenya to the Nairobi airport and on to their international destinations. Some farms even have on-line data links to obtain current market prices, rather like the foreign exchange traders in the financial markets. The pricing of agricultural exports must respond promptly to changing market conditions: a telecommunications failure can result in suboptimal pricing, mistakes in shipping produce, and disappointed customers. Farms and rural areas are not generally well served by Kenya's public network despite large investment expenditures by KP&TC in rural expansion in recent years. Such issues as cutoff calls and crossed lines (or "crosstalk"), in which a conversation on one circuit interferes with another circuit, are often cited as recurrent problems.

Other interviews provided insight into the specific ways in which inadequacies in public telecommunications services harm the effectiveness of horticultural export operations in regard to specific business functions, such as packaging, purchasing, shipping, maintaining customer relations, and pricing.

Product Packaging. A major Kenyan pineapple producer told us that it must communicate from its farm to the factory regarding changes in the size of the fruit being delivered, since this determines the setting of the plant's fourteen processing lines. When telecommunications problems prevent such communication, processing lines often have to be shut down unexpectedly and reset, which slows the plant to only 60 percent of its normal production capacity. Management thought that cellular telephone service in the future could greatly enhance plant productivity, enabling its export output to increase.

Buying. A Kenyan horticultural company that performs the marketing for a number of small farms described the significant problems it encountered in communicating with its buyers on the road. A buyer is authorized to buy a set number of boxes of beans at a certain price per box but has no authorization to buy at a higher price if an insufficient number is available at the lower price. When the buyer cannot find a working telephone to gain authorization from the head office to buy at the higher price, he or she returns without the required amount of beans and the company is unable to fill its export orders. Similarly, if the buyer's vehicle breaks down, it can take many hours for the driver to reach a public telephone (which may or may not be working) to request assistance and inform the office that the delivery will be delayed. These types of delays have a harmful effect on both freshness of products and customer satisfaction.

Freight Transport. The case study interviews showed that difficulties in communicating with farms when there are dockside delays, or when air freight space is canceled, resulted in inappropriate harvesting and delivery of perishable produce. If the docks or airport authorities can get through by telephone to the head office of an agricultural enterprise, the company can usually resolve the situation by sending a driver to the farm with a message. In extreme cases, however, a whole shipment may be ruined or sold at a lower than optimal price in the home market.

Customer Relations. Foreign customers of some Kenyan produce and flower growers like to be informed of the number of boxes that have been dispatched to them so they can make the necessary arrangements for delivery. If the telex is out of order for an extended period of time, as it was for ninety-three consecutive days for one flower grower in 1991, it is not possible to provide the level of service the customer expects.

Pricing. Other case study interviews were in the tea and coffee-growing industry. World markets are complex and volatile. Kenya's Tea Development Authority, a public-sector agricultural marketing organization, explained in a case study interview that it must obtain quotes from at least twenty-five agents. Telex congestion, however, often made this impossible in 1991 and earlier years, reducing the chances of getting the best prices. Kenya's coffee producers are also very dependent on up-to-date information on market movements and use Reuters Monitor, an international on-line information service providing commodity price data, for this purpose. When failed data lines prevent access to this critical information, Kenyan producers are at a major disadvantage in bargaining with buyers in Europe and elsewhere.

4.4.4 Manufacturing/Industrial Business

In 1991, Kenya's manufacturing sector contributed 12 percent of the country's GDP, but it has not yet become a significant export sector. Kenya's manufacturing industry contributed to the balance of payment by supplying products to the domestic market that otherwise would have to be imported. At the same time, this sector is one of the largest users of imported raw materials, parts, and equipment. Any savings that manufacturing companies can make in the cost of imported inputs will make finished goods more competitive, thus displacing imports, encouraging exports, and contributing to the growth of the economy.

According to our interviews, a variety of business functions of Kenya's manufacturing sector have been adversely affected by telecommunications problems, including purchasing, licensing, inventory control, production coordination, freight transport, and sales and pricing.

Purchasing. A petroleum company interviewed in one of our case studies told us that it buys six million barrels of crude oil per year on the often volatile world spot market, relying on good communications to buy at the best possible price. But difficulties in placing international calls from Nairobi have delayed the negotiation and closing of deals, resulting in higher costs. In one factory interviewed, the inability to get through to suppliers by telephone forced the factory engineer in several instances to drive into Nairobi to obtain new parts or quotations, with a consequent waste of management time and vehicle expense. The inability to get a sufficient number of quotes because of communications difficulties also caused the same company to pay a higher price for materials than it might otherwise have done.

Regulation and Import Licenses. In 1991, we found that a Kenyan chemical company was encountering problems in obtaining such licenses on a timely basis because of the restricted number of telephone lines into the Ministries of Agriculture, Livestock, and Commerce where the licenses were issued. Instead of being able to monitor the progress of a license via telephone, a company representative was often forced to travel to the ministry in person.

Inventory Control. A Kenyan importer of automobiles in "knockdown kit" form experienced problems with inventory control for spare parts in its fourteen sales outlets due to telecommunications failures. Since parts were often not available when they should have been, the company lost spare parts sales and its customers were dissatisfied.

Production Coordination. We found that a Kenyan battery manufacturer had to call the smelting manager in Athi River from Nairobi ten times a day. Because it frequently required thirty minutes to get through on the telephone, a considerable amount of management time was wasted.

Freight Transport. A Kenyan petroleum company experienced problems coordinating the transport of products because of Kenya Railways' frequent inability to reach the company by telephone to inform it of the daily loadings. Cargos can be shut out of the rail freight schedule because of the arrival of fertilizers or other priority goods; consequently, several days' loads may arrive at the same time, which incurs "demurrage" costs and overtime payments while the backlog is cleared.

Sales and Pricing. We found that a major Kenyan cement producer suffered considerable sales price penalties because poor communications made it difficult to contact buyers in such markets as Rwanda, Burundi, Tanzania, and Uganda where higher prices could be obtained.

4.4.5 Tourism

Tourism overtook agricultural exports as Kenya's most significant source of foreign exchange. In the late 1990s, however, the sector's performance continued to be constrained significantly by telecommunications problems in several important areas.

Customer Relations/Sales. For example, in one of our case studies, a Kenyan car rental company reported that it often lost business because of the time required to phone in to the main switchboard from sales desks in hotels and other locations.

Due to this congestion, it could take up to thirty minutes to confirm with the workshop whether or not a vehicle was available. A travel agent interviewed reported similar difficulties.

Similarly, a Kenyan touring company told us that it lost a large amount of business between November and April in its coastal office because the very limited number of exchange lines were constantly busy, and it was unable to react quickly enough to inquiries for railway/road tours and safaris. Travel agents had to send messengers in order to make bookings or confirm arrangements.

Reservations. Although there is a central reservations service for the hotel industry in Nairobi, hotel managers told us they lost significant amounts of business because of problems communicating with the service. Kenya's hotel companies were moving to computerize their reservation systems in the early 1990s, but by 1991 this had been accomplished only in Nairobi, with no leased-line links to other reservation centers because of the poor quality of the lines. Consequently, reservations were still communicated to the center by mail, courier, telephone, and fax.

Communication with Lodges. The remote location of tourist lodges in game parks makes communications very difficult. The problem has to some extent been resolved by the use of HF radio, but frequencies have to be shared, which leads to problems of congestion and confidentiality. Calls made by guests from the lodges could be an extra revenue source if direct lines were available. One travel operator estimated that it could justify a 1 percent price increase if telephone facilities were available for guests.

4.4.6 Economic Implications

These 1991 interviews confirmed the earlier findings, which showed that there are specific, quantifiable costs to Kenya's economy resulting from the inadequate availability and quality of telecommunications services. Furthermore, our results showed that these effects can be assessed and often quantified by detailed study of the impact on specific business functions and that losses of export revenues (foreign currency earnings) or additional import costs are often involved.

Specific effects of these telecommunications service problems include:

- Production stoppages due to machine breakdowns, spare parts shortages, raw material shortages, or other factors that could be avoided or mitigated with improved communications.
- Failure to obtain the best possible export price.
- Growth constraints as a result of poor communications restricting access to customers or suppliers.
- Wasted managerial time.
- Price penalties paid as a result of the inability to gain sufficient quotes for parts and raw materials.
- The cost of financing large inventory (stock) levels made higher because of reordering delays or unpredictable resupply exacerbated by poor communications.

- Increased transport costs because of the need to send messengers when phone contact is impossible: as much as 50 percent of the total transport cost to the business was an additional import cost (fuel and vehicle depreciation).

Data from the interviews were used to update the 1987 study's estimate of potential foreign exchange earnings forgone as a result of inadequate availability and quality of telecommunications services. The volume of earnings forgone, expressed as a percentage of Kenya's total exports, decreased from 0.75 percent to 0.65 percent over the four-year period 1987–91 as a result of the modest gains in service quality noted earlier. The trends in the availability and quality of service were clearly in the right direction. The improvement, however, was not yet sufficient to meet the full service requirements that arose from the explosive increase in exports and the increasing demand from local industries in that period. As a result, the total burden on Kenya's economy caused by telecommunications deficiencies had actually increased. We calculated the foreign exchange loss to Kenya's economy in 1991 resulting from the inadequacy of public telecommunications services at KSh271 million (U.S.$10 million) per year,[21] underlining the strong case for continued high levels of investment (of the right kind) in the public telecommunications infrastructure if the government aims to stimulate the expansion of the export economy.

4.5 Trends and Issues

4.5.1 Telecommunications and Electronics Manufacturing

Kenya has the small but promising beginnings of a manufacturing sector for telecommunications, electronics, and information technology. "Informatics policy" has been incorporated within the national development plan, and government policy favors local participation in the manufacturing of electronic equipment. In practice, manufacturing has so far mainly meant assembly of imported parts in "screwdriver plants." Sanyamco, for instance, started assembling radios in Kenya from imported parts in 1967. Although the employment and added value (and hence gross national product contribution) generated in Kenya per unit of sales is relatively small, local assembly using imported parts represents a first step on the ladder toward a more fully developed electronics manufacturing sector for Kenya.

High prices are commanded by imported computer hardware in a market protected by tariffs. This has created an attractive opportunity for local computer assembly, which Kenya Microcomputers has seized. The company began operations in 1986, and in the early 1990s offered a clone of IBM's PC XT with a matrix printer for KSh78,000 (about U.S.$2,900). Within fifteen months, 400 of these machines were installed in Kenya, with exports to Uganda, Tanzania, Rwanda, Ethiopia, and Zimbabwe. Other local computer industry companies have concentrated on backup support services while acting as agents for computer manufacturers in the United States and the Far East and also developing specialist sector-specific software packages.

The small scale of Kenya's electronics industry is mirrored by the small scale of technological education in the country. Approximately 20 to 30 technicians with "City and Guilds" diplomas leave Kenya Polytechnic every year, with a similar number graduating from Mombasa Polytechnic. In addition, 120 students graduate from the Electronics Faculty of the University of Nairobi and 40 from the Faculty of Technology at Moi University.

Kenya's only significant manufacturing activity for telecommunications equipment was a joint venture between KP&TC and AT&T's Irish subsidiary, Telectron, to manufacture cables and KP&TC's manufacturing complex at Gilgil, where telephone handsets, small PABXs (private automatic branch exchanges), and manual exchanges are assembled both for KP&TC and for export (mainly to other African countries). In the former arrangement, AT&T sent the raw materials for cable manufacture to Kenya—where labor costs are very small—and then bought back the completed cables; however, by 1993, AT&T phased out its participation in this line of business in Kenya.

In 1989, KP&TC's managing director suggested that neighboring countries in Africa should jointly develop local manufacturing: "Through regional cooperation, many African telecommunications administrations can succeed in manufacturing telecommunications equipment, whereby financial and human resources can be marshaled to achieve the required economies of scale."[22]

4.5.2 Service Innovation

One of the more intriguing aspects of telecommunications in Kenya in recent years has been the establishment of a variety of special-purpose networks making use of state-of-the-art technology to solve specific communications problems. The majority of these have been established by international public-sector organizations, by universities and other research institutions, by nonprofit organizations, and by the financial services sector.

4.5.2.1 Public-Sector International Organizations

In northern and eastern Kenya, UNICEF, the United Nations Children's Fund, and other United Nations agencies conducted a large-scale relief effort in the early 1990s to assist vast numbers of refugees from Sudan, Ethiopia, and Somalia. The project required immense logistical and communications capabilities. In April 1989, UNICEF installed Inmarsat-C mobile satellite communications terminals at its Copenhagen logistical base; in Khartoum, Sudan; and in project fieldwork areas in Kenya and Sudan. These terminals access the satellites operated by Inmarsat (the International Maritime Satellite Organization), which were originally intended for communications to ships at sea. The terminals can in practice be used on land for both fixed and mobile communications, and they have been used to coordinate the distribution of food and aid in Sudan and northern Kenya. The Inmarsat-C is a compact and lightweight terminal allowing two-way store-and-forward data messaging via satellite. It provides reliable service and uses very little power (15 to 50 watts).

The Food and Agriculture Organization of the United Nations (FAO) and the

European Space Agency (ESA) devised a satellite communications application—DIANA (Data and Information Available Now in Africa)—designed for rapid distribution of data obtained from remote sensing satellites on environmental and agricultural matters. The overall program, which was being implemented in the early 1990s, is known as ARTEMIS (Africa Real Time Environmental Monitoring using Imagery Satellite). User facilities were planned for Accra in Ghana, Nairobi in Kenya, Niamey in Niger, and Harare in Zimbabwe. The information is intended to help identify areas where severe crop failures may be expected due to adverse weather and to pinpoint locations where moisture and vegetation growth may favor an infestation of locusts and grasshoppers.

The FAO, World Bank, and United Nations Development Program (UNDP) also funded a worldwide agricultural information network, the Consultative Group Network of Agricultural Research (CGNET). This data network provides three organizations in Kenya with databases, travel reservation systems, information services, telex, and fax links to over 150 offices and organizations in other countries via packet-switched networks.

The Kenyan office of the African Regional Standards Authority is linked into the ARSONET information system funded by CIDA, the Canadian development agency.

4.5.2.2 Universities and Other Research Institutions

Five universities in Kenya, Uganda, Tanzania, Zimbabwe, and Zambia are collaborating through ESANET (East and Southern African Network), a data network, to experiment with network technology and to link researchers worldwide. The GreenNet conferencing system is available to users at the University of Nairobi and one other data network location in Kenya. GreenNet serves as the gateway to ESANET.

The International Centre for Research into Agroforestry has established a network similar to CGNET. Another new network, WedNet, links Kenya to several other countries and is dedicated to research on natural resources and women's issues (Godard 1994).

The University of Nairobi also participates in a five-country collaboration involving Tanzania, Uganda, Zambia, Zimbabwe, and Kenya for electronic text-message communications between health care professionals. This network uses a small satellite, HealthSat, launched in 1991, in low-earth orbit. Unresolved regulatory issues within Kenya prevented the University of Nairobi from acquiring a license to access the satellite for a while, but now this link is fully operational.

4.5.2.3 Nonprofit Organizations

In 1990, the Kenya Computer Institute, working with the African Medical Foundation, the Red Cross, St. John's Ambulance, and the Nairobi Hospital announced a prototype project in disaster management communications that would allow authorized institutions to broadcast messages and update information on resource availability. In an international communications program similar to those being pursued by the agricultural research centers, the Environmental Liaison Center in Nairobi was also planning a way to communicate with organizations via electronic mail.

These efforts are being assisted by the establishment of an Internet node by the Kenya Computer Institute and the local universities. Organization by UNESCO and funding by Korea and Italy contribute to the Regional Informatics Network for Africa (RINAF) in many countries of the continent. Kenya's Internet node is expected to be operational in the near future.

Currently, the widest international data access is gained via FidoNet, which is the least expensive way to connect to bulletin boards, news, and E-mail for many computer enthusiasts (usually through timed dial-up). FidoNet is a relatively informal worldwide network, developed initially by users themselves. Many of the new data communications networks emerging in Kenya are structured to accommodate FidoNet protocols (Godard 1994).

4.5.2.4 Financial Services and Commercial Sector

Kenya's membership in SWIFT, the global electronic payments network based on packet switching, was promoted by the Kenyan Bankers Association in 1994. Integration of Kenya's banking sector into SWIFT enhanced the telecommunications sector's performance and hence the competitiveness of Kenya's economy as a whole. Prior to the initiation of SWIFT, Visa card payments were settled by courier. This was costly, resulted in time lags, and posed the risk that information on stolen cards would not arrive in time.

Citicorp's internal worldwide satellite network, which includes links to its Kenya locations, is designed to achieve a competitive advantage by providing timely information and rapid execution of transactions.

With fifty-one members and eighty-one travel agency locations, SITA, the global affinity-group data network for airlines, also has a strong presence in Kenya. There were SITA nodes—at Nairobi, Mombasa, Kisumu, Nakuru, and Eldoret.

CompuServe, an information service that is offered commercially to users throughout the world, is available in Kenya via dial-up lines.

4.5.3 Business Challenges for KP&TC

In the early 1990s, KP&TC and its current and potential sources of capital, such as international lending agencies, faced a major dilemma. On the one hand, there was overwhelming evidence of vast excess demand for telecommunications services in Kenya, as demonstrated by huge waiting lists and by the congestion resulting from the many call attempts exceeding the capacity of the network.[23] On the other hand, there were indications that KP&TC would find it more and more difficult to finance a sufficient level of capital investment to catch up with demand, either by self-financing from retained profits (since profitability has been falling rapidly) or through outside loans or other forms of funding.

The financial deterioration of KP&TC in the late 1980s and early 1990s occurred in part because of a hugely expanded investment program, as we noted earlier. Commitments extended far beyond the projected level of investment for the Third Telecommunications Development Program that KT&PC agreed to with one of its major lenders, the World Bank, in 1985–86, and they appear to have strained the limits of KP&TC's financial ability to sustain its capital spending program.

Also, KT&PC's relations with the World Bank and other lenders and donors deteriorated in the early 1990s. In addition to relations with the World Bank, several bilateral relationships appeared to be troubled. The Danish aid program, Danida, as well as a similar Swedish program, both reduced their level of funding for telecommunications in Kenya. In Danida's case, the problem appears to have been (at least partly) a reflection of a controversy over the KP&TC pay phone program, in which KP&TC initially favored card phones using prepaid cards of relatively high denominations over the easier-to-use, lower denomination cards favored by Danida that would be more practical to low-income people. This issue was eventually overshadowed by a larger controversy with several aid donors over alleged corruption in the Kenyan government, which for a time resulted in the suspension of new aid from Denmark. Also, assistance from Norway ceased as a result of a disagreement over human rights (rather than telecommunications) issues.

The net result of these events was that KP&TC, with reduced access to finance on concessionary terms and with a weak financial position that limited its ability to self-finance its own capital investment program, appeared in the early 1990s to be depending increasingly on credits from equipment suppliers or their home-country governments to finance its telecommunications. Such credits can often be expensive and restricting.

Finding the right level and composition of investment in the network, balancing the need to catch up with high and growing demand against the requirements of financial solvency, continues to be the major challenge facing the management of KP&TC.

4.6 The Future

The future prospects for telecommunications in Kenya are poised between great promise and significant problems. The promise arises from the strong base for future growth provided by KP&TC's technical skills, the introduction of digital technology into its network, and the skills of its numerous sophisticated and demanding telecommunications users. The problems arise from the difficulty of lending to a large capital program at KP&TC on a financially sound basis. We believe that the telecommunications policy debate in Kenya over the next few years should, and probably will, focus on six key subjects: technology choices, the financing of the capital investment program, regulatory reform, "liberalization" (i.e., the policy option of allowing or even encouraging increased competition), "parastatal reform," and privatization. Following are our views on each of these areas.

"Technology choice" refers to the widening range of technology options available for expanding the network and especially the local loop. Issues of special importance are the following:

- What role should satellite communications linking small earth stations ("very small aperture terminals" or VSATs) play, especially in serving rural areas and large multipoint user applications?
- Is there a case to be made for providing direct optical fiber links to the largest users?
- Should the newer radio technologies, including digital radio, be used only for mobile services, or should KP&TC's strategy work from the outset toward a radio-based infrastructure ("wireless local loop") that can, in an evolutionary way, replace wired local loops as a means for providing user access to fixed services and mobile services?

In our judgment, the rate of advance in technical capabilities and cost reduction in all three areas is such that extensive use of these new technologies deserves serious consideration.

With respect to financing the capital program, notwithstanding KP&TC's financial difficulties of the early 1990s, the fundamentals that determine whether it is intrinsically feasible to finance a high level of capital investment in telecommunications continue to be favorable. Demand for service from users is strong and includes a large amount of demand from major business users who are willing to pay for substantially greater use of telecommunications (at profitable prices for KP&TC), especially if the quality of service is improved. The existence of a strong network infrastructure should enhance the return on additional investment in the network, provided that the investment program is planned and implemented appropriately. Finally, expanding access to the network and reducing congestion will generate additional international traffic, and hence foreign exchange revenues for KP&TC, as a result of the international settlements process, as long as KP&TC continues to experience a large surplus of incoming over outgoing traffic.[24] It may be possible to use settlements revenues as security for the loan financing of elements of the capital investment program, as has been done in other countries (notably the Philippines).

In the regulatory arena, Kenya clearly needs to adopt the principle, increasingly accepted in other countries, of separating responsibility for regulatory control of the telecommunications sector from the operational function of building and operating networks and providing services. An independent regulatory agency should be created that works at arm's length from KP&TC and the central government and is charged with both ensuring the economic viability of KP&TC and protecting user interests. Such an agency would be more likely than a centralized government agency to adopt a sound pricing policy rather than "squeezing" KP&TC with either no price increases or small increases at a time of inflation and major exchange rate depreciation that result in falling revenue in real terms per unit of traffic. Such independent regulatory supervision would be essential if any additional steps toward open markets and competition are to be undertaken.

Is there, in fact, a case for further movement toward competition? During our interview fieldwork, KP&TC's management argued strongly against such changes, primarily on the grounds that significant scale economies would be lost (because Kenya's national network is relatively small) and because "cream skimming" of profitable traffic by new entrants would undermine KP&TC's ability to bear the high costs and limited revenues derived from serving rural areas. On the other side of the issue are the benefits of innovation, customer responsiveness, business discipline, and cost/price reduction that have clearly arisen from the transition to competition in telecommunications in other countries. Although the concerns expressed by those on the anticompetition side of the debate have some merit, policies can be developed that satisfy these concerns and still permit competition to develop and yield large benefits.

Whether or not competition in the telecommunications services business arrives in Kenya in coming years (as it already has in the CPE market), there is no doubt that KP&TC will continue to be by far the biggest provider of telecommunications services in Kenya well into the next century. Its performance is therefore a matter of critical importance for Kenya's economy. A wide variety of proposals for improved management of parastatal bodies such as KP&TC are under discussion worldwide under the heading of "parastatal reform."[25] The general idea behind such reform is to introduce into parastatal enterprises, such as KP&TC, a more commercial orientation, with greater cost consciousness, innovation, and responsiveness to users.

The World Bank is increasingly making parastatal reform a condition for many of its investments in developing countries. In the case of KP&TC, we understand that the World Bank has made the recommendation that the posts enterprise be separated from telecommunications; that the regulatory framework be restructured and regulatory ground rules made more favorable to competition; and that KP&TC be run as a commercial entity. The possibility of privatizing KP&TC may also have been discussed. Kenya's treasury, however, was arguing that certain parastatals should be excluded from this reform program because they are of strategic importance to the state. The Kenya Post and Telecommunications Corporation would be considered one of these. In 1995, Kenya's Parliament reviewed legislation to separate KP&TC postal and telecommunication divisions. Formal separation of these two divisions took place in 1996.

Leaving aside the issue of privatization, there is a pressing need for parastatal reform to stimulate KP&TC to improved economic performance through more commercial management. At the same time, however, the corporation needs to maintain its social obligations with respect to, for example, the provision of service in rural areas.

The option of privatization returned for debate despite strong opposition. Even more than the successful privatization of British Telecom and of NTT in Japan, the decisions by developing countries such as Mexico to privatize their PTTs, and the successful results, are likely to have a major impact. If Mexico, which has a long tradition of sensitivity to issues of economic independence and national control of the economy, finds that it has been beneficial to privatize Telmex (subject to certain continuing limitations on foreign shareholdings), it will be difficult for policy makers in other developing countries to ignore the precedent.

In short, telecommunications planners and managers in Kenya, and national policy makers considering policy toward the sector, are likely to have a very full agenda. Fortunately, they can draw on a substantial reservoir of Kenyan expertise and experience developed over the last several decades.

Notes

1. The BPO's telecommunications business subsequently became British Telecom.

2. Note that between 1985 and 1991, the value of Kenya's exports, as measured in U.S. dollars, grew 5 percent annually. This rate of growth for export value is lower than the rate of export volume growth due to the impact of the falling exchange rate of the Kenya shilling against the U.S. dollar.

3. Statements from documents made available by KP&TC and interviews conducted during our fieldwork in Kenya.

4. To ensure that all met the approval and licensing standards, KP&TC announced courses on a fee basis for staffs of private contractors.

5. During mid-1991, when this information was gathered, the exchange rate of the Kenya shilling was approximately 27 Kenya shillings to 1 U.S. dollar.

6. The increase in working lines in 1993 over 1983 was 122 percent (International Telecommunications Union 1994; Mulugua 1994).

7. These data were provided by KP&TC in an interview conducted by H. Renfrew in 1991; additional data on call completion are contained in the KP&TC annual reports. This measure includes call failures for all reasons, including "called party does not answer." Nevertheless, the very low figure for long distance call completion certainly reflects a significant problem of congestion.

8. High traffic growth is evident. Between 1983 and 1987–88, telephone traffic more than doubled: automatic dialed units grew from 565 million in 1983 to 1.2 billion in 1987–88. More recent comparable data on domestic traffic (i.e., traffic stated in automatic dialed units) are not available. Statistics showing growth in operator-dialed calls, despite a dramatic increase in automatically dialed calls, are another indication of total traffic increases in Kenya: the annual volume of operator-controlled national calls is 42 percent greater in 1990 than in 1983—an indication of high traffic growth despite increased automation on the domestic network. International traffic has also increased, assisted by network automation: the annual volume of outgoing international traffic in 1989 was 107 percent more than traffic in 1983. The percentage of international outbound direct-dial traffic from Kenya rose from 68 percent in 1985 to 97 percent in 1992 (International Telecommunications Union 1993).

9. This is compared with 62 percent of all rural exchanges in 1987 (Mulugua 1994).

10. In Kenya, this problem of interpretation is compounded by discrepancies between the data available from KP&TC and those reported in the ITU's *Yearbook of Public Telecommunications Statistics,* the other standard source.

11. See chapter 12 in Saunders, Warford, and Wellenius 1983.

12. In 1978, there were only 1,490 such phones in Kenya, including rented coin phones (coin boxes) on private premises such as hotels and restaurants.

13. It is difficult to reconcile data from different sources on Kenya's international circuit capacity. For example, the ITU *Yearbook of Common Carrier Statistics* cites 680 international circuits serving Kenya in 1992.

14. In 1980, the VHF system had twenty-four subscribers, with capacity for ninety-six. By the end of June 1988, the system still had the same capacity but with forty subscribers sharing six frequency channels.

15. Inferences on total leased-line growth cannot be derived simply from comparing the 1980 leased-line figures and the 1988 modem figures, because domestic leased lines used for data require two modems (i.e., one at each end of the line), while international leased lines require only one modem for the Kenya end of the link.

16. Dial-up X.28 and X.32 and dedicated X.25 access are available at speeds between

300 and 9,600 bps. Where access is via dial-up network or analog leased lines, KP&TC advises that type-approved modems be bought from Racal or NCR and installed by KP&TC personnel or their authorized agents. There are two switches in Nairobi and nodes in Kisumu, Nyeri, Mombasa, and Nakuru. Gateways give X.75 access to IPSS in the United Kingdom and Telenet in the United States. We were told by KP&TC that Kenpac will support a wide variety of data communications protocols: X.28, 3270 SNA DSP, SDLC, 3270 Bisynchronous DSP, and X780 Bisynchronous (2780/3780) after conversion to the standard interface protocol, X.25. During our interview fieldwork in 1991, however, some users questioned whether the full range of planned protocol-support capabilities was fully effective.

17. The World Bank loan for the Third Telecommunications Program was approved by the World Bank board in June 1985: the loan agreement was signed in May 1986. The original loan amount was U.S.$32.6 million.

18. For another example beyond the research discussed in detail here, see Cleevely and Walsham 1980.

19. The methodology and findings are described in Tyler and Jonscher (1982).

20. By "full demand" we mean the volume of traffic that would occur if all customers wanting to rent exchange lines and make calls *at the currently prevailing prices* were accommodated.

21. Based on the exchange rate of the Kenyan shilling at 27 KSh to U.S.$1 at the time this field research was conducted in 1991.

22. From documents made available during our interview fieldwork in Kenya.

23. Strictly speaking, the evidence shows that there is large excess demand at KP&TC's current levels of pricing, which are probably below economically efficient levels because of government restrictions. However, in our judgment there would continue to be a large amount of excess demand at any reasonable level of pricing.

24. In 1990, there were 26.1 million minutes of incoming telephone traffic to Kenya and only 21.2 million minutes of outgoing traffic. By 1992, outgoing international traffic was 22.3 million minutes, showing a modest 5 percent rate of growth.

25. See, for example, Wellenius, Stern, Nulty, and Stern 1989.

Bibliography

Cleevely, David D., and Geoffrey Walsham. 1980. Interim Report on Modelling the Role of Telecommunications within Regions of Kenya. Working paper, Department of Electrical Engineering, University of Cambridge, England, May.

Godard, Paul. 1994. Africa and Science: The Availability of Computer Communications. In *Telecommunications and Development in Africa,* edited by B. A. Kiplagat and M. C. M. Werner. Amsterdam: IOS Press.

International Telecommunications Union. 1993. *African Telecommunications Indicators.* Geneva: International Telecommunications Union.

______. 1994. *Yearbook of Common Carrier Telecommunication Statistics.* 21st ed. Geneva: International Telecommunications Union.

Kenya Post and Telecommunications Corporation. 1990. *Annual Reports and Accounts.* Nairobi: Kenya Post and Telecommunications Corporation.

Mulugua, Rufus. 1994. Financing the Development of Rural Telecommunications: The Kenyan Experience. In *Telecommunications and Development in Africa,* edited by B. A. Kiplagat and M. C. M. Werner. Amsterdam: IOS Press.

Okundi, Philip O., and B. G. Evans. 1975. Afrosat: Proposals for an African Domestic

Satellite Communications System. Paper presented at the International Conference on Satellite Communications Systems Technology, London, April.

Okundi, Philip O., Arthur W. Ogwayo, and James K. Kibombo. 1977. Rural Telecommunications Development in East Africa. In *Proceedings of the International Telecommunications Exposition INTELCOM 77,* Vol. 1. Dedham, Mass.: Horizon House International.

Saunders, Robert J., Jeremy J. Warford, and Bjorn Wellenius. 1983. *Telecommunications and Economic Development.* Baltimore, Md.: The Johns Hopkins University Press.

Tyler, Michael and Charles Jonscher. 1982. The Impact of Telecommunications on the Performance of a Sample of Business Enterprises in Kenya. Case Study no. 18, ITU-OECD Project "Telecommunications for Development." Geneva: International Telecommunications Union.

Wellenius, Bjorn, Peter A. Stern, Timothy E. Nulty, and Richard D. Stern. 1989. *Restructuring and Managing the Telecommunications Sector: A World Bank Symposium.* Washington, D.C.: The World Bank.

5
Tanzania

M. L. LUHANGA

5.1 Physiography

The United Republic of Tanzania is a littoral state of the Indian Ocean located between longitudes 29 and 40 degrees east and latitudes 1 and 11 degrees south. The United Republic of Tanzania is made up of a union of two formerly independent countries—Tanganyika (on the mainland) and the islands of Zanzibar and Pemba in the Indian Ocean. Tanzania mainland (formerly Tanganyika) has an area of 94,000 square kilometers of which 20,000 square kilometers is covered by lakes. The islands of Zanzibar and Pemba have a combined area of 2,500 square kilometers.

The population of Tanzania in 1995 was estimated to be 30 million growing at an average annual rate of 2.8 percent. The overwhelming majority of the rural population consists of small holder farmers and herders.

5.2 Economics and Politics in Tanzania

The primary economic activity in Tanzania is agriculture, which in recent years has accounted for about 40 percent of the GDP and about 70 percent of annual export earnings. Tanzania also has a small industrial sector and a small manufacturing sector, both accounting for less than 20 percent of the GDP.

Between 1967 and 1991 Tanzania followed a socialist economic philosophy. A large parastatal sector was built up to carry out activities in the areas of industry, agriculture, transport and communications, mining, education, health, fisheries, and so on. But starting in 1991, Tanzania has been following a free-market, capitalist philosophy. The government has started divesting itself of all production parastatals and is encouraging private-sector participation in the economy.

For about the past thirty years, Tanzania has been ruled by one political party. In 1992, however, Tanzania adopted a multiparty political system. The first multiparty parliamentary and presidential elections were held in October 1995.

Economic and political liberalization has gone hand in hand with the liberalization of policies governing the news media. Tanzania has gone from a state where all news media were government controlled to a state where there is vibrant private press, private radio stations, and private television companies.

5.3 Recent History of Tanzania

In April 1964, Tanganyika united with the Indian Ocean islands of Zanzibar and Pemba to form the United Republic of Tanzania.

Tanganyika was a German colony up to 1918, the end of the First World War. From 1918 to 1961 Tanganyika was ruled by Great Britain as a Trustee Territory of the League of Nations (1918–45) and of the United Nations (1945–61). Tanganyika obtained her independence from Great Britain in 1961 and became a republic in 1962.

5.4 Telecommunications in Tanzania Today

Telecommunications services in Tanzania were provided by one public-switched telephone network (PSTN) operator, the Tanzania Telecommunications Company Limited (TTCL). Private telecommunications networks are also operated by the railways, the harbors, the electric utility, the military and security establishment, and private and public mobile and nonmobile radio communications systems.

5.4.1 Private Telecommunications Networks

Although the military and other security services operate private radio and telecommunications networks, no data on these networks are available in the public domain, and therefore, these networks will not be covered in this chapter. Of the remaining private networks, the large ones belong to the Tanzania Electric Supply Company (TANESCO), SITA (the airline reservation system), and the railway signaling and telecommunications systems of the Tanzania Railways Corporation (TRC) and the Tanzania Zambia Railway Authority (TAZARA).

The TAZARA telecommunications system links all stations along the railway line. The system has automatic exchanges in all electrified towns and magneto exchanges in nonelectrified towns. Some magneto exchanges have been replaced with automatic ones by using photovoltaic power systems in nonelectrified stations.

The TRC telecommunications system also links all stations along the railway network. The TRC network has automatic exchanges in electrified towns and in stations with photovoltaic power systems. It uses magneto exchanges in other stations. Maintenance of the TRC network is done under contract by TTCL.

The SITA airline reservation system links major airline offices in Dar es Salaam with other offices worldwide via leased voice grade circuits through Nairobi to London. Although the SITA network is a packet-switched network, it is operated in a circuit-switched mode from Tanzania to London. Within Tanzania, the Air Tanzania Corporation (ATC) operates terminals in major up-country towns it serves and in Zanzibar. These terminals access the SITA network through a gateway in Dar es Salaam. The ATC network uses a star topology with messages between any two terminals in the network having to pass through the node in Dar es Salaam. The SITA network enables any ATC terminal to access, store, and/or retrieve data from host computers belonging to SITA that are situated in the state of Georgia in the United States.

The TANESCO telecommunications system provides voice communications services needed to control the operation of its interconnected electricity generation, transmission, and distribution system. In order to improve its reliability in operating this interconnected network, TANESCO has installed a system for supervision, control, and data acquisition (SCADA) that is built up as follows:

- A control center in Dar es Salaam.
- A SCADA system connected to existing power plants and substations with possibilities of up to fifty connections.

The SCADA system transmits data over voice grade circuits using several voice frequency telegraph (VFT) channels that employ frequency modulation (FM) in accordance with international telecom recommendations: R.35 (50 baud), R.37 (100 baud), and R.38A (200 baud). Separate VFT channels are also provided for telephony and telex systems.

The TANESCO telephone system deployed fifteen automatic exchanges with a capacity of 288 connections, out of which 75 connections were used as trunk lines. This telephone system has more sophisticated traffic handling facilities than those available on the TTCL network. For example, TANESCO's system offers a conference facility, multiaddress signaling, and prioritized customer categories. The telex system operated by TANESCO links ten substations and offers full duplex communication between the substations.

5.4.2 The Telecommunications Network of TTCL

5.4.2.1 Basic Services

Voice telephony and telex services are offered by the Tanzania Telecommunications Company Limited, the only PSTN operator so far licensed. The PSTN consists of switching, transmission, and subscriber loop facilities. In switching there is a diversity of equipment ranging from old step-by-step exchanges through crossbar exchanges to digital switches.

Tanzania had a total of 218 exchanges in 1995, out of which 34 were digital with 39,000 direct exchange lines (DELs). Table 5.1 presents data on these exchanges, including their number, their suppliers, and their capacity. All manual exchanges are from Ericsson of Sweden. The total installed capacity in 1995 was 130,000 lines. The average exchange fill was about 74 percent, and 99,000 DELs were connected. Of the DELs connected in 1995, 38 percent were served by thirty-two analog exchanges and 52 percent were served by thirty-four digital automatic exchanges. The remainder of the lines were served by 152 small manual exchanges. Seventy percent of the exchanges, serving over 40 percent of the subscribers, are more than twenty years old and in poor condition.

In 1995, the telephone density in Tanzania at 0.33 DELs per 100 population was among the lowest in the world. Service is concentrated on Dar es Salaam,

Table 5.1. Exchanges and Suppliers

Switch Type	Supplier/ Manufacturer	Country	Quantity	Capacity (no. of lines)
		Digital Exchanges		
GX 5000	Mitel	Canada	17	17,700
NEAX61E	NEC	Japan	6	26,000
FETEX F150	Fujitsu	Japan	4	10,000
HDX 10	Hitachi	Japan	2	6,000
S 12	Alcatel	Belgium	2	3,000
FEDEX 100	Fujitsu	Japan	1	2,000
CDOT 128	W. S. Telesystems	India	1	80
ELTEX V	Sagem	France	1	20,000
Total			**34**	**66,780**
		Analog Exchanges		
C 400	Hitachi	Japan	10	22,200
C 23	Hitachi	Japan	13	8,200
S × S	Plessey	U.K.	4	13,100
C5	Hitachi	Japan	3	2,502
X-Bar BTM	Bell	Belgium	2	550
	Total		**32**	**46,442**

connecting it with all major towns in Tanzania and with the neighboring countries of Malawi, Zambia, Kenya, Rwanda, and Uganda. International voice traffic is transmitted via earth satellite stations at Dar es Salaam. By 1998, the telephone density increased to 0.62 DELs per 100 population.

There is an automatic international digital telex exchange in Dar es Salaam. Telex service is also available to fifteen other major towns in Tanzania served by three analog and one digital exchange. Total telex capacity is about 4,000 subscribers.

Although the only problem with telex service is inadequacy, telephone service is not only inadequate but also congested and unreliable. The inadequacy of the telephone service is so severe that the number of customers waiting to be provided with service countrywide is about 150,000, whereas the number of DELs is 99,000. The real pent-up demand is estimated to be 400,000. Congestion on the network is caused mainly by poor maintenance and inadequate capacity (both switching and transmission capacity) on the trunk network. The poor maintenance results in high fault rates, with an average of about 70 percent of DELs being in working condition at any one time.

The traffic on the network of the Tanzania Telecommunications Company Limited (TTCL) is 60 percent local and 30 percent trunk, with the balance being international traffic.

A Memorandum of Understanding signed by TTCL and the government of Tanzania set up specific performance targets on network performance, revenue generation, and staffing that TTCL has to meet. An independent consultant is engaged annually by the regulatory authority to carry out a performance audit of TTCL to verify whether or not TTCL is meeting the agreement and license targets.

Some of TTCL's customers have started using callback (or call reorigination) equipment to take advantage of the low telecommunications tariffs available in the United States. The use of the callback facility has had a negative impact on the revenue that TTCL obtains from international traffic.

Tanzania is one of the twenty-five poorest countries in the world. As such it receives most of its aid from developed countries in the form of grants rather than loans. The telecommunications sector has not been an exception. Prior to the inauguration of the Telecommunications Restructuring Program (TRP), some developed countries provided grants that enabled TTCL and its predecessor TPTC to rehabilitate and expand the telecommunications network in Tanzania.

Some of the donor countries that have provided grants for investments in the telecommunications network in Tanzania have not been keen to see TTCL privatized. These countries called for a feasibility study on the privatization of TTCL, specifically taking into account investments they had made in TTCL through grants to the government of Tanzania.

5.4.2.2 Mobile Cellular Telephony

Tanzania licensed a private company to offer mobile cellular telephone service starting in 1993. The company used analog technology to provide the service to customers in Dar es Salaam and Zanzibar.

In 1994, another private company was licensed to offer mobile telephone services in Dar es Salaam and Zanzibar, using digital GSM technology. Subsequently, TTCL began to offer mobile cellular telephone service to some of the other large towns in Tanzania.

5.4.2.3 Data and Paging Services

Data services were provided in Tanzania after 1996 through a joint venture between TTCL and a French company. TTCL also offered paging services.

5.4.2.4 E-Mail and Internet Services

Electronic mail services are at their infancy in Tanzania with less than ten nodes connected to the FIDONET network in 1995. The largest number of E-mail users was at the University of Dar es Salaam. The computing center at the university, which operates the E-mail system, also operates terminals where the public can send and receive E-mail messages.

Starting in January 1996, full Internet services became available for the first time in Tanzania via a leased circuit to Uninet (South Africa). The circuit had 9.6 Kbps of capacity. The University of Dar es Salaam was the first institution to offer Internet services in Tanzania.

5.4.2.5 Taxes

The government requires TTCL to pay the following taxes:

1. Sales tax on telephone and telex services at the rate of 15 percent of the revenue earned from the provision of the service.
2. A corporation tax at the rate of 35 percent of the profits.
3. A payroll levy at the rate of 4 percent of the gross emoluments paid to staff.
4. A stamp duty at the rate of 1 percent of the receipts of the company.

5.4.2.6 Non-Core Business Activities

The operating license of TTCL permits the company to engage in businesses other than telecommunications. The policy of the company is, however, that of divesting itself of all non-core businesses.

5.4.3 The Regulatory Environment

5.4.3.1 The Tanzania Communications Commission

The government of Tanzania established the Tanzania Communications Commission in 1994 to regulate, among other things, the provision of telecommunications services in Tanzania. In that regard, major functions of the commission are as follows:

1. To license providers of telecommunications services.
2. To license and regulate the allocation and use of satellite orbits and the radio frequency spectrum.
3. To approve telecommunications tariffs.
4. To promote competition in the provision of telecommunications services.
5. To approve equipment to be used in the provision of telecommunications services.

Prior to the formation of the Tanzania Communications Commission, regulatory functions and the provision of postal and telecommunications services were being carried out by the same organization, the Tanzania Post and Telecommunications Corporation (TTPC)—an organization that was created in 1977 to replace the East African Posts and Telecommunications Corporation following the breakup of the East African community in 1977.

The Tanzania Communications Commission is mandated to promote the development of rural telecommunications in Tanzania. To achieve this objective, the commission plans to set up a Rural Telecommunications Development Fund to which all licensed operators of telecommunications networks in Tanzania must contribute.

5.4.3.2 The License Granted to TTCL

The Tanzania Communications Commission (TCC) granted a license to TTCL in 1994 allowing the company to provide telecommunications services in Tanzania. The main highlights of the license are as follows:

1. It does not give TTCL a monopoly status. The commission is at liberty to license other providers of telecommunications services.
2. The duration of the license is twenty-five years, and it cannot be revoked in the first fifteen years.
3. Cross-subsidies in TTCL's tariffs are not allowed.
4. Once every year TTCL must undergo an efficiency audit to verify operational performance vis-à-vis targets set in the license.
5. Interconnection service must be provided by TTCL, at rates approved by

the commission, to all other parties wishing to offer telecommunications services in Tanzania.

The terminal equipment market is fully liberalized, and TTCL has allowed private individuals and companies to perform internal wiring of customer premises.

The license granted to TTCL commits it to the provision of universal service. Since most people in Tanzania live in rural areas where there is little business to pay for telecommunications facilities, provision of universal service is to be made possible by requiring all telecommunications licensees to contribute to a Rural Telecommunications Development Fund. This fund will then provide for the expansion of telecommunications services to the rural areas.

5.5 The Future of Telecommunications in Tanzania

5.5.1 Constraints on Efficiency

Future activities in the telecommunications sector will have to address the following problems currently afflicting the telecommunications network in Tanzania:

1. Severe congestion at the local and trunk exchange levels.
2. A high percentage of faulty equipment and deteriorated external plants.
3. A lack of spare parts and qualified staff.
4. A low level of successful interurban calls (about 25 percent).
5. The length of time it takes to repair a fault (an average of ten days).
6. Incorrect and late billing.

5.5.2 Strategies for Removal of the Constraints

Prior to 1991, the approach to network expansion and to solving the constraints just mentioned was ad hoc and without any clear strategy. Starting in 1991, TTCL has reviewed the existing quality of service and broadly defined the following areas for improvement:

1. Institutional development to ensure efficient and effective management and maintenance.
2. Major rehabilitation and expansion of the telecommunications network.
3. Alleviation of traffic congestion.

A Telecommunications Restructuring Program (TRP) has been conceived to address these issues. The program addresses network expansion and rehabilitation, staff training, and the stocking of spare parts. The program raised revenues from U.S.$56 million in 1991 to U.S.$166 million in 1995.

5.5.3 Competition

The licensing of two mobile cellular telephone operators in Dar es Salaam and Zanzibar allowed a private company to break ground for a satellite station and to put up exchanges that will enable the company to offer telecommunications services in Zanzibar.

Initially, one company offered mobile cellular communications services. The second operator commenced operations later in 1995. The initial tariff was U.S.$0.4 per minute for local calls. The quality of service was good, although there was congestion between about 9:30 A.M. and noon.

A private company, ZANTEL, was the first to apply for and be granted a license to provide basic telecommunications services in Tanzania. This company provides basic telecommunications services in Zanzibar. It will also provide international telecommunications services using its own satellite earth station. The Tanzania Communications Commission granted ZANTEL its license.

Set up as a joint venture company between the government of Zanzibar and a private company based in Ireland, ZANTEL is responsible for obtaining financing for the implementation of the telecommunications project in Zanzibar. The government of Zanzibar is not required to contribute funding, but it will be able to be an equity shareholder in the joint venture company.

The funds for the project, estimated at U.S.$25 million, was raised via a combination of the following:

- Medium-term suppliers' credits.
- Support from the Overseas Private Investment Corporation, an organization of the U.S. government.
- Support from the United States Agency for International Development.
- Other participants, comprising local, private, and foreign investment groups.

The ZANTEL network includes the following:

1. An earth station.
2. Digital switching equipment, comprising an international transit exchange and local exchange equipment (10,000 lines) to provide telephone service in Zanzibar.
3. Customer local telephone distribution system using wireless communications technology.
4. Cellular mobile telephone service.
5. Interface and interconnection equipment to enable the ZANTEL network to work with the TTCL network.

The license granted to ZANTEL is more restrictive than TTCL's. It contains the same provisions as TTCL's license with the following special issues:

1. It limits ZANTEL to providing only basic fixed telephone services, meaning that the company cannot provide mobile cellular services or use wireless local loop technology.
2. It confines ZANTEL to offering telecommunications services to customers in Zanzibar and Pemba islands only.
3. International traffic carried by ZANTEL must be from and to customers only in Zanzibar and Pemba.
4. It prevents ZANTEL from offering value-added services.

5.6 Conclusion

Telecommunications services in Tanzania are provided to only a few individuals. The services currently provided are inadequate and poor. A program for an expansion of the service, rehabilitation of the network, and training of staff that has been embarked upon may improve the situation. The liberalization of the telecommunications sector, allowing more operators and service providers, augurs well for the future of telecommunications in Tanzania.

Bibliography

Licence Granted by the Tanzania Communications Commission to the Tanzania Telecommunications Company Limited under Sections 9, 11, and 12 of the Tanzania Communications Act 1993. 1994. Tanzania Communications Commission, December 30.

Tanzania Communications Act No. 17. 1993. The Government of Tanzania.

Telecommunications Restructuring Programme—TRP. 1994. Tanzania Telecommunications Company Ltd.

Report of the Directors. 1994. Tanzania Telecommunications Company Ltd.

6
The Congo

E. BISIMWA GANYWA AND BUKASA TSHILOMBO

The Republic of Congo (formerly known as Zaire) is the third largest country on the African continent with 2,345,000 square kilometers. It is also one of the most populous countries with more than 41 million inhabitants. For over a decade, the country has faced several serious economic and political challenges that have affected the prospects of this otherwise resource-rich country.

This chapter discusses the development of the Congo's telecommunications sector. It presents the state of telecommunications in the Congo as of the mid-1990s and assesses the obstacles confronting further development. The conclusion provides broad recommendations for improving the current state of telecommunications and for moving the Congo's telecommunications sector into the twenty-first century.

6.1 The Colonial Period

Telecommunications technology was introduced into the country during the colonial period solely for the colonial power and its administration. The Belgian Congo was a region of intense economic activity, and because the productive units were scattered, considerable resources were placed into telecommunications (relative to other African colonies). The telecommunications system linked the several production units within the Belgian Congo and the capital Leopoldville, which is now called Kinshasa. The Belgians invested in infrastructure expecting the country to remain a colony. The events of 1959–60 went against all Belgian expectations.

6.2 Postindependence

The Congo gained its independence in 1960. Telecommunications was placed under the administration of the National Office of Zaire for Post and Telecommunications (the ONPTZ). Having had no role in the development of the system, however, and no experience in technology or management, the newly independent nation and its citizens did not realize the importance of telecommunications for social and economic development. During the 1960s and the 1970s, for example, children were still making toys out of the wires used for telecommunications. Further, the lack of political stability during the postindependence era resulted in a discontinuity of telecommunications development. Subsequently, there has been no strategy for thirty years; each administration has simply restarted the same telecommunications projects. Since independence, ONPTZ has continuously blamed the poor service on the antiquated and obsolete equipment it is forced to use. While this is partially true, other questions must be answered.

When the telecommunications data from 1980 to 1988 is examined, it is evident that while the network was antiquated, it was still underutilized. In 1985, the number of telephones was 38,845. By 1988, these figures had dropped dramatically, except the number of major lines, which grew from 29,010 to 29,186. With respect to traffic volume, 1985 was the year with the greatest volume: for national calls, volume changed 786.3 percent (1,604,000 more calls) over 1980; 598 percent (or 1,549,000 calls) over 1982; and 343 percent (or 1,400,000 calls) over 1988. Thus, there is considerable room for increased utilization of the current network. The government invested an estimated U.S.$33.1 million in telecommunications infrastructure in 1988 and employed 6.75 (±0.025) percent of the population in the telecommunications industry. According to the ITU, the number of main lines in Zaire increased from 26,600 main lines in 1980 to 29,186 in 1988. Meanwhile, the amount of customer premises equipment (CPE) connected jumped from an estimated 27,300 in 1980 to 38,845 in 1985, only to fall back down to 32,116 in 1988.

In the 90s, the following telecommunications development projects were under way:

1. Installation of 8,000 lines for the town of Kinshasa, divided into four digital telephone exchanges using system S 1240. (Financed by Belgium for a total cost of 350 million Belgian francs. Supplier: Alcatel-Bell telephone.)
2. Installation of two central telephone exchanges with 9,000 lines, one in Kinshasa with 6,000 lines and the other in Kisangani with 3,000 lines. (Financed by Italy with a U.S.$9.3 million fund. Supplier: Alcatel-Siette.)
3. Installation of a digital microwave system with large capacity (34 megabits per second) linking Kinshasa to the port of Matadi and serving the towns of Kasangulu, Inkisi, Mbanza Ngungu, Lufutoto, Tmba, Lukala, Kimpese, and Songololo. (Financed by France for a cost of 20.5 million French francs. Supplier: Alcatel-ATFH.)

The following telecommunications projects were announced:

1. The construction of a trunk network associated with the four telephone exchanges just mentioned. (Financing by Belgium for 368 million Belgian francs, of which 100 million was agreed on for the first phase. Supplier: Belca.)
2. Acquisition of exchanges and trunk networks in Bukavu, Gbadolite, and Goma. (Financing by Italy for U.S.$15.4 million. Supplier: Alcatel-Siette.)

3. Study of the reorganization of telecommunications at the national level. (Financing by the World Bank for U.S.$750,000. Execution agent: Tracte-Bell.)
4. Implementation of the completed studies for the restoration of the telecommunications network of the town of Kinshasa and the restructuring of the ONPTZ at a cost of U.S.$80 million.

In 1991, ONPTZ began using Intelsat's Planned Domestic Service to extend service to the rural areas via a 36 Mhz bandwidth leased transponder and fourteen standard Z earth stations supplied by Telespace of France. Zaire uses microwave relay links for national trunks and international links to neighboring countries. Microwave relay links were utilized along the following routes:

Kinshasa–Mbanza Ngungu–Matadi
Lubumbashi–Likazi Kolwezi (in good condition)
Goma–Bukavu–Uvira
Bukavu-Cyangugu (Rwanda)
Lubumbashi-Tshingolo
Kinshasa-Brazzaville (Congo)
Kananga–Mbuji Mayi (in good condition)

Thus, by the early 1990s, ONPTZ was making some headway in improving and expanding telecommunications service. The political troubles of the 1990s, leading to the eventual ouster of President Mobuto, imperiled these projects. Demand was clearly shown: in 1993, for example, 10,420 were on the waiting list even as line capacity increased by over 70 percent from 30,000 lines in 1988 to 51,000 lines in 1991. But the collapse of the economy and of the Mobuto regime put telecommunications development on hold.

6.3 The Present

6.3.1. Domestic Telecommunications

It is extremely difficult in the mid-1990s to make a call from Kinshasa to Bukavu—a distance of 2,000 kilometers, and telephone calling between Bukavu and Goma—only 200 kilometers away—is often impossible.

Further, telecommunications continue to be plagued by the same problems that cripple the rest of the country's public sector: inflation and nonindexed salaries lead government employees to personal entrepreneurship. For example, communications with Europe are often sold by individuals with access to a business line who charge others for the use of their phones. Similarly, some ONPTZ employees sold minutes or hours of international calls to private individuals. These calls were then undertaxed and the money divided among the participating employees.

In 1993, Zaire's minister of Telecommunications presented an urgent priority plan that concerned the city of Kinshasa, the ten regional capitals, the town of Gbadolite, some other important towns, and the Panaftel (Pan-African Telecommunications) project. In this program, the qualitative and quantitative objectives of the restoration and modernization plan for Zaire's telecommunications were defined as follows:

1. Attain 100,000 telephone lines by the end of 1995, compared with 34,000 at the end of 1993.[1]
2. Attain 4,000 telex lines by the same date.
3. Link Kinshasa, the capital, with all the regional capitals through an efficient transmission system.
4. Link Zaire to all the neighboring countries by microwave relay links.
5. Improve the quality of service by installing new central telephone exchanges and some new cable networks to replace the antiquated ones.
6. Reorganize the satellite telecommunications service and build several terrestrial stations including a type A international station.

6.3.2 International Telecommunications

The state of Zaire's international telephone communications is little better than its domestic system. In the mid-1990s, for example, it had become extremely difficult to reach Kinshasa directly from Europe, sometimes requiring repeated attempts over several days to make a single connection. Officially, this situation is attributed to network congestion, but the actual problem is efficient routing. The only effective route for foreign calls is through an operator in Kinshasa. Even then, however, it is practically impossible for an international caller to reach Bukavu via Kinshasa. One way to contact Bukavu from Paris or Brussels is to make a connection with a town like Cyangugu, on the border between Zaire and Rwanda, and hope that someone in Cyangugu will relay the message to Bukavu.[2]

The ONPTZ had practically no control over receipts, and the big departments of the state, which are the largest consumers of telecommunications, did not pay their bills. This, combined with the free fall of the local currency in the 1990s, meant that Zaire could no longer pay the bills for international communications. This situation reached drastic portions in 1993 when Intelsat threatened to disconnect Zaire because of failure to settle its accounts. In 1994, Zaire paid U.S.$1.2 billion to Intelsat and averted having its international link to the world severed.

6.3.3 Solar-Powered Telephone Stations

A project of the Belgian-Zaire Development Cooperation for the restoration and installation of eighty-six solar-powered telephone stations was in the execution phase in late 1993. Thirty-six stations were being installed before the breakup of the Development Cooperation. Providing service to rural locations was one of the objectives of these stations.

Solar-powered technology is useful in a country like the Congo where distances are great and intense solar energy exists in abundance. However, the technology is expensive when photocells are used. In addition, the infrastructure can be stolen and/or vandalized easily. An intermediate solution to these issues is to locate the solar-powered communications devices in protected areas.

6.3.4 Cellular Telephony

The ONPTZ was dealt a hard blow in 1989 when TELECEL, a mobile carrier operating its own telecommunications system, was authorized by the government. Launched in Kinshasa, TELECEL has extended service to Lubumbashi, Goma, and other locations. The ministries and the major companies who were clients of the ONPTZ were first to subscribe to TELECEL. In 1992, TELECEL had 4,200 subscribers. It offered a better quality of service than the ONPTZ, but only the elite could afford to subscribe to it. Subsequently, some argued that TELECEL's success was merely another manifestation of the inequity and centralization that has characterized Zaire's society. Further, it is ironic that much of the ONPTZ's low service quality resulted from nonpayment of telephone receipts by the state departments, exactly those who migrated to the new TELECEL: in 1994, government ministers and officials owned 1,000 TELECEL accounts.[3] As of 1993, TELECEL maintained a virtual monopoly over telecommunications, as the ONPTZ's service was completely unreliable. In 1995, the cellular operator had an estimated 6,500 subscribers using the service throughout the country.

Originally, TELECEL was allotted both the A and B bands; however, in 1994, there was new entry into the cellular market—Comcell. Subsequently, TELECEL was ordered to cease transmission over the B band. The company refused to move out of the B band, stating that it would require U.S.$30 million to restructure its services. In April 1994, armed men raided TELECEL transmitters, cutting cables and causing the network to go down for thirty-six hours.

Comcell eventually completed construction of its own earth station and began building its central exchange. In 1994, it had 1,000 subscribers and a capacity for 3,000 simultaneous connections. Another firm, Transglobal Telecom, was poised to enter the wireless market. Thus, by 1995, we find a scenario in which the PTT had ground to a halt and private companies were moving in to provide adequate communications to those who could afford the service—the elite.

The introduction of competition in the Congolese market is expected to result in lower airtime rates in the short run. In a country where cellular telephony is used as a substitute for wireline telephony and where airtime rates are as high as U.S.$6 per minute for international calls, competition is bound to have a positive effect on the market. In the future, as cellular phone service continues to spread to the mass market, other wireless technologies will be increasingly used to provide communication to both the business and consumer markets in major cities like Kinshasa, Lubumbashi, Mbuji Mayi, and Matadi. These include mobile satellite technology and wireless local loop technology, all of which need a commitment from the government to allocate spectrum. There is evidence of a growing trend toward this necessary development. For example, the IDM Satellite

Division has become a key telecommunications provider for Congo, signing a $12 million deal to set up a wireless network. The deal includes a long-distance communications project.

6.4 The Future

A question remains about the type of technology that should be implemented in the Congo in order to address the bottom line: better quality, reliable, low cost, and profitable communications throughout the country. When social and business factors are taken into account, it is clear that there is a need for immediate voice and fax communications, and later data and Internet-type communications. When the geographic and economic factors are taken into account, it becomes evident that the cost- and time-effective technology to be used in upgrading and extending the Congo's telecommunications network is wireless communications.

Specific wireless solutions include:

- *Cellular and personal communications services (PCS) technologies.* These could be implemented in a multiple carrier environment allowing the entire country to be covered and leveraging the entrepreneurial spirit of private parties. The government could set directions by allocating the frequency spectrum and by suggesting a standard such as the Advanced Mobile Phone System (AMPS), which is already used by TELECEL and Comcell. A digital standard like the Global System for Mobile Communications (GSM) should not be disregarded, but some consistency with existing systems is also desirable.
- *Wireless local loop technologies.* These are at the intersection of wireless and wireline technologies and could be used to rapidly bolster the existing public telecommunications network. Alternatively, licenses could be given to a private organization who would then compete with the PTT. Motorola, among other equipment vendors, has introduced its Wireless Local Loop system (WiLL) in several African countries already. The vendor could push its system through the existing channels in the Congo where it has a long-standing presence.
- *Satellite technology.* Already used in the Congo to a large extent in public telephony and television, the technology could be extended in the form of mobile satellite services provided by private parties. COMSAT's Planet 1 system offers competitive, ubiquitous communications. Iridium, a 66 Low Earth Orbit (LEO) satellite system, has made agreements with local service providers throughout Africa, including the Congo. And the IDM Satellite Division has already become a key telecommunications provider for the Congo, signing a $12 million deal to set up a wireless network. The deal includes a long-distance satellite communications project, and the company has drafted a trio of contracts covering wireless communications, billing systems, and long-distance calls.

- *Private or Specialized Mobile Radio (PMR or SMR) systems.* These trunking systems, which were already widely used in the Congo, could be upgraded with the advanced technologies, including digital and encryption.

5.0 Conclusion

In 1994, the Congo's telephone penetration was estimated at 0.09 DELs per 100 inhabitants. It was estimated that in order to reach the goal of 1 data exchange line (DEL) per 100, a total of U.S.$600 million would have to be invested. Despite the tremendous needs of the telecommunications sector, by 1994, the ONPTZ had ground to a halt. It had more than 100 employees per 1,000 main lines, which was among the highest in the world. As a result, salaries and fringe benefits accounted for 80 percent of the ONPTZ's earnings. Additional expenses including "cream skimming" that occurred with the licensing of the several wireless providers and the shock of a U.S.$1.2 billion retroactive payment to Intelsat that cut into funding that the ONPTZ could have used to improve service.

It is imperative to extend the network beyond Kinshasa, whose centralized role is in any case slowly wearing away. The political changes put into place after Mobuto's fall already give a larger degree of autonomy to the Congo's regions.

The third goal for the sector is to define a more precise administrative framework by separating the postal service and telecommunications. These two sectors face enormous challenges that in many cases can no longer be encompassed by a single ministry or a single national office. Also, the two sectors are managed differently and have different human resources requirements.

The fourth goal of the Congo's future telecommunications policy will be to clarify issues at the organizational level: the sector's operational structures; the responsibilities at different levels, as well as the responsibilities assigned to the regions and to Kinshasa; and the delegation of powers.

The country's telecommunications accounting and financial system also needs to be revised. The tariff system, the collection methods, and the internal controls structure must be reviewed and adjusted. Billing must be automated, and information technology employed to facilitate management.

Finally, the Congo must develop a campaign for training and informing its population in the use and advantages of telecommunications. In a country in which oral communication has historically been important, the widespread adoption and effective use of the telephone and other telecommunications services would translate into good business.

Notes

1. Because of numerous factors, including the failure of twisted pair wiring and attempts to transition to new technology, the number of lines decreased from 51,000 in 1991 to 34,000 in 1993.

2. Seeking an alternative to such haphazard methods of international communication, the authorities of the Catholic University of Bukavu opted in the 1990s for a private tele-

phone line in Rwanda and a fax—avoiding the Congo's international telephone system altogether.

3. The telecommunications sector, however, was not the only victim of the spirit of privatization that pervaded the Congo's power structure in the late 1980s and early 1990s. The national airline, Air Zaire, was also sacrificed for Scibe-Zaire, a new airline whose fares were inaccessible to the average citizen and whose planes were said by some to be simply Air Zaire craft with a new coat of paint.

Bibliography

African Research Bulletin. 1994. "Communications and Transport." *African Research Bulletin: Economic Financial and Technical Series.* September 16–October 15.

Datapro. 1992. Zaire: The Commercial and Regulatory Environment. In *Telecommunications in Africa.* Delran: Datapro Information Services Group.

International Telecommunications Union. 1990. African Conference for the Development of Telecommunications (CAD 5–90). Working paper, vol. 2, Harare, Zimbabwe, December 6–11.

______. *African Telecommunications Indicators.* Geneva: International Telecommunications Union.

7
Rwanda

ALPHONSE KABAYIZA

Rwanda is a landlocked country located at the heart of the African continent, 1,700 kilometers from the Indian Ocean and 2,000 kilometers from the Atlantic, and surrounded by the Sahara, Zaire, Burundi, Tanzania, and Uganda. The topography of its 26,338 square kilometer area is dominated by mountainous regions and large lakes: Lake Kivu to the west, Lake Tanganyika to the south, and Lake Victoria to the east. The country has an average annual temperature of 18 degrees Celsius.

Rwanda, whose capital is Kigali, has a population of 8.3 million inhabitants, the majority of whom are under sixteen years of age. Rwanda's three ethnic groups are the Hutu (90 percent of the population), the Tutsi (9 percent of the population), and the Twa (1 percent). In the early 1990s, Rwanda's gross national product was U.S.$1.813 billion, or U.S.$250 per capita. Coffee exports served as its main source of hard currency, but the economy also relied on exports of tea, pyrethrum, tin ores and concentrates, and wolframite.

Until 1991, Rwanda's political regime was based on single-party rule. However, since then Rwanda has had sixteen political parties, of which five were part of the transitional coalition government of the early 1990s. The Rwandan Patriotic Front (FPR), a Tutsi-dominated rebel organization, invaded from Uganda in 1990, and the conflict intensified as the coalition government was formed. A peace treaty was signed in August of 1993, but before the peace could be implemented, the head of state, Major General Juvénal Habyarimana, was killed in a plane crash. Almost immediately, the Hutu government forces and militia slaughtered between 500,000 and 1,000,000 Tutsi civilians in an orgy of blood that shocked the world. The FPR resumed its status in an active rebellion and eventually won in July of 1994. They then set up a new Tutsi-dominated government headed by a moderate Hutu. Many other Hutus fled the country, and in the process the entire region's stability suffered. All this affected the telecommunications system of the country in a serious way.

7.1 The Past

Originally a monarchy ruled by the Tutsi tribes in 1899, Rwanda was part of a German protectorate until 1916. It then became a Belgian trust territory in 1916, as part of Ruanda-Urundi, and gained its independence on July 1, 1962. Its first republic lasted for eleven years, until July 5, 1973, when the military deposed President Gré goire Kayibanda, a civilian-military government was set up, and Major General Juvé nal Habyarimana became head of state. In 1991, a multiparty system was declared legal, and elections were announced. Unfortunately, in 1990 a civil war broke out between the FPR and the government, and it was not until 1993 that an accord was reached. This proved to be a temporary peace, however, and the FPR eventually won the civil war in July 1994.

7.1.1 The Colonial Period

It was only in 1924 that the first automobiles began crossing Rwanda's few passable roads, which had been completed just two years earlier. Mail that previously could be delivered only by messengers walking at an average speed of six miles an hour could now be transported by motor vehicle within a day to its final destination—but only in locations served by Rwanda's scant road system.

By 1930, the first wireless telegraphy station (TSF) was established linking Kigali to Bujumbura, the capital city of the territories under Belgium's guardianship (now the capital of Burundi). Its primary purpose was the processing or forwarding of official correspondence. Within the next quarter century, each prefecture's capital city was equipped with similar TSF stations. Each TSF station communicated exclusively with Bujumbura by telegraph, using manual transmission and audio reception of Morse code signals. The connection linking Bujumbura and Bukavu (The Congo) was the only outlet for international traffic. International-bound messages had to be retransmitted from Bukavu to Leopoldville (Kinshasa) before accessing international connections.

In 1956, Rwanda and Burundi set up their own provincial administration at Bujumbura-Kigali, and the first local telephone networks—served by small manual telephone exchanges—were established in Gisenyi, Butare, and Kigali. In 1958, local telephone networks were created for Cyangugu, Nyanza (Nyabisindu), and Gitarma. Gisenyi, Cyangugu, and Butare maintained their direct connection with Bujumbura while the other inland stations were linked to Kigali, forming the embryo of a truly Rwandan national telecommunications network. The forwarding of international traffic was enhanced by the opening of a connection between Bujumbura and Kinshasa. In 1959, an aerial line was installed between Kigali and Butare. Rwanda's telecommunications system experienced radical changes between 1956 and 1962, from the first successful attempt to forward telephone messages directly without routing them through Zaire to acquiring modern Philips-brand telecommunications equipment from Holland.

7.1.2 The Postindependence Period

At this stage of telecommunications development—July 1962—Rwanda gained national sovereignty. During its first transitional year of independence, Rwanda's telecommunications system operated within the Common Telecommunications Agency of Burundi and Rwanda (ATCBR), and the country focused on creating new national and international direct telecommunications lines. For example, by the end of 1962 telephone services between Kigali and Bujumbura had been opened, and in February 1963 a Morse code link between Kigali and Kampala was established. By March–April 1963, telegraph, telephone, and telex services were opened between Kigali and Brussels, and by mid-year the first radiotelegraph operators trained in Rwanda were graduating in Kigali. Also in 1963 an international telex network and a manual telephone switchboard center with seventy-five numbers were installed at Kigali, and Rwanda began serving its first manual table telex customers.

In 1965, telegraph services (by teleprinter) were opened between Kigali and Nairobi, and the Morse code manual radio connection between Kigali and Kampala was ended. Several milestones occurred in 1969: TSF stations at Gikongoro and Gitarama were opened, the telex connection between Bujumbura and Kigali was established, automated telephone switchboard centers at Cyangugu were installed, and construction work for a Hertzian frequency band network was begun.

Expansion continued in the 1970s. In 1970, the automated telephone switchboard center at Kigali was extended, radiotelephone connections between Kigali and Cyangugu, Butare, and Ruhengeri were inaugurated, and automated telephone switchboard centers were installed at Butare, Gisenyi, and Ruhengeri. A radiotelephone connection was established between Kigali, Abibjan, and Brazzaville the following year, and in 1972 telex connections between Kigali, Frankfurt, and Kampal were established and a radiotelephone connection was opened with Paris. In the next four years, five more TSF stations were opened (1973), a telex was installed at four locations (1974), and a radio transmission center was created at Nyanza-Kicukiro (1975).

The late 1970s and early 1980s witnessed further expansions and changes in Rwanda's telecommunications network. In 1977, the switchboard center at Kigali was extended from 2,000 to 3,000 customers and then in 1979 from 3,000 to 5,000 customers. The year 1979 also saw the installation of an automated telex center and the inauguration of the Ecole Nationale Mixte des Postes et Télécommunications at Kigali. In 1980, fourteen linkages of a rural telecommunications network were established, and in November of the following year, the high-frequency connection between Kigali and Ruhengeri was replaced by a connection in the one Hertz frequency band. In 1982, the first Intelsat tests on the Nyanza-Kicukiro Terrestrial Station were performed. A new TSF station was opened at Karengeras, and the new automated switchboard at Kigali was established.

7.2 The Present

When the civil war of the 1990s stopped many projects, Rwanda's domestic telecommunications network consisted of the following facilities:

- Twenty-six telegraph stations, which aside from the Central Radio-Telegraph Bureau of Kigali, include stations in Gitamara, Butare, Gikongoro, Cyangugu, Kibuye, Gisenyi, Ruhengeri, Byumba, Kibungo, Nyabisindu, Rwamagana, Kansenze, Ruheango, Gatsibo, Nyamasheke, Nyange, Kabaya, Vunga, Kirambo, Rushashi, Karengera, Kiyuhba, Garenke, Rubengera, Kaduha, and Masango.

- Twelve automated telephone switchboard centers, six of which were launched in February 1987 for Kigali and its satellites, with the remaining six launched in March 1987 for the capital cities of the prefectures: Butare, Nyabisindu, Gikongoro, Cyangugu, Ruhengeri, and Gisenyi.
- The telex center of Kigali.
- Connections between Rwanda and other countries—either by telex, telegraph, or telephone: Kigali/Brussels, Paris, Frankfurt, Amsterdam, Nairobi, and Bujumbura.
- Two earth stations for satellite communications, one Intelsat and one Symphonie, both located in Kigali.

7.2.1 The Domestic Network

Rwanda has four major telecommunications centers—in Butare, Cyangugu, Gisenyi, and Ruhengeri—and seven local centers in Gitarama, Nyabisindu, Gikongoro, Kibuye, Byumba, Rwamagana, and Kibungo. Each major telecommunications center is headed by a center manager who is responsible for the center's administration, and each local center is under the authority of a local center manager who controls production and management.

The infrastructure of Rwanda's telecommunications network consists of the following elements: a telephone exchange, a telegraph switchboard, rural telephony, transmission by microwave links, an earth station, and local networks. The telephone exchange consists of modern digital switches. In Kigali, a telephone exchange operates as an international transit center, a national transit center, a grouping center, and a manual center facility with operators' transmitters.

In the other regions, four grouping centers were installed in Gisenyi, Butare, Cyangugu, and Ruhengeri. These automatic switchboards are modern digital systems and serve 500 to 5,000 subscribers. Through the regional exchanges and the Kigali grouping center, with its distant concentrators, twelve major towns in Rwanda have telephone networks. All the regional telephone exchanges are connected to the Kigali center with interurban lines.

Rwanda's telex network initially consisted of a Siemens TWKN-type autoconverter installed in 1979 and was equipped with sixty-five channels. In 1988, Rwanda's only telex exchange was in Kigali. This switchboard served Kigali's subscribers and twenty-one consumers spread over different regions of the country. However, because the exchange was old and had reached the saturation point, the Rwandan administration decided in 1987 to replace it with a Japanese NEC automatic switchboard with a capacity of 200 local channels and 120 international channels. Additionally, a French Sagem Eltex U Alpha switchboard was installed in 1987.

The microwave system, which is mostly digital, is used to support the interurban communications. The Kigali exchange employs transmitters that can function manually for both regional and international communications. Rwanda's basic transmission network has a star-shaped structure for both analog and digital connections. The central nerve of this star is in Kigali, and connections link Kigali to other important towns. Rwanda's analog microwave system interconnects the following centers: Kigali-Butare, Kigali-Gisenyi, and Kigali-Cyangugu. The total length of the connections is 290 kilometers, and the total number of microwave links is six.

7.2.2 The International Network

Before the launching of the earth station in 1982, Rwanda used to process international traffic through high-frequency connections to Brussels (one phone line plus one graph line), Paris (one phone line plus one graph line), and Frankfurt (one phone line). A high-frequency connection with Nairobi, Kenya, was created. In addition, two international connections by Hertzian frequency bands were established to link Bujumbura and Kampala.

There are two types of international microwave links in Rwanda: ground communication microwaves and microwaves forwarded by the two earth stations. In the early 1990s, the Kigali-Butare connections were being used to forward the Kigali international channels to Bujumbura (Burundi) and were also intended for use as a means of emergency communication between Butare and Gisenyi.

The ground traffic centers on Uganda and Burundi, although communication with other countries is possible through the earth station and satellite. International traffic, however, is of less economic importance because there are simply not enough channels.

7.2.3 Organizational Structure

The organizational structure of Rwanda's Telecommunications General Administration, which dates back to 1984, is as follows: at the top of the hierarchy is the General Management Board, followed by two management boards—one technical, the other operational—beneath which are five departments (commutation, transmission, terrestrial station, research, and operations), and then nine bureaus (accounting, consumer service, cables, telephone and telex switchboards, transmission, radiotelegraph, maintenance, internal taxation service, and logistics and supplies).

A general directorate of telecommunications is in charge of the country's telecommunications system. This general directorate is in turn controlled by the Ministry of Transports and Telecommunications. Through the early 1990s, Rwanda's telecommunications had neither financial autonomy nor self-sufficiency, although the World Bank was sponsoring a program meant to put national telecommunications under the control of a limited company called Rwandatel.

In the early 1990s, an appropriate national law or set of laws relative to telecommunications was nonexistent in Rwanda. For relevant legislation one had to refer to the laws and recommendations of both the International Telecommunications Union (CCITT-IFRB and the like) and Intelsat. All potential investors in Rwanda's telecommunications must sign a contract with the general directorate of telecommunications.

The general directorate is structured as follows:

1. General Directorate.
2. Directorate of Equipment and Maintenance.
3. Directorate of Exploitation.
4. Seven departments: administration and finances; inspection and control; planning and programming; exchanges and cable network; energy and transmission; earth station; and data processing.

Each department is in turn divided into sections. In the early 1990s, Rwanda's telecommunications personnel consisted of 615 agents, with 400 in Kigali and the remainder spread throughout the other regions.

7.2.4 Tariff Structures

7.2.4.1 Telecommunications Tariffs

Local communications in Rwanda are not charged by time—only a fixed rate is applied. For interurban communications, charges depend on the amount of time used, but both daily and night tariffs are the same. In the early 1990s, interurban communications were charged 3 Rwandan francs, and the charging system was divided into fifteen-second periods.

Charges for international communications depend on both the length of communication and the distance between countries. Communication with certain specific services are not charged (these include national and international registrations, the state police force, and the fire brigade).

In 1992, the efficiency of Kigali's local traffic was only 49 percent, which is below expected performance levels based on the general efficiency trend of telecommunications (65 percent). Of the total number of local calls not completed, 74 percent received busy signals and 17 percent received a "no response" from the call's intended recipient. Rwanda's international traffic also shows a significant level of inefficiency.

To solve Rwanda's traffic flow problems, tariff arrangements were planned to motivate subscribers to adjust their telephone usage with respect to the importance and type of traffic.

Eighty-one percent of Rwanda's telephone traffic is consumed during the busiest hours of the day. In particular, 51 percent is consumed between 7 A.M. and noon. Thirty percent of the traffic is consumed during the working hours in the afternoon (2 P.M. to 5 P.M.), but between 1 P.M. and 3 P.M. the traffic decreases to only 4 percent of total daily traffic. Only 5 percent of Rwanda's telephone traffic occurs after 8 P.M. Saturday mornings represent an important traffic period, but such traffic occurs only at long intervals. The average duration of telephone calls in Rwanda depends on the distance of the call.

Rwanda uses a lump tariff system for local calls, which constitute nearly 80 percent of the calls made by the system's consumers. However, because local calls represent only 13 percent of the incoming traffic, they are undercharged.

Incoming international traffic dominates the total incoming traffic in Rwanda's telecommunications system, just as in most African countries. This reflects the fragile character of the management of this service. Generally speaking, the quality of Rwanda's international service is unsatisfactory, resulting in dissatisfied customers and losses of incoming traffic, with a resultant loss of revenue for the general directorate of telecommunications. The reasons for the inefficiency of the international service include busyness of lines, nonresponse, and line congestion.

A close study of the international tariffs to and from Rwanda in the early 1990s revealed an important asymmetry: there was a tendency for traffic to be reversed because subscribers preferred to be called from abroad rather to call from Rwanda. Tariffs therefore need to be adjusted gradually to compensate for the asymmetric network traffic.

7.2.4.2 Tariffs for Other Services

The use of telegraph services for international traffic decreased in Rwanda in the early 1990s due to the growing use of the telephone and fax services. The international traffic, however, brings substantial incoming revenues. Because the telegraph remains the only means of communication for most people in Rwanda, the institution of a flexible tariff seemed called for.

Although telex services are often used within Rwanda, gross budgets were decreasing in the early 1990s. However, the international telex traffic has increased in recent years—accounting in the early 1990s for about 90 percent of the incoming telex traffic. Telex services may face stiffer competition from the increased use of faxes in the near future. Although fax service is quite new in Rwanda, by the early 1990s it was in full use, primarily by the business sector. It is therefore likely to generate an important level of telephone traffic and should continue to attract great interest.

7.2.5 Quality of Service

Rwanda's rate of efficiency in 1992 was 58 percent call completion for domestic calls and 35 percent for international calls.

With a telephone density of 0.11 main lines per 100 inhabitants in 1988, Rwanda ranked last among African nations. However, the telephone exchanges installed in 1987 were all digital and constitute a strong framework for the network's future. Nevertheless, it should be pointed out that the distribution of telephones between Kigali and the rest of the country is not proportional to the breakdown of the population by location. Kigali, for example, represents 4 percent of Rwanda's population but possesses 58 percent of its telephone lines. In addition, several telephone exchanges have already reached the point of saturation (Butare, Gisenyi, Nyabisindu, Byumba, and Gitarama).

In the early 1990s, Rwanda's telephone system faced high installation costs, which are calculated according to the length of the line from the applicant's property limits to the connecting box of the cable (20 meters or more). At the same time, the low number of functioning lines per month discourages potential subscribers.

Most of Rwanda's telex consumers live in Kigali. By the early 1990s, the telex system, with 100 teleprinters, had reached the point of saturation. The initiation of the NEDIX exchange in 1988 allowed for an extension of the park, and another telex exchange, ELTEX, was put into service in 1989. However, low traffic prohibits the installation of an interurban telex exchange, and the increased interest in new services such as telecopy has reduced the importance of telex in the system.

Rwanda's telephone network is small, but it does possess modern digital exchanges that make possible the evaluation of traffic. In the early 1990s, Rwanda's telecommunications authorities acknowledged that increases in demand were possible and that the network might therefore be extended.

In Rwanda, consumers may buy both the receivers and their accessories themselves and may also purchase telex and fax machines together with their accessories provided that they inform the general directorate of telecommunications. Private companies in Rwanda are also entitled to purchase private exchanges under three types of arrangements in which: (1) the equipment belongs to the subscriber but is maintained by the general directorate of telecommunications, (2) the equipment belongs to the subscriber but is maintained by the general directorate of consumers, or (3) the equipment belongs to the subscriber and is maintained by the subscriber.

7.2.6 Government Control

Rwanda's telecommunications are under the total monopoly control of the state, and the government is the only operator and service supplier. The general directorate of telecommunications is in charge of both international and domestic telecommunications. Except for the circuit leased by SITA (an international company for aeronautical telecommunications), leased circuits do not actually exist in Rwanda. Furthermore, as part of a public telecommunications administration, the general directorate of telecommunications has no relationships with specific national groups such as enterprises, syndicates, and businesspeople. The only relationships so far have been those between customers and suppliers.

The general directorate of telecommunications, however, maintains close relationships with such international institutions as the International Telecommunications Union (ITU), Intelsat, and Inmarsat. Because the Rwandan government has sought to free itself from telecommunications policy, in the early 1990s a limited company, Rwandatel, was expected to eventually assume control of national and international telecommunications with the assistance of various economic operators (investors, businesspeople, and national security agencies).

Insofar as the general directorate of telecommunications is under the control of the Ministry of Transports and Communications, the Rwandan government is the only investor in the country's telecommunications sector. Telecommunications bills are paid via the account of the Ministry of Finances and contribute toward the state's gross budget. A budget totaling approximately U.S.$2.5 million is approved by vote each year. Although the government is aware that this amount is insufficient and requires adjustment, the telecommunications budget is not likely to increase the national budget substantially, which itself consistently shows a chronic deficit. Moreover, the development of Rwanda's telecommunications sector could have proceeded at a faster pace if the civil war had not intervened.

7.2.7 Telecommunications Modernization Programs

7.2.7.1 The Terrestrial Station of Kicukiro-Nyanza

To eliminate Rwanda's telecommunications bottlenecks, it became urgent to improve its intercontinental telecommunications, and the direct connections with correspondent countries that were needed could only be obtained via satellite. In contrast to the high-frequency circuits used previously, satellite circuits would improve the quality and reliability of the international service and make possible connections with nonadjacent African nations.

Rwanda thus presented the Japanese government with a general program that consisted of a terrestrial station, an international transit center, an international telex center, and Hertzian frequency band connections between various cities. Following negotiations, Japan agreed to finance only the international telecommunications portion of the project and granted funds of 1.35 billion yen. The project was awarded to joint affiliate of the Sumitomo Corporation and NEC Ltd. (Nippon Electric Company Limited), and on July 15, 1980, Sumitomo and the Rwandan government signed the project agreement. The funds were divided into two categories: 160 million yen were allocated for financing the installation and 1.183 billion yen for purchasing the equipment. For its part, Rwanda was to provide the construction of facilities made to house the new equipment, as well as the customs clearance and shipping of the equipment to the project sites.

7.2.7.2 Modernization Projects Financed by France

Rwanda's telecommunications modernization projects financed by France included the installation of new switches, a billing and invoicing center, a technical facility at Kigali, two other facilities in the provinces, and the replacement of the Philips UR 49 A central of Kigali with an electronic central (the E-10B).

Rwanda's old communications network was equipped with rotary telephone switchboard centers of very small capacity and was hampered by technical problems related to spare parts, maintenance, and the like. Moreover, it simply did not represent a harmonious fit in Rwanda's modernization program. As a result, the rotary telephone switchboard centers were replaced by telephone switchboard centers of a more recent design.

The establishment of a modern billing and invoicing center using recording tapes made it possible to handle delays and larger traffic volumes. Its construction cost (12 million French francs) was more than twice that of the technical facility built in Kigali (5 million French francs or 65 million Rwandan francs). Between 1978 and 1986, the number of telephone customers in the urban center of Kigali grew at a rate of 6 to 15 percent, necessitating the replacement of the Philips UR 49 A central with a larger electronic central—at a cost of 5 million French francs.

Overall, of the 75 million French francs financed by France, 60 million came from a loan from the Caisse Centrale (CCCE) and the remaining 15 million from a subsidy of the French Aid and Cooperation Fund (FAC). Of this 75 million total, 45 million francs were allocated for supplies, installation, and maintenance, as well as miscellaneous or unexpected costs.

7.2.8 Optimizing Rwanda's Telecommunications System

Some of the major objectives—or aspirations—of Rwanda's policy to establish a modern and reliable national and international telecommunications network were the following:

- To draw up a plan—recruiting and training system—for highly qualified individuals capable of utilizing and maintaining the system's technical equipment.
- To restructure the system's organization to eliminate the confusion of tasks and the absence of functional separations between the managerial and operational services.
- To introduce high quality and large capacity into the telecommunications service in order to make possible the substantial expansion of the networks and the automation of national and international connections.
- To overcome the impediments to the functioning of the system caused by the fact that management and its budgetary decisions must still answer to the ministry.

7.2.8.1 Independence of the Telecommunications Service

In spite of the financial and political difficulties and the demands of other projects, the Rwandan government was forced in the 1990s to consider giving financial and administrative independence to the telecommunications sector in acknowledgment of the growing importance of telecommunications to the future development of the country as a whole. Without financial and managerial independence, the telecommunications sector will not be able to perform its tasks within the new infrastructures emerging in Rwanda. Moreover, independence in and of itself will lead to increased income and improved customer service. Thus, starting in 1996, the government gave serious consideration to privatize the national system of Rwandatel.

7.2.8.2 Extension of the Rural Network

As we have seen, the continued expansion of Rwanda's rural telecommunications system—particularly the communal administrative centers and the socioeconomic, industrial, and health centers—is vital to the country's economic future. The goal of the third phase of the rural development program was to establish a target structure by transforming the local centers into rural zones. The equipment ordered in this phase was expected to satisfy Rwanda's rural telecommunications needs for approximately five years. Unfortunately, the civil war delayed or stopped much of the planned development.

7.3 The Future

Since 1987, Rwanda's telecommunications sector has experienced considerable shocks as well as progress, including the installation of new modern telephone exchanges and digital telex systems that have made possible automatic international communications. In the 1990s, Rwandan telecommunications remained a state property, and the absence of judicial, financial, and management autonomy left it with no particular way to develop meaningfully. The creation of the limited telecommunications company Rwandatel should thus result in greater power with which to exploit and control the entire field of Rwandan telecommunications. Such restructuring can only proceed, however, with a clear definition of the roles of both the government and the limited telecommunications company.

7.4 Conclusion

Most of the planned changes were negatively affected by the country's political situation. Since the civil war intensified in 1992, the situation in Rwanda deteriorated. The telecommunications system was destroyed in many places, repairs to the system have been slow. The telecommunications system has been a low priority for the government as it struggles to hold together the country. With stability returning, telecommunications will again become a focus for economic development efforts.

8
The Ivory Coast (Côte d'Ivoire)

HUGUES KONE

As in other countries, the roots of the Ivory Coast's modern telecommunications system stretch back to the end of the nineteenth century. In the decades since, the country's system has developed considerably, both quantitatively and qualitatively, and has experienced—particularly in the early 1990s—notable technological, structural, and regulatory transformations. This chapter describes the development of the Ivory Coast's telecommunications infrastructure and outlines its current structure, the problems it faces, and its various development efforts.

Located in Africa's humid tropical zone between 5 and 10 degrees north latitude, the Ivory Coast (officially, Côte d'Ivoire) encompasses an area of 322,463 square kilometers. In 1994, its population was about 13.5 million (up from 3,230,000 in 1960) and was dispersed among 8,500 cities and villages and 75,000 camps. These communities are grouped into 10 administrative regions, 50 departments, and 184 subprefectures. Historically, the Ivory Coast's annual population growth rate has been high—for example, 3.74 percent between 1975 and 1988—and the country's population has experienced rapid urbanization, from 15.4 percent in 1958 to 39 percent in 1988. Abidjan, the Ivory Coast's major city, for example, had 2.5 million inhabitants in the early 1990s. At the same time, the country's population distribution reveals a lack of regional equilibrium, with three out of four people living in the coastal or forest region in the south, which constitutes slightly less than half the area of the country.

8.1 Overview

8.1.1 Economic Development, 1950–1994

Between 1950 and 1980, the Ivory Coast's economy enjoyed a long period of prosperity, with a GDP (gross domestic product) increase rate of 7 to 8 percent annually in real terms. This allowed the country to develop a remarkable infrastructure (roads, ports, schools, electricity, hydraulics, and the like) relative to the rest of the African continent. However, because of the historical lack of a certain class of national businesspeople, the state became the first investor, employer, and main holder of entreprise capital through state-owned companies (or commercial or market sector). Moreover, most companies were unfortunately plagued by bad management and deficits, resulting in financial difficulty for the state.

In the 1980s, the Ivory Coast experienced a recession that was provoked by a drop in agricultural raw material prices on the world market as well as excessive public debt. As a result, between 1986 and 1990, the country's GDP suffered a yearly drop of 1.8 percent on average. The Ivory Coast subsequently initiated a series of Structural Adjustment Programs. Through the early 1990s, however, these programs had not resulted in any noticeable success. A devaluation of the local currency (the CFA franc) was decided on January 1994.

Confronted with its first economic crisis since its independence in 1960, the Ivory Coast was forced to do its accounts, revise its priorities, redirect its field of intervention, restructure its services, and reconsider role distributions among the different players in the national economy. The country had to find its way back to growth and confront its economic deterioration and the growing poverty of its population. It was in this context that the privatization of several sectors and state companies was decided upon.

8.1.2 Electronic Communications

With respect to its electronics communications infrastructure, the Ivory Coast is one of the best-equipped countries in West Africa. The country had a national radio station as early as 1951, a national television station by 1963, and a regional radio station in the heart of the country by 1964. By 1986, 70 percent of the Ivory Coast's households had at least one radio and 32 percent had a television set.

In the early 1990s, the Ivory Coast's television and radio sectors were in the process of being deregulated. In fact, after long being run as a state agency, they were transformed into a national office in 1991 and then, in 1992, into a public company under the name RTI (Radio Télévision Ivoirienne, or Ivory Coast Radio and Television). The state owns 98 percent of RTI's capital, with the balance set aside for its personnel. Furthermore, in December 1991, a law allowing the creation of private radio and television stations was adopted, and five private radio stations and one television channel were authorized at the beginning of 1993: four radio stations and the television channel are operating now along with three Catholic radio stations and a few local radio stations without a clear status. In 1992, the government decided to install four rural radio stations in the remote areas of the country by 1995. It signed an agreement with COMSAT, an American company, to cover the country with national television and radio signals.

Computers entered the Ivory Coast in 1962, initially for use by the Ministry of Economics and Finance but soon by other sectors. In 1983, the coordination of computer development was given to the National Council of Computer Systems (CNI), headed by a state minister. Unfortunately, the economic crisis of the 1980s put a stop to large public computerization projects, such as the introduction of computers in the Ivory Coast's primary and secondary education systems.

8.2 The Past: Telecommunications in the Ivory Coast

The history of modern telecommunications in the Ivory Coast began in 1887 with the introduction of the telegraph and ended in 1990 with the decision to privatize the telecommunications administration. There were two major phases: the colonial period, which ended in 1960, and the period after independence.

8.2.1 The Colonial Period

Until the end of the nineteenth century, the Ivory Coast was a territory occupied by a mosaic of ethnic groups organized into a few kingdoms and several autonomous local law enforcement units. France then demonstrated its will to colonize the territory by establishing its presence on the coast and beginning the exploitation of wood, coffee, and cacao—the products upon which the Ivory Coast would build its economic reputation. Today, the Ivory Coast is the leading world producer of cacao and the fourth leading producer of coffee.

The first school in the country was erected in 1887, and in 1893 the colony of the Ivory Coast was officially constituted, with Grand-Bassam as its capital. The country was ruled by a French governor and was part of French West Africa (Afrique Occidentale Française, or AOF), which had Dakar as its capital city and a governor general as its head. The country's conquest, however, did not take place until 1915 following the surrender of the resisting Samory Toure (1898) and the repression of several local revolts. The participation of the Ivory Coast's native population in internal politics began in 1946 with the creation of the country's first political party, the PDCI-RDA, and the election of an African deputy to the National French Assembly. This was followed in 1956 by the application of a law giving higher internal autonomy to the various French colonies in Africa, the proclamation of the Republic of the Ivory Coast in 1958, and the constitution of the first government in April 1959.

8.2.1.1 Telecommunications Development to 1960

The development of the Ivory Coast's infrastructure really began with the construction of the country's first wharf, which was built in Grand-Bassam in 1901. This was followed by the initiation of railroad "Abidjan-Niger" construction in 1904; the opening of the first road between Grand-Bassam, Bingerville (the country's capital from 1900 to 1933), and Abidjan in 1912; and the creation of the country's first industrial units (sawmills and oil works). The Ivory Coast's development increased sharply in the 1950s with the opening of a deep seaport and Abidjan's international airport, the development of its roadways, the equipping of its urban centers, and the growth of export crops.

Telecommunications in the territory that became the Ivory Coast actually preceded the birth of the country: the first telegraphic link was completed in 1887 after the French government authorized the British firm Western African Telegraph Company to open a telegraph office in Grand-Bassam, thus linking the country to the underwater cable connecting Dakar (Senegal), Freetown (Sierra Leone), Monrovia (Liberia), Accra (Ghana), Cotonou (Benin), and Libreville

(Gabon). The first national telegraphic connection between Grand-Bassam, Jacqueville, and Grand-Lahou along the coast was opened in October 1894.

During the next ten years, the Ivory Coast's post and telegraph system spread inside the country and along the coast to the borders of Liberia and Ghana, following the pace and trails of conquest, as well as the country's exploitation by the colonial force. The north of the country was linked by telegraph to Dakar via Bamako (Mali), then connected to the rest of the country in 1902. In 1909 an agreement was reached to set up a telephone line between Grand-Bassam and Accra.

By December 1905, there were forty-seven post offices, thirty-three of which offered full services—postal and telegraphic—and six of which offered telephone service only. At that time, the Ivory Coast's telegraphic cable measured 3,260 kilometers. The first radiotelegraphic links with the neighboring territories were opened in 1930, with the Bamako connection used for traffic flow to France. Until 1945, all the Ivory Coast's internal telecommunications were made by telegraph. The Akouedo-Bingerville center for radioelectric transmissions was installed in 1943, and in 1957 the Abidjan-Marcory center for radioelectric reception was put in service. These two centers provided almost all of the Ivory Coast's international telecommunications.

In 1895, the Ivory Coast's first telephone service was installed between Grand-Bassam and Assinie (the location of the country's first post office), using an aerial cable 50 kilometers long. That same year, the two cities were connected to Alepe and Jacqueville. The country's first urban phone lines appeared in Grand-Bassam and Bingerville in 1903 and in Abidjan in 1910—which benefited from the country's first urban telephone exchange (of ten subscribers)—and then in Bouaké. It was not until the end of the Second World War and the adoption of new policies for colony valuation, notably the creation of FIDES (Investment Funds for Economic and Social Development) in 1946, that the provision of telephone equipment resumed and extended into the heart of the country.

In the postwar period, the following telephone connections were established: Abidjan to Grand-Bassam (1949), Abidjan to Dabou (1951), Abidjan to Agboville (1952), and Dabou to Tiassalé, Gagnoa to Divo, Dimbokro to Bongouanou, and Agboville to Abengourou between 1954 and 1959. The urban lines of Dimbokro and Agboville were built in 1954, and in 1955 the first automatic telephone exchange with a capacity of 2,000 lines was put in service in Abidjan. In 1958, an electromagnetic wave was installed between Abidjan and Aboisso.

During the 1950s, telephone connections were made with the Ivory Coast's neighboring countries: Senegal, Sudan (Mali at present), Haute-Volta (Burkina Faso at present), Dahomey (Benin at present), Togo, Ghana, Nigeria, Guinea, and Mauritania. A direct radioelectric link was opened between Abidjan and Paris in 1959. At the time of independence in 1960, the Ivory Coast's national telephone system was composed of 100 kilometers of electromagnetic waves, 1,325 kilometers of aerial wires, and 125 kilometers of underground cables. On the international scene, there were twelve African connections and one connection with Europe via Paris. The number of telephone subscribers grew from 591 in 1950 to 3,667 in 1960, which represented 0.11 telephones per 100 inhabitants. Of the total number of subscribers, 69.8 percent were residents of Abidjan.

8.2.1.2 Administration of Telecommunications

During the colonial period, the Ivory Coast's telecommunications system was administered by the French PTT (Post, Telephone, and Telegraph) but coordinated at the level of the AOF at Dakar. The PTT services of the Ivory Coast were allowed neither to build nor equip themselves, and their revenues were directly transferred to the public treasury to fund the general state budget, which, in turn, ensured the PTT's operational and investment expenses.

The Ivory Coast's first government appointed in 1959 included a secretary of state for the postal and telecommunications services. On April 1 of that same year, an office for those services was created, and international telecommunications was given to the French company France Câbles et Radio (FCR).

In short, the administration of telecommunications in the Ivory Coast clearly emerged in the context of the French conquest, organization, and control of the territory, which started in the coastal area and continued with the objective of linking the territory to the AOF—notably at Dakar—and to the "mother country" (i.e., France). The telecommunications infrastructure was closely tied to the economic exploitation of the country.

In the 1950s, the need to further develop the country guided the installation of the Ivory Coast's internal telephone and telegraph network. The network was structured from Abidjan, the country's economic and political capital city, with the aim of satisfying the administration's needs as well as those of the colonial economy. In most cases, telecommunications cables (ways) were planted along the railways and roads, the basis of economic development. The areas producing cacao, coffee, and timber were first equipped. The north was not to be reached until after the independence. However, the telephone remained largely inaccessible to the local population.

In general terms, telecommunications in the Ivory Coast has historically been seen mainly as a public service, and its management has thus been given to an administrative body. Therefore, financial profitability was seen as a secondary issue.

8.2.2 Telecommunications after Independence

In August 1960, the Ivory Coast gained its independence and in the following month was admitted to the ITU (International Telecommunication Union). It continued to equip its telecommunications infrastructure emphasizing international connections in particular, which clearly conveyed its intention to insert itself in the international exchange systems. The result of this emphasis was the emergence of an economy that is still considered by some to be extroverted in nature. Consequently, national telecommunications is not considered a domestic priority in the Ivory Coast: the decision was made long ago to develop the country's roadways in order to move its raw materials toward the port of Abidjan.

8.2.2.1 Telecommunications Development Projects

Historically, there has been no explicit national policy for telecommunications or for other economic activity sectors in the Ivory Coast. Nevertheless there was a consensus on the necessity of making the telephone available to a great number of people, of supplying enterprises with new communications techniques, and of linking the country to the rest of the world, particularly the industrial countries. Despite the general absence of a commitment to development in the past, efforts have been made in recent years to systematically upgrade the country's telecommunications network.

The development of the Ivory Coast's telecommunications network was scheduled to take place over several years and was later divided into five-year projects: 1971–75, 1976–80, 1981–85, and 1986–90 (not completed until 1992 due to a late start). These were in turn to be followed by a three-year plan (1992–95) and another five-year project (for 1996–2000). Moreover, there exists a general plan for telecommunications development that spans the period 1986–2005. Given such long-range schedules, the need for financial forecasting seems obvious. But in reality, important investments were made outside the projects for political reasons. For example, to commemorate the country's independence in 1960, significant investments in infrastructure were made in the designated host city with no regard for profitability. Such investments were also made during the equipping of Yamoussoukro, the country's official capital since 1983. Politically motivated investments like these have resulted in a poor distribution of the country's telecommunications equipment and an overbuilt infrastructure in some places.

The main providers of exterior funding for the Ivory Coast's telecommunications development projects have been the FED (European Funds for Development), the FAC (Fonds d'Aide et de Coopération, or French Aid and Cooperation Fund), the EIB (European Investment Bank), the ADB (African Development Bank), the CCCE (Caisse Centrale de Coopération Economique), and the United States. For all this external aid, however, the Ivory Coast has made a significant effort at self-financing: for example, roughly 62 percent of the 1.8 billion French francs spent on the 1986–90 period project was financed by internal funds provided by the country's National Office of Telecommunications (ONT).

8.2.2.2 Telephone Services

In 1970, there were ninety-seven urban telephone networks in the Ivory Coast (of which fourteen were automated) and 17,000 phone subscribers, 75 percent of which resided in Abidjan. At that time, the country was linked to almost all the African countries and more than sixty other countries worldwide. In 1975, the number of subscribers in the country increased to 28,000, in other words, 0.4 phones per 100 residents. The 1976–80 plan aimed at having 1 phone for every 100 residents, a goal that had still not been reached by the mid-1990s. This same plan envisaged the digitizing of the country's network, which eventually took place in 1979 with the installation of the first temporary electronic central telephone exchange (an E-10 by Alcatel).

In 1985, 54,675 subscribers were registered, making the country's penetration rate a little less than 0.6 phones per 100 residents. There were 68,380 subscribers in 1990, or 0.57 phones per 100 residents. The geographic distribution of the subscribers accurately reflected the distribution of the country's economic activities and the urban concentration of the Ivory Coast's population, especially the importance of Abidjan. In 1977, for example, 74.7 percent of the country's phone subscribers resided in Abidjan whereas Bouaké, the second-largest city, only had 3.3 percent and the other cities accounted for less than 2 percent each. Abidjan's importance remained constant through the early 1990s: in late 1992, the proportion of the Ivory Coast's telephone subscribers who lived in Abidjan reached 75.2 percent while the city itself holds only 20 percent of the country's population.

Even inside Abidjan, phone coverage is unequally distributed and reflects the spatial repartition of the economic activities and social classes. In fact, high-density telephone areas include embassies, industries, services, and administrations, as well as populations with high income (Europeans, Americans, and Africans who are well off). This is why a county township such as Cocody has 14,517 subscribers for 120,000 residents, whereas Yopougon and Abobo have only 7,682 subscribers for more than 1 million residents.

Until the 1986–90 development project called for enhancing the phone coverage of the Ivory Coast's rural areas, those regions were almost entirely without phones. The goal was to open up certain semirural subprefectures and large villages that were particularly active on the agricultural and commercial levels by offering them twenty lines linked by numerical electromagnetic waves to the automatic telephone exchanges of a large city. The subscription was billed at one-third the price of service in an urban area. The enthusiasm for the project surpassed the forecast of the ONT and demonstrated that a real need had existed.

The telephone is an important social issue in the Ivory Coast, a fact reflected in the number of telephone subscription requests processed and in the installed network's use. As a former minister of the Ivory Coast's Post and Telegraph stated: "The instant influx of traffic on the lines as soon as a connection is automated says a lot about people's wishes to be linked to the network."[1]

In 1982, there were 27,000 unmet requests in Abidjan alone, which had a total number of subscribers of only 33,000. Such demand pressures led to illegal actions, fraud, bribes, and the use of influences and connections to gain access to a telephone.

In order to increase access to the telephone, coin-operated public cabins appeared in Abidjan in 1974. Unfortunately, they were soon destroyed by vandals and in 1978 were replaced by telecommunications offices that remained open until 8 P.M. and were managed by agents of the ONT. Some phone cabins for telex service were also made available to the public. As a result of the high level of demand for telephone service, some people even transformed their private lines into public phones, charging their clients higher tariffs. In 1988, the ONT installed 206 Swiss-made phone cabins operated by magnetic cards in Abidjan and other locations.

Over time, the traffic for international calls in the Ivory Coast has increased dramatically, growing from 161,253 minutes in 1961 to 975,392 in 1970 and finally to 4,848,663 minutes in 1975. The main destinations called were Western Europe, the United States, Lebanon, the Francophone Africa, and Morocco. The number of outgoing calls has always been greater than those coming into the country which serves again to indicate the extroverted nature of the Ivory Coast's economy. In 1965, for example, the ratio of outgoing to incoming calls was 204,304 to 152,338; in 1970, it widened to 621,622 (outgoing) to 353,770 (incoming); and continued to widen further still in 1975, with 3,360,801 outgoing calls versus 1,487,862 incoming calls.

8.2.2.3 Telex and Telegraph Services

Traffic for the telegraph in the Ivory Coast increased from 7,782,836 words in 1960 to 22,406,491 in 1975. In 1965, the country had seventy-seven telegraph centers, of which twenty-seven were used by radiotelephony in Morse code and fifty by telegraphy over wire.

The evolution of the Ivory Coast's international telegraphic traffic results in tangible fluctuations from one year to the next. These fluctuations seem to coincide with the evolution of the telephone and telex: as they grow their extensions and capabilities through new installations, the traffic for the telegraph decreases and later starts to progressively increase. This was the case in 1966–67, in 1972 (the year in which the Ivory Coast's first international telecommunications station was serviced by satellite), and again in 1976 and 1977.

With respect to the telex, the country's network is linked with the telex center of Abidjan, which is equipped with an automatic switch of 1,000 directions and has been in operation since 1961. After 1977, the center offered eight African connections, six European (France, Germany, the United Kingdom, Belgium, Pays-Bas, and Switzerland), one connection with the United States, and twenty-one other connections (including Kuwait, Saudi Arabia, Brazil, Canada, Mexico, South Africa, Sweden, and the former Soviet Union). The number of telex subscribers in the Ivory Coast grew from 54 in 1961 to 320 in 1970 to 736 in 1975, and the traffic grew from 117,459 minutes in 1961 to 1,4767,020 in 1975.

8.2.2.4 Other Services

In the 70s, the spectacular growth of computerization in the country created an insistent demand for specialized telephone circuits for data transmission. In order to address this demand, which had been assessed through surveys with actual and potential users of telecomputerization services, the decision was made to develop a national network for telecomputerization called SYTRAN (Transactional Systems), which began operation in March 1978. It consisted of the subnetwork of Abidjan, the national subnetwork (linking the six regional capitals and the port city of San Pedro), and the international subnetwork. The SYTRANPAC telecomputerization network was placed into service in 1989, linked with the rest of the world by NTI (Node for International Transit) in Paris. Finally, the Ivory Coast's telecopying service became available to the public in 1987.

8.2.2.5 International Transmission

In 1971, the Ivory Coast became a member of Intelsat. The country installed its first land station at Akakro in November 1972 with sixty circuits for seven connections. Another land station followed in 1978, giving the country the capacity to receive and transmit two television programs simultaneously.

In 1978, the Ivory Coast participated in the realization of the underwater cable linking France, Morocco, Senegal, and the Ivory Coast with a capacity of 4,800 circuits of 4 kilohertz. It was then extended to Nigeria in 1981. The cable was connected to the Atlantis cable system link in South America, Africa, and Europe and then was associated with the regional project Panaftel, which sought to link the African countries with each other by radio waves. As a consequence of these projects, the Ivory Coast was linked to two-thirds of the ECOWAS (Economic Community of West African States) countries.

On the level of regional and inter-African cooperation, the Ivory Coast is a member of several organizations such as the PATU (Pan-African Telecommunications Union), the UAPT (African Union for Postal Services and Telecommunications), and the CAPTEAO (Administrative Conference for Postal and Telecommunications for West African States). The country also participates in telecommunications-related activities through the ECA (United Nations Economic Commission for Africa) and ECOWAS. As a result of its various international transmission projects and associations, by 1980 the Ivory Coast was linked to 113 other countries by automatic telephone.

8.2.2.6 Institutional and Regulatory Organization

The Postal and Telecommunications Ministry. Since 1959, the Ivory Coast's telecommunications had been under the administrative supervision of a ministerial department, a secretary of state, and then a ministry, while its management has been left with a public office, always under the supervision and technical control of the ministry. The ministry (Ministère des Postes et Télécommunications) is in charge of the design and implementation of the policy and the regulations of the telecommunications and postal sector. Toward the end of the 1970s, the ministry included directorates such as general inspection, OPT (Postal and Telecommunications Office), INTELCI (International Telecommunications Company of the Ivory Coast), BIPT, external relations, teaching services of the PT (Post and Telecommunications), and telecomputerization and specialized networks. In addition, it had consulting organizations such as the Superior Council for PT, the National Committee for the Coordination of Telecommunications, and the Surveillance Committee of the OPT.

The Operational Structures. From 1959 to 1964, the Ivory Coast's post and telecommunications sector consisted of a public establishment with an industrial and commercial character that was equipped with a guiding set of principles and benefited from financial autonomy. On January 1, 1965, it became a public organization with an administrative character and was equipped with a budget separate from that of the state's general budget, following the example of similar projects like the RTI.

In 1975, the OPT came to life with two autonomous directorates: the General Postal Directorate (DGP) and the Directorate General of Telecommunications (DGT), as well as an administration council headed by the minister of PT. The PT code of 1976 gave the OPT the state monopoly for the exploitation of the public services for the post and telecommunications.

The Ivory Coast's international telecommunications remained in the hands of the French company France Cables et Radio (FCR) until 1969, when the Ivory Coast's international telecommunications company INTELCI was created after several years of negotiations with FCR. The company's capital was composed of 10 million French francs, of which 52 percent was held by the Ivory Coast government and the other 48 percent by FCR.

The government of the Ivory Coast increased its participation to 80 percent in 1976 when the capital was raised to 400 million French francs. In 1981, INTELCI became wholly owned by the government prior to being liquidated in 1984, with international telecommunications given to the ONT.

In view of INTELCI's commercial success stemming from its administrative autonomy, the Ivory Coast's government had two options: transform the DGT into a state-owned company or integrate INTELCI with the OPT. In 1984, the government decided to liquidate INTELCI and split OPT into two structures, the ONP (National Office of Postal Services) and ONT. The latter became a public establishment with an industrial and commercial character, applying a state monopoly to all national and international telecommunications services with the exclusion of image distribution and the transport of sound guaranteed by the RTI. The ONT benefited from greater managerial autonomy than the previous structure, which was purely administrative. Nevertheless, it remained subject to the managerial rules of public finance: it could not make deals without the consent of the national directorate of public deals of the Ministry of Economics and Finance. Furthermore, investment decisions were not always subject to return criteria: political considerations sometimes became a factor. In short, the ONT had neither the flexibility nor the administrative independence necessary to confront the demands of modern services and telecommunications performance.

It was at this point in the development of the Ivory Coast's telecommunications sector that a study performed by the West German PTT research counsulting arm DETECON (Deutsche Telepost Consulting GmbH) in the early 1990s concluded that an evolution of the legal status of the ONT was necessary. Three years later, this legal structure was put into place.

8.3 The Present

After almost forty years of independence and continued economic development, what is the current state of telecommunications in the Ivory Coast and what transformations are expected to occur through the end of the century? These issues are addressed in the following sections.

8.3.1 Telecommunications Services and Equipment

In 1997, telecommunications customers in the Ivory Coast had access to a diversified range of equipment that included the telephone, telex, transmission equipment, telecomputerization and telematics equipment, and other services such as mobile radiotelephones, maritime radio service, and facsimile service.

8.3.1.1 Telephone Service

In 1996, 144 cities and villages are connected to the telephone network thanks to 188 telephone exchanges, satellites, and rural telephone systems, although 25.7 percent of them are served by manual exchanges. There are 149,572 lines available, of which 76 percent operate on the digital network (compared to 95 percent in France). Subscribers number 103,456, representing 0.77 telephones per 100 inhabitants (against 0.67 in 1992), with a roughly equivalent number of secondary telephones. Of these subscribers, 74.3 percent are living in Abidjan.

There are two international transit centers and one national transit center (CTN) located in Abidjan, seven regional transit centers (Abengourou, Abidjan, Bouaké, Daloa, Korhogo, Man, and San Pedro), and one regional and national transit center (Yamoussoukro), as well as two operator services.

The regional distribution of telephone services in the Ivory Coast, shown in Table 8.1, indicates that there is a strong concentration in Abidjan, where the major modern economic activities take place.

In addition to the subscriber lines, public telephone is available, although not widely. The types of telephone cabins available to the public in the early 1990s included taxi phones, hotel-type cabin phones, and authorized operators phones. Taxi phones work with either magnetic cards or coins and are installed in places that are self-monitored, such as hotels and post offices. Cabin phones can be found in hotels and comparable establishments. The legal tariffs for such phones are 1.25 French francs (FF) for a local call and a tax of 1.00 FF for international calls.

Authorized operator phones work through telephone lines provided to individuals or commercial firms. In 1992, there were forty-two authorized operator phones.

8.3.1.2 Telegraph and Telex Service

In the early 1990s, the Ivory Coast's equipment for telegraphic communication consisted of an automatic switch for telegraphic messages with fifty-two circuits. The country's telex system consisted of a central electronic exchange for national and international communications with a capacity of 3,558 circuits, of which 1,813 were in service in 1994. There were 1,109 recorded subscribers to the telex system, against 1,395 two years before. The decline is due to competition with facsimile.

Table 8.1. Distribution of Telephone Subscribers by Region (in 1992)

Region	Subscribers	Percentage
Abidjan	63,253	77.0
Bouaké	4,401	5.4
Yamoussoukro	3,201	3.9
San Pedro	2,526	3.1
Daloa	1,816	2.2
Agboville	1,734	2.1
Abengourou	1,603	2.0
Man	1,542	1.9
Korhogo	1,379	1.7
Odienne	673	0.8
Total	**82,128**	**100.0**

8.3.1.3 Transmission Equipment

The Ivory Coast's national grid in 1994 was made up of 10,817 kilometers of arteries divided into 2,822 kilometers of aerial lines; 4,028 kilometers of connections by analogous electromagnetic waves (with a capacity of 4,499 circuits); 3,004 kilometers of connections by digitized electromagnetic waves (with a capacity of 10,268 circuits); 528 kilometers of connections by fiber-optic cables between Abidjan, Yamoussoukro, and Bouaké (with 8,130 circuits); 35 kilometers of coaxial pair cables (with 3,360 circuits); 400 kilometers of VHF connections (with a capacity of less than twenty-four channels); 263 transmission centers; and one satellite connection (DOMSAT) between Abidjan and Yamoussoukro (with 72 circuits).

The Ivory Coast's international transmissions are handled by an earth station (located at Akakro, not far from Abidjan), submarine cable connections, and a maritime radio service. The earth station is equipped with two antennas and 360 circuits, of which 326 were in service in 1994. The station links the Ivory Coast to the Intelsat network through two satellites (Atlantic and Indian Ocean), and it can receive or transmit simultaneously two television programs coming from or going to the two satellites. The Ivory Coast's international submarine cable connection includes the West African–Europe cable offering 960 telephone circuits (358 of them in use in the early 1990s) and the Atlantis cable linking South America (Brazil), Africa (Senegal), and Europe (Portugal) and offering 2,000 circuits. The international transmission system includes a maritime radio service that operates through the emitter station in Akouedo and the receiving station of Marcory.

8.3.1.4 Telematics

Telematics services in the Ivory Coast are offered through two networks: the SYTRAN network and the SYTRANPAC network. The SYTRAN network can transmit and receive data from terminals at speeds of 300, 1,200, 2,400, and 9,600 bits per second. It allows the establishment of point-to-point or multipoint connections at the national level as well as some specialized data transmission links between the Ivory Coast and the rest of the world. In total, it offered 582 connections in 1994. There were 774 subscribers to the SYTRAN network in 1994 (only 60 in 1992)—mainly airlines, banks, maritime companies, distribution companies, and large public-sector enterprises (water, electricity, and the like), as well as some temporary subscribers.

The Ivory Coast's linked SYTRANPAC network consists of ten nodes, four of which are in Abidjan; the rest are in the interior of the country (Abengourou, Bouaké, Daloa, Man, San Pedro, and Yamoussoukro). The Abidjan-Plateau node is linked to NTI (International Transit Node) in Paris, and it allows the other nodes in the network to access the foreign networks working with the packet commutation system. All networks or equipment conform to the X.25 protocol of the CCITT. All videotex terminals conform to the CEPT 2 Minitel standard. The SYTRANPAC network offers 453 access gates and 1,500 subscriber lines. French Minitel technology is used by some organizations such as the National University (for documentation purposes), banks, ports, and pharmacies. There are around fifteen data banks in the country. The sector is growing because there is a strong demand despite the high tariffs (twenty times higher than in France for the same consumption) and lack of commercial promotion.

On the whole, however, users deemed the quality of data services to be unsatisfactory and to delay development. According to Citelcom managers, SYTRAN and SYTRANPAC were more profitable than telephony, all things being equal. A new tariff was planned, with prices reduced considerably.

8.3.1.5 Other Telecommunications Services

Other telecommunications services in the Ivory Coast included the mobile radiotelephone, radio and television broadcasting services for the Ministry of Communications, maritime radio service, private radio with 6,785 authorized stations and facsimile services.

8.3.1.6 The Ivory Coast's Electronics Industry

Although the Ivory Coast's local electronics industry in the early 1990s was an uneven one, it showed the potential for substantial development. In fact, the Ivory Coast's master plan for industrialization (established in 1988) identified the electronics industry as one of the country's twelve priority sectors. Moreover, by the early 1990s, the Ivory Coast as a whole was beginning to have a real industrial culture, and some very well-equipped training centers for its electronics industry were in place.

In the short term, however, the government had no plans to develop a local telecommunications industry. It argued that the technology evolves extremely quickly, research and development is very expensive and requires highly qualified personnel, and no economies of scale appeared likely to emerge in the near future.

Although the UAPT had at one point conducted a study to determine the feasibility of a cable manufacturing factory to serve the local market, the project was rendered obsolete before its completion due to the development of optical fibers and radioelectrical technology. In the 1990s, there were nevertheless a few small firms in the local sector: Tonfack Telefon (TTI), which manufactures small products such as telephone exchanges for companies, telephone consumption control systems, and some coin telephones ("an Ivory Coast technology with a Japanese finish"); SARITEL, which mounts telephone sets without artificial intelligence technology; and SICABLE, which manufactures cables.

In the area of heavy and sophisticated equipment, the PT uses the services of large companies such as Alcatel (especially for telephone exchanges), Siemens (optical fibers and radio waves, for example), COMSAT (for the domestic satellite link), or Câbles de Lyon. Major supplies are from France, Germany, the United States, and Italy. There was also a large number of local installers authorized by CI-TELCOM.

In computer technology, US-AITC-CI assembles microcomputers in the Ivory Coast, and a subsidiary of IBM France was the leading company in the very active software and equipment distribution sector. In the audiovisual sector, National Electric Côte d'Ivoire (NELCI), a subsidiary of the Japanese group Matsushita, assembles television and radio sets with the "National" brand name for the local and regional market.

8.3.2 Institutional and Regulatory Organization

Although the previous organization of the Ivory Coast's telecommunications sector was put into place during 1991, during the mid-1990s the sector was in a transitional phase characterized by a great degree of confusion over future goals and projects. As a result, a new organizational structure for the sector was introduced in 1995, together with a new telecommunications code.

8.3.2.1 The Postal and Telecommunications Ministry

The mission of the ministry in charge of the Ivory Coast's Post and Telecommunications (PT) in the 1990s is the elaboration and establishment of the country's postal and telecommunications policies, including defining the sector's general objectives, organization, and regulation; representing the sector internationally; and putting into place and controlling the entities in charge of the PT and training personnel. To accomplish these objectives, the ministry was organized into the following sections:

- A general inspectorate in charge of inspecting and controlling the services of the ministry.
- A consultative committee for PT responsible for assisting the minister in the definition of and the political choices entailed in the post and telecommunications sectors. This committee grouped together the people responsible for the PT, including private and public partners, as well as certain consumer organizations as they are created.
- A general directorate for regulation in charge of defining the legal framework applicable to the sector as a whole. This body has several missions: ensuring that the Ivory Coast's general policy guidelines and international engagements are respected; drawing up a list of authorized public-sector users; dealing with all legal matters of the PT; looking after the national interests in the radio communications sector and the distribution, attribution, and use of the country's radio frequencies.
- A directorate for planning and development in charge of conducting all the general studies necessary to define the general policy of the PT.

The ONT and CI-TELCOM. After 1991, two entities participated in the development and management of telecommunications in the Ivory Coast. The first was a public body named ONT (National Office of Telecommunications), serving as the owner of the Ivory Coast's telecommunications assets. ONT is in reality an empty shell, employing less than ten people in the early 1990s. Its mission was to control the assets granted to the Ivory Coast Telecommunications Company (CI-TELCOM), and it operated under the economic and financial supervision of the Economics and Finance Ministry.

The second entity was the CI-TELCOM, created on May 14, 1991. It was a mixed company with the legal status of a private company; owned by the state (98 percent) and its personnel (2 percent). By the terms of its concession, CI-TELCOM's mission is to operate the Ivory Coast's telecommunications sector, which in the early 1990s was a monopoly. The country's general telecommunications policy, including the pricing policy and all regulatory aspects, were to be defined by the government. However, the transfer contract was delayed for a long time, and as a result, CI-TELCOM was essentially using the state's assets illegally. This situation rendered the company fragile from a financial point of view because the banks were hesitant to lend to a company without assets and legal guarantees.

The creation of CI-TELCOM was the product of the various studies performed in the 1980s on the best way to enable the country's communications sector to respond efficiently and rapidly to the needs of its customers. The establishment of CI-TELCOM was thus part of the global trend toward deregulation that began in the United States at the beginning of the 1980s. This trend in turn was a reaction to the diversification of services offered to a growing and more demanding set of customers; to the competition between manufacturers, suppliers, and equipment installers; and to the need for a more efficient and profitable management.

After 1994, CI-TELCOM focused its efforts on three major goals: changing the mentality of its personnel to increase the quality of the services; improving productivity and working conditions; and investing in the companies that produce telecommunications technology.[2]

A board of directors leads CI-TELCOM and nominates the chief executive officer (director general) who manages the company. The company employs a staff of over 4,000 agents (including engineers, administrators, inspectors, controllers, workers, and other positions) of which 12 percent are managers and 23 percent master agents. The number of employees per 1,000 lines decreased from 58 in 1989 to 45 in 1992 and 36 in 1994.

The training of personnel has always been one of the central preoccupations of those responsible for telecommunications in the Ivory Coast. In fact, as early as 1909 a school for the post and telegraph was opened in Bingerville for the training of subordinate jobs, while the training for superior jobs was performed in France or in Algeria. In 1959, the National School for Post and Telecommunications (ENPT) was created for the training of the sector's subordinate and ordinary personnel. In 1975, it was divided into two structures: the National Higher School for Post and Telecommunications (ENSPT) for the training of the superior administrative and technical managers, and the Center for Professional Training for Post and Telecommunications (CFPPT) for the training of the operational and control personnel. Since September 1992, the ENPT has been a private institution belonging to CI-TELCOM and the Ivory Coast Company for Post and Savings (SIPE) under the name Higher African Institute for Post and Telecommunications (ISAPT).

8.3.2.2 Regulation

Generally speaking, the Ivory Coast applies the international regulatory texts adopted by the ITU and Intelsat for the use of satellites. Excluding the documents that create the various regulatory entities and determine frequency assignment, the standard texts in force in the early 1990s were as follows:

- Law No. 76–501 of August 1976, called Code of the PT, which reserves the monopoly on the Ivory Coast's telecommunications to the state (with the state, however, having the right to concede it if it wishes).
- The decrees and decisions relative to the pricing of telecommunications, which date most recently from March 1989.
- The decision of December 1977 requiring that projects for new urban quar-

ters and buildings plan for the installation of telephone and mail distribution networks in addition to the networks for water pipes and electricity. This decision, however, has been little respected.

- The decisions that fix the conditions of authorization and intervention for the telecommunications equipment suppliers and the private installers. In January 1995, a new text for this area was adopted by the government.

On the whole, these texts do not sufficiently punish such offenses as the misappropriation of a private line for commercial use (i.e., using it as a public phone cabin) and fraud.

8.3.3 Administration of Telecommunications

8.3.3.1 Technical Administration

The sector's maintenance problems, outside those common to all tropical countries, involve problems related to the dilapidation of the equipment; the equipment's diversity, which requires large stocks of replacement parts; and problems to create an efficient maintenance system.

From the point of view of human resources, CI-TELCOM has stated that it has the technical potential to run the Ivory Coast's telecommunications satisfactorily on its own, with the intervention of foreign technical experts considered seldom necessary.

8.3.3.2 Financial Administration

CI-TELCOM does not receive subsidies from the state, and must in addition pay import duties and value-added tax. The company also had to deal with clients' unpaid bills and the ONT's debt. The sector was nevertheless profitable, with a rate of return estimated at 20 percent per year. This profitability has been one of the reasons why some supported the sector's privatization and others opposed it.

In 1991, CI-TELCOM's sales increased to 1.03 billion FF, with the telephone alone accounting for 95 percent of sales, and international service accounting for 65 to 70 percent of telephone sales. Major sources of expenses included labor costs (30 percent) and maintenance (10 percent). The transformation of CI-TELCOM into a private company was followed by an increase in wages, with no employees laid off. As a private company, its personnel is no longer part of the Ivory Coast's civil service, which made the personnel's union fear eventual layoffs. In 1994, CI-TELCOM's sales were 630 millions FF. This was a drop from 1991, due to the FCFA devaluation in January 1994 (1 FCFA = 100 FF, against 50 FF before the devaluation).

8.3.3.3 Commercial Administration

Customer Administration. The Ivory Coast's telephone customers are divided into three categories: large subscribers, official subscribers, and ordinary private subscribers.

Large subscribers are large consumers from the private and public sectors with a minimum of ten lines and an average consumption of more than 20,000 FF every two months. There were 200 such subscribers with 11,000 lines in 1992, accounting for slightly more than one out of every seven lines. In order of importance, these subscribers included the African Development Bank (ADB), the Ivory Coast Electricity Company (CIE), the Port of Abidjan, Abidjan Transportation Company (SOTRA), and Air Afrique. The embassies, town halls, and national public institutions also belong to this category.

These large subscribers enjoy certain benefits: their bills are delivered and their payments are picked up; their lines are not cut off as quickly as the lines of ordinary subscribers; and the very largest customers (more than 200 lines) are given supplementary services.

Official subscribers belong to the public administration and number 8,000—roughly 10 percent of the Ivory Coast's total customer lines. Official subscribers account for numerous unpaid bills (360 million FF in the early 1990s), but CI-TELCOM rarely takes the risk of disconnecting them. Instead, it tries to make sure they do not consume more than their budgeted amount for the year and helps the state to control and slow down the agencies' telephone consumption.

Ordinary private subscribers represent the remainder of the country's telephone customers. It is with this group that CI-TELCOM has the largest number of disputes over bills (about 1,200 per year, of which more than 60 percent is for international communications). The company planned to improve its invoicing system in the 1990s in response to this problem.

Private radio brings in 60 million FF per year to CI-TELCOM, and the service is given in concession by a committee made up of representatives from CI-TELCOM, the presidency of the republic, and the Ministry of Defense. It is granted to certain individuals, companies, or administrations such as the CIE, the Ivory Coast Company for the Distribution of Water (SODECI), Sodesucre, Palmindustrie, the Civil Aviation, the police, the Interior Ministry, the armed forces, the presidency of the republic, RTI, and others. The current regulation provides that the authorization to use private radio can be taken away from the beneficiary as soon as the telephone network allows it

Private telecommunications networks authorized and supervised by the Ivory Coast's telecommunications administration also existed in the early 1990s. These included the networks of certain large commercial banks—such as the Société Générale de Banques en Côte d'Ivoire (SGBCI), whose agencies are linked together by a special telephone network financed by the bank itself—and of the Ivory Coast Railway Company (SICF), whose stations are linked by its own telephone and telegraph network.

Pricing of Services. In the early 1994, a telephone line in the Ivory Coast cost 420 FF in an urban center and 220 FF in a rural zone (against 1,760 FF prior to 1990). The cost structure broke down as follows: 400 FF for the connection and 20 FF for a fiscal tax. The rental charge is 52.20 FF every two months.

Subscription to the Ivory Coast's telex service in 1994 cost 1,920 FF if the subscriber owned his or her own terminal, 5,484 FF if he or she did not, and 3,484 FF if the terminal belonged to the state. A subscription for the maritime radiotelephone cost 1,420 FF of which 1,000 represented advance payment for consumption. A mobile radiotelephone subscription cost 3,000 FF, with the price set high to encourage the use of the first maritime radiotelephone. Finally, the establishing of a SYTRAN link cost 1,800 FF.

Since 1989, there has been a 50 percent price reduction on national calls made during off-peak hours.

Problems in the Commercial Administration. According to CI-TELCOM:

- The rate of traffic continuity—or rate of availability (number of cutoff calls per unit of time)—was 98.00 percent for domestic calls and 99.99 percent for international calls, compared with the 90.00 percent international standard.
- The efficiency rate (the number of calls reaching their destination) for local trunk calls was 30 to 40 percent in the early 1990s and 45 percent for international calls. The minimum international standard for efficiency rate was 60 percent and 40 percent for local trunk calls and international calls, respectively.
- The speed with which CI-TELCOM attended to service problems was 90 percent within three days (against 76 percent in 1990), and 97 percent within eight days.
- The number of service disruptions per year was 0.8 per subscriber, compared with the international standard of 0.3.
- The delay for a new connection was about twenty-five to thirty days, which is still long in comparison to international standards.

Fraud is another major problem that has been facing CI-TELCOM for many years. It generally takes two forms: internal, when employees perform clandestine installations or redirect telephone lines, and external, when it is performed by consumers. Fraud primarily affects international communications, which are relatively expensive, and represents an average loss of 12 million FF per year. In the early 1990s, CI-TELCOM began to address the problem aggressively by identifying and imprisoning individuals who had committed significant fraud.

Unpaid bills represent a third problem area for CI-TELCOM. In the early 1990s, the rate of unpaid bills in the Ivory Coast was 12 to 13 percent for ordinary private subscribers, 14 percent for embassies, 5.1 percent for large private-sector subscribers, 61 percent for government establishments, and 77 percent for the country's municipalities. Because this created financial problems for CI-TELCOM, the company began suspending the lines of the individuals in the public institutions who were responsible for the accumulation of too many unpaid bills.

Relationships with special partners were a fourth source of problems for CI-TELCOM in the 1990s. For example, RTI is a special client of CI-TELCOM because it in effect uses the country's telecommunications circuits to retransmit events of international significance by satellite. However, CI-TELCOM is the only entity authorized to deal with Intelsat: it pays the rent fees and needs to be reimbursed by RTI. Unfortunately, by the early 1990s, RTI's payment arrears had reached 40 million FF.

The relationship between CI-TELCOM and RTI suffers from other problems as well. For example, the telecommunications administration tries to exercise a monopoly of the country's telecommunications infrastructure whereas RTI wants to preserve its autonomy in the transmitting of sound and images. Similarly, the communications and PT ministries could not agree on who would be responsible for controlling the domestic reception of satellite transmitted programs (DBS), and a committee created by the two ministries to regulate this reception could not come to an agreement.

Although telecommunications in the Ivory Coast is well established, its management, as we have seen, is characterized by difficulties. The attempts to solve the Ivory Coast's telecommunications problems are described in the following sections.

8.3.3.4 The New Telecommunications Law and Privatization of CI-TELCOM

In January 1995, the government offered a new telecommunications bill to replace the law of 1976. This new law was inspired by the global movement toward deregulation. The bill entailed the following:

- The existence of a ministry responsible for defining a national telecommunications policy—determining the general objectives and acting as an arbitrator for telecommunications issues.
- The creation of a national public body responsible for the regulation and selection of private companies to exploit various new telecommunications services (such as data transmission, message services, cellular radiotelephones, and the like), as well as the allocation of the frequencies for these services. Its financing will be secured by dues paid by the users of the frequencies and user companies.

  With respect to radio and television, the Ministry of Communications, specifically, le Conseil National de la Communication Audiovisuelle, would select the television and radio channels while the new institution would assign the frequencies. This new institution was said be inspired by the American Federal Communications Commission (FCC), but it was to remain under the administrative supervision of the telecommunications minister insofar as the texts that regulate public bodies in the Ivory Coast would not permit it to be completely autonomous.
- The permission of greater competition at the level of telephone service. This choice was justified by (1) the state's decision to pull out of productive sectors and privatize them and (2) the acknowledgment that telecommunications evolve too quickly for one company alone to satisfy the needs of the population. Under the new law, however, the monopoly of the Ivory Coast's telephone service would continue and be given to CI-TELCOM to assure the amortization of the existing equipment. All other private-sector companies, however, could be called on to build local networks if CI-TELCOM were unable to perform the task. Moreover, the supply of terminals to the system's subscribers would be liberalized with the authorizations to be delivered to the telecommunications companies for a definite amount of time (for example, fifteen years).

The transfer of CI-TELCOM to the private sector has also been at the center of a heated debate since 1990. The former prime minister (1990–93) wanted to transfer the state's share of CI-TELECOM's capital (98 percent) to domestic companies and/or a foreign partner with solid experience in telecommunications. The PT minister, the chairman of the board of directors of CI-TELCOM, and the union of telecommunications personnel (SYNAPOSTEL) approved the transformation of CI-TELCOM into a private company, but they wanted it to remain in the hands of the state. This position led to the dismissal of the minister and caused the union to fear layoffs following the state's pullout. Management, in turn, argued that CI-TELCOM did not need private or foreign partners to be run well, and some opposition parties supported by a part of the press disapproved of the state's pullout and warned against the involvement of a foreign partner in a sector so strategic to the country. Nevertheless, the privatization process continued. In 1998, it led to the selection of France Télécom's subsidiary FCR as the majority holder, with 51 percent of CI-TELCOM, for about $200. About 15 percent were sold to private investors on the stock exchange. The remaining 35 percent was kept by the state, with the possibility of subsequent sale to private investors.

8.4 The Future: Telecommunications Development Plans

The master development plan for telecommunications, which covers the period 1985–2005, anticipated 500,000 telephone subscribers (i.e., about 2.3 telephones per 100 inhabitants) and 13,000 telex connections by the year 2005, as well as the provision of complete satellite coverage for the country. Investments for the ten-year period 1991–2000 alone were estimated to be 5 billion FF.

Between 1993 and 2002, the Ivory Coast's analogous system was to be transformed into a digital network thanks mostly to the use of optical fibers and the introduction of large new public-access services such as videotex, cellular radiotelephone, and cable distribution. The Ivory Coast also placed great hope in the Regional African Satellite Communications project (RASCOM)—whose headquarters it hosts—to improve its telecommunications system.

The private sector and foreign corporations will be widely involved in the future. For instance, Teleglobe (Canada), France Telecom (France), and Comstar (United States) were authorized by the government to develop cellular radiotelephone networks in the country.

8.5 Conclusion

It has been a little more than a century since telecommunications services, beginning with the telegraph, first appeared in the Ivory Coast. During the colonial period, the network developed to facilitate the exploitation of the country and to link it to the "mother country" (France). Telecommunications equipment was installed in the areas where raw material and crops were available. After independence, the Ivory Coast's telecommunications equipment was expanded, diversified, and modernized while its management operated under state monopoly conditions and according to the principles of the national administration. In the nearly four decades following independence in 1960, the number of telephone subscribers in the Ivory Coast multiplied by a factor of twenty-eight while the population increased only by a factor of four. For technology and the supply of telecommunications material, however, the country is still almost totally dependent on Western countries, especially France.

Considering the quality of services and access to those services, there is need of a vigorous policy in order to accomplish the following:

- Improve quality and viability of the domestic network.
- Minimize errors of invoicing.
- Fight fraud and nonpayment of bills.
- Improve the delays of connecting a new subscriber and take care of subscriber's line disruptions.
- Promote public phone cabins, including rural settings.
- Improve the quality of the telematics network and reduce tariffs.
- Promote a better access to telephones across society and geography.

Faced with the rapid evolution of telecommunications technology and a telecommunications management system at the international level, together with the incapacity of the national telecommunications body to address the quantitative and qualitative needs of economic agents and the general public, the government made the decision to privatize the sector against the position of the telecommunications trade union and leftist political parties.

Despite its weaknesses, the Ivory Coast's telecommunications system is considered one of the most effective and advanced in sub-Saharan Africa and a significant number of companies and international organizations have set up their regional or African headquarters in Abidjan.

Abbreviations

AOF	French West Africa
CAPTEAO	Administrative Conference for Postal and Telecommunications for West African States
CEDEAO	Economic Community for West African States
CI-TELCOM	Ivory Coast Telecommunications
DGT	Directorate General of Telecommunications
FCR	France Câbles et Radio
FF	French francs
INTELCI	International Telecommunications Company of the Ivory Coast
ISAPT	Higher African Institute for Post and Telecommunications
ITU	International Telecommuncation Union
MPT	Postal and Telecommunications Ministry
ONP	National Office of Postal Services

ONT	National Office of Telecommunications
OPT	Postal and Telecommunications Office
PATU	Pan-African Telecommunications Union
PDCI–RDA	Parti Démocratique de Côte d'Ivoire, Rassemblement Démocratique Africain (Ivory Coast Democratic Party–African Democratic Union)
PT	Post and Telecommunications
RTI	Ivory Coast Radio and Television Broadcasting
SIPE	Ivory Coast Company for Savings and Postal Services
SYTRANPAC	Information Transmission System
SYTRAN	Transactional Systems
UAPT	African Union for Postal Services and Telecommunications
UPAT	Pan-African Union for Telecommunications

Notes

1. *Les Echos,* no. 14 (1983): 6–7.
2. *Ivoire Télécom,* no. 1 (December 1991): 1–4.
3. *Soir Info,* no. 196 (February 17–19, 1995).

Bibliography

"Les postes et télécommunications en Côte d'Ivoire." *Revue de l'UAMPT* no. 3 (August 1965): 4–21.
"Côte d'Ivoire: les P et T." *Jeune Afrique Economie* no. 133 (July 1990): 150–257.
La Côte d'Ivoire en chiffres. Dakar: SAE, 1978.
Journet, B. "Les télécommunications en Côte d'Ivoire." Paris: Industrie et travaux Outre-Mer, 1960.
Plan quinquennal 1976–80 des télécommunications. Abidjan: Ministère des PT, 1976.
Project ONIDOCUSTE. Abidjan: Ministère des P et T, June 1992 (unpublished).
"Les télécommunications." *Afrique Industrie Infrastructures* no. 110 (March 1976): 14–63.
Système Régional Africain de Communication (RASCOM), vol. 1. UIT(ITU), April 1992.
Plan de développement des télécommunications 1971–75. Perspectives 1980. Document de travail. Paris: SOFRECOM/STET, 1970.
Ivoire Télécom, published by CI-TELCOM in Abidjan since December 1991, has replaced *Les Ondes*, published by ONT.

9

Nigeria: After a Century of Telecommunications Development, What Next?

G. O. AJAYI, R. I. SALAWU, AND T. I. RAJI

Since its inception a little over a century ago, Nigeria's telecommunications system has progressed through various stages of development from the primitive communications equipment in its colonial days to the great variety of technologies available today. In this chapter, the processes of Nigeria's telecommunications development and its progress, problems, and prospects are examined and discussed from its emergence to the expansion and modernization efforts of the 1990s.

9.1 The Past

Preindependence Era

The development of telecommunications in Nigeria began in 1886 when a cable connection was established between Lagos and the colonial office in London. By 1893, government offices in Lagos were provided with telephone service, which was later extended to Ilorin and Jebba in the hinterland. A slow but steady process of development in the years that followed led to the gradual formation of the nucleus of a national telecommunications network.

In 1923, the first commercial trunk telephone service between Itu and Calabar was established. Between 1946 and 1952, a three-channel line carrier system was commissioned between Lagos and Ibadan and was later extended to Oshogbo, Kaduna, Kano, Benin, and Enugu, thus connecting the colonial office in London with Lagos and the commercial centers in the country with local authority offices.

The main transmission medium during the preindependence era was unshielded twisted pairs of copper wire. This evolved later from rural carrier systems on high gauge lines to line carrier systems of twelve-channel capacity. Small- to medium-capacity systems employing VHF and UHF radio were introduced around 1955. The first serious attempt at planning telecommunications services in the country was the 1955–62 Development Program. It provided for the expansion of the trunk using a VHF Multichannel Radio System on a nationwide basis and a short microwave link between Lagos and Ibadan.

In the early days, the primitive coordinate pegboard switching system was used. This progressed through manual switchboards of different sizes, shapes, and capacities until Strowger exchanges were installed into the national network at Lagos Island, Ikeja, Ebute Metta, Apapa, and Port Harcourt between 1955 and 1960, along with 116 manual exchanges. The installation of the Strowger exchanges marked the beginning of automatic telephone switching in Nigeria. By the time of independence, automatic exchanges were established at the main centers and a subscriber trunk dialing system (STD) was introduced between Lagos and Ibadan.

The telegraph service also witnessed a parallel development, from telegraph delivery by way of manual coordinate pegboard switching to the use of Morse code for telex switching. By about 1960, a manual telex exchange of sixty subscriber lines was in service in Lagos. While all the efforts just described were essentially aimed at improving internal telephone services in Nigeria, external telephone services in the preindependence period were wholly owned by Cable & Wireless of the United Kingdom, which was a colonial private company.

9.1.2 The Postcolonial Era

With the attainment of independence in 1960, Nigeria embarked on a periodic national development plan. Telecommunications development was featured in each of these plans, which were usually of a five-year duration. It is more meaningful, however, to discuss the development of Nigeria's telecommunications since independence—its objectives, achievements, and features—on a decade-by-decade basis.

9.1.2.1 The 1960s

The focus of attention in this period was the expansion of the network to meet the needs of the fledgling commercial and industrial sectors. The specific objectives included:

- Installation of an additional 60,000 telephone lines to bring the total number of lines to 90,000 by the end of the decade.
- Expansion of trunk dialing facilities to link the major urban centers that were then springing up.
- Establishment of the Nigerian External Telecommunications (NET) Limited.

Unfortunately, these objectives could not be completely realized by the end of the plan period. For example, only about 26,000 lines (just over 40 percent of the planned target) were added to the existing network, partly because of underfunding and partly because of the disruption caused to the economy by the Nigerian Civil War (1967–70). Nevertheless, some the decade's major achievements included the installation of a microwave radio transmission system to link the cities of Lagos, Ibadan, Enugu, Benin, and Port Harcourt, all of which are in the southern part of Nigeria's transmission system. Preparatory work toward the establishment of NET as a limited liability company started during this period as well.

9.2.1.2 *The 1970s*

The second decade of independence incorporated two five-year plan periods—Nigeria's second and third development plans.

The 1970–1975 Plan Period. During the first half of the decade efforts were concentrated on the reconstruction and rehabilitation of the telephone equipment and other infrastructure damaged during the civil war. In order to achieve the objectives of the second plan period, developments in five major areas were considered. In telephony, new automatic exchanges were to be constructed and existing automatic exchanges expanded. New manual telephone exchanges were considered for construction as well. In telegraph communication, the torn-tape system was to be replaced with teleprinter automatic switching systems, and key and sounder circuits were to be converted to teleprinters.

For Nigeria's transmission system, the plan considered construction of subsidiary radio routes to provide trunk services from toll centers to end offices, construction of twisted pair carrier systems to provide links to rural areas, and provision of additional coaxial routes for Lagos-Ibadan-Ilorin-Kaduna. The plans for Nigeria's external line plant included construction of a local line plant network for new subscribers and an increase in existing line plant to achieve the objectives of the expansion. Finally, the 1970–75 plan stipulated the establishment of a Nigerian Satellite Communications earth station at Lanlate in the southwestern part of Nigeria.

Due to underfunding, however, the only objective of the 1970–75 plan period that was completed was the national telex network.

The 1975–1980 Plan Period. The third National Development Plan Period (1975–80) was the most ambitious. It aimed at increasing the telephone facilities from 50,000 lines to 750,000 lines—an increase of about 1,400 percent. In the area of switching, three contracts were awarded to add over 340,000 lines to Nigeria's networks. In the first contract—for the contingency plan—forty-five locations were to receive exchanges with a total installation capacity of 162,000 lines, and twelve other exchanges were to be expanded by 48,000 lines. In the second contract, covering turnkey projects, 147 locations were to receive external line plant and switching equipment to add 121,000 lines. The third contract, covering mobile exchange, provided for installation of twenty-nine mobile exchanges with 11,300 lines.

In the area of transmission, the following projects were considered: introduction of the Nigerian Domestic Satellite (DOMSAT) to provide television and sound broadcasting (later modified to accommodate telephony and teletype services between the states); introduction of the aerostat (balloon) system, which was intended for television and sound broadcasting and telex and telephone services; provision of coaxial cable between Lagos and Kaduna; expansion of the existing microwave radio link system intended for telephone services as part of the contingency plan exchanges; and provision of new transmission links for the exchanges in the contingency plan not covered by existing radio links.

There were at least six achievements associated with the 1975–80 plan. First,

177 locations were provided with telephone exchanges, as well as twenty-nine mobile exchanges, increasing the number of lines in the network from 52,000 to 241,000 and the number of telex lines from 874 to 4,950. Second, the DOMSAT earth station project was completed. Third, work started on the aerostat balloon (which turned out to be a disaster). Fourth, a second satellite antenna was built at Lanlate, which increased the global coverage of the external services. Fifth, an International Telephone Switching Center (ITSC) was installed at the Nigerian External Communications (NECOM) house in Lagos. And finally, a new microwave link was provided between Lagos and Cotonou (Benin Republic) and computerized telex, telegraph, and data switching centers were provided at NECOM house.

The 1975–80 plan period was not a complete success, however. The aerostat balloon project, which was abandoned, was a colossal waste of money. The proliferation of different technologies in the network made spare parts procurement difficult and complicated manpower training by limiting the number of personnel who could be switched from one part of the network to another. There was also a shortage of technical manpower to operate and maintain the additional facilities and a lack of adequate levels of finance to execute the projects.

The most serious problem, however, was bad planning. There was not adequate coordination between project management and implementation. Buildings were not available for the installation of purchased equipment, and vital links—such as external line plants—were omitted in the contract awarded.

9.1.2.3 The 1980s

The first half of the decade covered the fourth National Development Plan period (1980–85), which was essentially aimed at completing all outstanding projects from the previous plans. In addition to the primary objective, the development plan was designed to provide a total of 370,550 additional telephone lines, a terrestrial toll and trunk transmission network to link all switching centers throughout the country, and Telex/Gentex exchanges for about 9,000 telex lines with external line plant and teleprinter machines. It was envisaged that the total number of installed telephone lines in Nigeria at the end of the plan period would increase to 612,000.

During this period, the telecommunications arm of the Department of Posts and Telecommunications was merged with the Nigerian External Telecommunications to form, in 1985, the Nigerian Telecommunications Ltd. (NITEL), a limited liability company that today administers both internal and external telecommunications services in Nigeria.

9.2 The Present

9.2.1 The Existing Network

The Nigerian national telecommunications network of the 1990s is made up of the following elements:

1. Telephone services
   a. Total capacity has risen to over 800,000 lines.
   b. Other services include: over 10,000 cellular mobile telephones and 15,000 voice mail lines.
2. Telex services: Fourteen telex exchanges with a total installed capacity of 12,800 lines.
3. Transmission systems
   a. Microwave.
      There are 264 terminal stations and 172 unmanned repeater stations. The channel capacities range from 300 to 1,800 channels with capability to accommodate one color television on the 960- and 1,800-channel systems.
   b. Coaxial.
      This system provides 960 channels between Lagos and Kaduna.
   c. Optical fiber cable.
      Optical fiber cables were developed to link primary/secondary exchanges in the Lagos zone.
   d. DOMSAT.
      This system consists of nineteen standard B earth stations and operates on the leased transponders from Intelsat. The 36 MH2 bandwidth transponders were initially grossly underutilized. They were originally used mainly for television transmission for only a few hours daily. However, they are now utilized for both television and telephone.
4. International services
   a. International satellite system.
      This originally had two gateways at Lanlate and Kujama, which were linked to switching centers at Lagos (NECOM) and Kaduna, respectively. In 1992, two additional gateways were commissioned—one at Victoria Island, Lagos, to cater to the ever-increasing traffic in the Southwest, and the other in Enugu to cater to the traffic in the eastern part of the country.
   b. Submarine cable.
      This provides a transmission system from Lagos through Abidjan, Dakar, Casablanca, and on to Europe.

9.2.1.1 Extent of Services

The services offered by NITEL are telephony, telex and telex delivery services, telegraphy and registered telegraphic addresses, pay phones and public coin telephones, transmission and reception of real-time television for network services, private leased telephone and telex service, private wire, leased telephone and telegraph services, alternate voice data (AVD) circuits, voice-cast and press reception, international public counter services, facsimile service, switched data, electronic mail, and cellular.

Maritime Services. Nigeria's shore-to-shore and ship-to-shore maritime communications services are provided via the high frequency radio. Its limitations are poor transmission quality, low reliability, and lack of automatic access to the national telecommunications network. In 1988, Nigeria joined the International Maritime Satellite Organization (Inmarsat), which operates a system of satellites to provide mobile communications for the world's shipping and offshore industries. Through the Inmarsat system, NITEL offers Maritime Mobile Service (MMS) as well as satellite mobile communications.

The Scope of National Service. Nigeria's telephone penetration rate was still low, in 1994, at about 0.8 direct exchange lines (DELs) per hundred inhabitants. Its major challenge in extending its facilities continued to be the provision of telecommunications services in the rural areas, where there is little or no penetration. In the mid-1990s, however, the telecommunications facilities in the urban areas continued to be inadequate as well. Nigeria's telecommunications services—especially telephony—are not sufficient to meet the needs of all those who require them, especially in the big cities like Lagos, Ibadan, Enugu, Kano, and so on. This has led to long waiting periods for obtaining facilities (which was above ten years in 1993) and congestion of existing exchanges. Finally, new telecommunications facilities such as facsimile, international business services, and high-rate data transmission are not readily available.

Nevertheless, in recent years, attempts were made by the Nigerian authorities to spread the telecommunications facilities throughout the country, and in the mid-1990s all the twenty-one state capitals, as well as Abuja and many of the 589 local government headquarters, were connected to the national network. The government attempted to ease the wait for phone lines by increasing the penetration of public telephones. In recent years, the government licensed seven companies to provision and operate public pay phones in different regions of the country. The major obstacle hampering the extension of the system was lack of funds and an absence of the engineering infrastructure needed for the development and production of spares and components.

Cellular Telephony. Cellular telephony was first introduced in 1992 with the formation of Mobile Telecomunications Service (MTS). This company is a joint venture between NITEL and Digital telecommunications of Atlanta; in 1994, MTS had a nationwide monopoly over cellular service. It began with a capacity of 10,000 lines, and due to the high level of unmet telecommunications demand, the system was filled to capacity within one year. Subsequently, in 1994, MTS added 20,000 lines, with plans to add 25,000 more lines. Even with this additional capacity, it is common to receive a fast busy signal during peak hours due to network congestion.

Recognizing the need to meet the growing demand for cellular services, the government licensed four additional companies: M-Tel, Wireless Systems Nigeria, Ltd., Tele-Africa Nigeria Ltd., and Motophone Nigeria Ltd. Additionally, fourteen companies were awarded licenses to operate paging services.

Nigeria's cellular market is expected to grow at a rate of 25 percent annually through the 1990s. The major markets are in Lagos (among the young urban elite), the oil-based communities of Port Harcourt and Warri, and the cities of

Kano and Abuja. Over half of the cellular equipment was supplied by U.S. companies, mostly Motorola.

9.2.2 Institutional Structure

In May 1992, NITEL, which was fully owned by the Nigerian government, was corporatized, with the ultimate objective being full privatization. The present system seeks to protect the sovereignty and security of the country by keeping NITEL under government control while, at the same time, making the telecommunications service less dependent on the government. The motivating force behind the decision to corporatize rather than privatize was the fear that control of the national network might be lost to foreign companies.

Although the Nigerian government still provides funds for major capital projects, the conditions of service of telecommunications workers have been moved out of the civil service structure. The corporatization has also brought about increases in the cost of telecommunications services in the country.

The administrative and policy matters of telecommunications remain with the Ministry of Communications, which represents the country at the International Telecommunications Union (ITU) and other international telecommunications organizations.

9.2.2.1 Management Structure of NITEL

In the mid-1990s, NITEL continued to operate under a board appointed by the government to serve various interests. The structure of the company is based on the three-tier system of territorial administration, zonal administration, and headquarters in order to decentralize functions and optimize operational efficiency. Abuja, the federal capital, and each of Nigeria's twenty-one states constitute a single territory, while Lagos constitutes two territories. There are five zones: the northwest, northeast, southwest, southeast, and Lagos, and each zone is made up of between four and six states or territories. The zones are semiautonomous in their operations; however, the functions of the zones are coordinated at the center through the office of the managing director, who is the chief executive. The headquarters operates as six divisions.

9.2.3 Telecommunications Regulation

The Cable and Wireless Act of 1962 established the Ministry of Communications as the regulatory body for telecommunications in Nigeria. The ministry regulated NITEL until a decree in 1992 established the Nigerian Communication Commission (NCC), which was charged with the duty of regulating the telecommunications sector. The NCC became operational in September 1993. It is responsible for the following:

1. Licensing telecommunications operators.
2. Facilitating private-sector participation and investment in the telecommunications sector.

3. Ensuring the improvement of Nigerian telecommunications penetration.
4. Establishing and supervising technical and operational standards and practices for network operators.
5. Overseeing the quality of service provided by operators.
6. Setting terms for the interconnection of carrier networks.
7. Ensuring that the interests of telecommunications consumers are protected by promoting competitive pricing and guarding against abuse of market power.

Since becoming operational, the NCC has taken aggressive steps to open the telecommunications sector to private investment and enterprise. In June 1994, the following services were open to private-sector participation:

1. Customer premises equipment (CPE).
2. The provision and operation of public pay phones.
3. The provision and operation of private network links.
4. The provision and operation of community telephones for rural areas and industrial parks.
5. The provision and operation of value-added network services for the banking and airline sectors, including packet-switched networks.
6. The repair and maintenance of telecommunications facilities.
7. Telephone cabling.

In 1995, NITEL still maintained a monopoly over the following services: (1) the provision and operation of public switches and trunks and their associated infrastructure; and (2) the provision and operation of international network links.

To participate in Nigeria's telecommunications sector, a company must be either owned by a Nigerian citizen or be registered in Nigeria; the NCC hoped that this requirement would encourage joint ventures between foreign companies and Nigerian companies. The NCC also required that a licensee submit technical and organizational plans that demonstrate a commitment to sustained service and that the equipment be compatible with the existing infrastructure. Since the NCC was inaugurated, over one hundred licenses have been granted:

1. Twenty-one companies in sales and installation of CPE.
2. Seven companies in the provision and operation of pay phones.
3. Ten companies in private network links—including domestic satellite links.
4. Eighteen companies in mobile communications—including cellular telephony and paging.
5. Two companies in local community telephony.
6. Thirteen companies in value-added network services.
7. Two cabling companies.
8. Seventeen companies in fixed telephony services.
9. Twenty-two companies in Internet services.
10. Four companies in equipment repair and maintenance.

Telecommunications concerns have to pay 5 percent of the capital value of a system as a license fee and 2.5 percent of the turnover as an annual concession fee. Additionally, there is a 2.5 percent surcharge for special licenses; for example, companies with extensive networks that use over five pairs of frequencies and are not willing to share their infrastructure.

In 1997, the NCC made a landmark decision by authorizing a second national carrier to compete with NITEL in all services, effectively ending NITEL's monopoly and creating a duopoly. While the second national carrier is currently concentrating on offering doomestic and international satelite-based services for business telephony services for the general population.

The National Broadcasting Commission (NBC) was established in 1992 to regulate and control radio and television broadcasting in the country. The NBC is vested with powers to handle matters connected with mass communication and technical aspects as they relate to the broadcast bands for radio and television. The NBC has licensed many radio and television stations, including the first international satellite television broadcasters. Minaj, based in Obosi, and Daar communications, based in Lagos, were authorized to transmit their programming internationally via their own satellite teleports. Many companies seized the opportunity to enter into the fixed satellite services (VSAT) sector once the NCC authorized competition in value added network services and private network links. The primary customers for such services are in the banking and oil industries.

9.2.4 Telecommunications Financing

Almost all of Nigeria's telecommunications development plans since independence have suffered from underfunding. For example, during the first National Development Plan (1962–68), only 35 percent of the expected expenditure was provided, and consequently only 40 percent of the expected 60,000 lines were added to the network. Similarly, most of the objectives of the ambitious program under the third National Development Plan (1975–80) could not be achieved, partly because of underfunding.

Inadequate funding has continued to inhibit the rapid development of Nigeria's telecommunications since the creation of NITEL in 1985. The situation has worsened in recent years because of the large-scale devaluation of the national currency and the shortage of foreign exchange with which to prosecute many projects. Worldwide inflation has also led to high prices for telecommunications equipment.

Nigeria's revenue from the provision of telecommunications is also still comparatively low because telecommunications service is inexpensive there compared to the United States and other parts of the world. The initial telephone charge, for example, is about U.S.$0.10 for the first three minutes for local calls. Similarly, for international calls, the first three minutes cost about U.S.$6.00.

9.2.5 Manufacturing

The basic support industries for telecommunications manufacture and assembly in Nigeria belong mainly to the private sector. The country's telecommunications and electronics industries include electronic rewinding factories; teleprinter machine and telephone set assembling factories; radio, television, cassette, and cartridge and record player assembly plants; intercom, mini-EPBX (private branch exchange), and key telephone assembly plants; telecommunications components factories; TVRO (TeleVision Receive Only) earth station assembly plants; cable and wire factories; and plastic extrusion and injection industries. Production levels for some of these products in 1991 were: telephone handsets, 5,000 units per year; intercoms, 30,000 units per year; key telephones, 500 units per year; and mini-PABX (private automatic branch exchange), 500 units per year.

Both Nigeria's telecommunications and electronics subsectors are in their infancy stages. Because domestic manufacturing input to telecommunications development is very small, large amounts of foreign funds are required for telecommunications projects. Export of electronic and telecommunications products is virtually nonexistent in Nigeria, and importation still continues on a large scale in sophisticated consumer electronics; telecommunications; and defense, computer, medical, and industrial electronics. Domestic assembly also cannot fulfill the demand for simple consumer items, and importation is used to supplement the local production. It is worth noting that in Nigeria there is no industry in the electronic components manufacturing subsector.

9.2.6 Trends in Technology Adoption

In recognition of the fact that telecommunications is an infrastructure that may aid a country's economic development, many African countries, including Nigeria, have embarked on programs of modernization and rehabilitation of their outdated and failing national telecommunications network. For example, in Kenya, the telecommunications administration has put into service a digital microwave system to service the eastern and southeastern parts of the country using solar technology as its power supply. Malawi, Zambia, and Tanzania also began constructing microwave links and satellite earth stations some years ago to ensure efficient and reliable telecommunications services.

The telecommunications administration in Nigeria decided to adopt digital technology for the national network with a view to improving services for the existing customers as well as meeting new demands. At the time of this decision, the network was experiencing three categories of problems that hindered the provision of fast and reliable telecommunications service and inhibited the desired rate of telecommunications development in the country:

- *Inadequate capacity.* This hindered considerably the rapid and cost-effective expansion of telephone service. Considering the size of the population and the level of economic development in the country, the number of installed telephone lines was seen to be grossly inadequate to meet demand, and the resulting inadequate capacity was responsible for poor call completion rates, subscriber dissatisfaction, and hence loss of revenue.
- *Poor maintenance.* This had contributed in no small way to the inefficient utilization of Nigeria's existing network. The maintenance problem itself was attributed to such factors as lack of or inadequate supply of tools, test equipment, and materials for maintenance; government policy on procurement of spare parts; poor maintenance organization; and the poor work attitude of maintenance personnel.

- *Low revenue generation.* The revenue being generated from the existing public telephone service was rather low in comparison to the cost of providing the service. This was attributed partly to inefficiencies in management, partly to unproductive use of capital, and partly to an inefficient billing system.

NITEL's management believed that by adopting appropriate technology, most of the technical aspects of these three problem areas could be solved. Accordingly, the administration opted for digital technology and sought suitable strategies for its introduction and implementation in the existing national network, even though the vast majority of the network was analog, personnel requirements for the new technology were significant, and methods of financing the digitalization projects had to be found.

NITEL decided that the digitalization of the network would commence with the switches, to be followed later by the transmission aspects. Due to financing issues, the implementation of the digitalization was divided into three phases with the priority areas—mostly multiexchange areas and the international gateways—being digitalized during the first phase. It was also decided that Abuja, the nation's new capital, should be made a "digital island"; that all existing analog switches should be gradually phased out, to be replaced at the end of their lifetime by a digital switch; and that all new exchanges would be digital. In addition, all further telecommunications expansion and development in Nigeria was to be digitalized in order to (1) surmount the problems associated with maintaining the old analog network and (2) meet the increasing demand by customers for such services as facsimile, telex, data transmission, and the like. Most of these decisions were being implemented in the mid-1990s.

In addition to the existing telecommunications network administered by NITEL, the Nigerian National Petroleum Corporation (NNPC) has also installed a cross-country high-capacity digital communications system for its pipeline operations. This system combines two advanced or high-level technologies—microwave digital radio and optical fiber—in a complementary rather than competing manner. The system consists of terrestrial microwave links in the riverine areas, like the delta region of the country, and densely populated areas such as Lagos, and fiber-optic links in the larger part of the network. In the mid-1990s, the system was highly underutilized given the low level of traffic being generated for the pipeline operations. It was therefore suggested that NITEL could use part of the system to supplement its own backbone system. In 1995, the major foreign oil companies—Shell, Chevron, AGIP, and ELF—were leasing fiber-optic lines from NNPC in order to connect their oil fields with regional bases and computer centers.

9.2.7 Regional and Continental Collaboration

As a member of the Economic Community of West African States (ECOWAS) and the Organization of African Unity (OAU), Nigeria has been collaborating with the member nations of these organizations to develop telecommunications services at the subregional as well as continental levels. This collaboration has taken the form of meetings among the telecommunications engineers and planners of the various member countries to discuss the technical issues involved in planning, operating, and designing telecommunications systems suitable for use in Africa's environment.

The segment of the Pan-African Telecommunications Network (Panaftel) linking the eastern part of the continent with the western part passes through Nigeria and uses portions of its domestic network. The objective of the Panaftel is to provide the African continent with reliable and effective telecommunications systems that will enable telephone and telex circuits to be set up readily between any two African countries without the need to transmit through extra-African centers.

Convinced that the development of telecommunications is one of the essential requirements for the building of a meaningful ECOWAS organization, this subregional group has made the intertelecommunications (INTERCOM) program one of its priority actions since it was initiated in 1979. The objective of the program is to link the capital towns of the member states through earth networks and microwaves and provide them all with an international transit center to facilitate automatic telephone communication among them. Nigeria continued to be an active participant in this endeavor in the mid-1990s.

The feasibility study of the Regional African Satellite Communications System (RASCOM) addressed the need for developing an African satellite for a comprehensive telecommunications network that would ensure reliable communication between African countries. The findings of the study complemented the objectives of Panaftel and provided a viable option for achieving an effective and efficient telecommunications interlink within the continent. Nigeria embraced the study and has continued to participate actively and fully in the work.

One effect of these regional collaborative activities was demand for a high level of performance on the part of the domestic network. Beyond that, however, the "big brother" role that Nigeria plays in African affairs makes it even more imperative that she develop her telecommunications infrastructure for an efficient and reliable domestic service.

9.2.8 The Pattern of Traffic between Nigeria and Other Countries

One study of Nigeria's international traffic indicated that Nigeria generates significant traffic volume (in engineering terms, as much as 290 erlangs). Forty percent of this large traffic volume terminates in the United Kingdom, which has had a long-standing historical and economic bond with Nigeria. About 30 percent of the traffic from Nigeria is now directed toward North America, a reflection of the growing trade relations between the two. The increasing flow of traffic into North America can also be attributed to the rising number of Internet connections from Nigeria to major Internet backbones located in the United States.

Nigeria is geographically bounded in the south by the Atlantic Ocean and in the north, east, and west by French-speaking countries. By virtue of their different colonial experience, these Francophone neighbors have a trade and social outlook that is quite different from that of Nigeria. Although there are terrestrial networks linking Nigeria with neighbors, most of these networks are either not operational or rarely available.

9.3. The Process of Change

The emergence of an electronics industry in Nigeria is gradually introducing new dimensions into the system. Now data transmission and electronic mail delivery are possible, although few have been added to the services. Computer communication has also been introduced, and most establishments have now computerized their services and systems. The banking sector especially is now partially computerized. There is a move to have intrabank computer communications services for the clearance of checks, but that has not yet taken off.

9.4 Future Trends

In the mid-1990s, NITEL continued to hold a monopoly over basic telecommunications services. And in 1997, the NCC licensed a second national carrier, creating a duopoly. It was hoped that an additional national carrier would improve telecommunications penetration and hasten the expansion of service. Possibly spurred by the imminent licensing of a second carrier, NITEL in 1994 announced plans to invest naira 40 billion over a five-year period. The money is to come from both internal and external (unilateral lending institutions and the private sector) sources.

The NCC planned to expand the telecommunications services in the country, increasing penetration to about 8.5 per 100 inhabitants by the turn of the century. To that purpose, it planned to provide 1.2 million new lines. It also intended to license several new cellular operators in order to reach a goal of 100,000 cellular lines by the year 2000.

To accomplish this expansion, the government planned to have manufacturers set up plants in Nigeria, consequently requiring the selection of technologies for adoption in the country. However, this plan was not finalized. The government also decided that for security reasons the telecommunications industry should be corporatized rather than privatized.

The present structure of thirty states and 589 local government areas in Nigeria has increased the pressure to provide telecommunications services to the many administrative headquarters. Since about 70 percent of Nigerians reside in the rural areas, there is an urgent need for these areas to have telecommunications facilities.

The digitalization of both the exchanges and transmission systems will continue at a rate determined by the availability of funds. In effect, the analog systems will be gradually phased out. Several new technologies (e.g., the wireless system) are available for the rapid development of telecommunications in Nigeria and developing countries in general. The wired conventional system has led to long waiting periods for telecommunications to be made available to both urban areas such as Lagos and the rural areas. The use of wireless communications systems, both land based and satellite based, will continue to play an increasing role in future telecommunications development. For example, the Celluphone (CELLUlar telePHONE), making use of a cellular terminal with the necessary interface, has been adopted for the provision of rural telecommunications.

Nigeria may also take advantage of the introduction of Low Earth Orbit (LEO) and Medium Earth Orbit (MEO) mobile satellite systems such as IRIDIUM, Globalstar, and so on, to provide wireless telecommunications services in the country for both urban and rural areas.

9.5 Conclusion

This chapter has considered the development of telecommunications in Nigeria for a period of over one hundred years following the introduction of the technology into the country. The discussion has covered the state of telecommunications during the colonial era (before 1960) and the postindependence period. The objectives, achievements, and failures of telecommunications during the five National Development Plans of these three decades have been highlighted, as have other aspects of the telecommunications sector: regional collaboration in telecommunications, local manufacturing facilities, telecommunications policy, adoption of new technology, the financing of facilities, and the future prospects of the sector's development.

Currently telephone penetration in Nigeria is low (about 0.8 DELs per 100 inhabitants) and the available facilities exist mainly in the urban centers. The government is consequently now focusing on expanding telecommunications facilities to the rural areas.

There has been a gradual introduction of digital switches, digital radio, and optical fiber transmission into the network. The local manufacturing of telecommunications components and equipment is at present low; however, the government has started to take the necessary actions to correct this imbalance by entering into joint ventures with foreign companies to establish telecommunications industries in Nigeria.

In order to achieve the target of 8.5 DEL per 100 inhabitants by the turn of the century, the country needs a large amount of investment, including foreign financing. Options for this funding include subventions from the government to NITEL, contractor financing, and foreign loans from international agencies such as the World Bank, the African Development Bank, and so on.

The telecommunications scenario in Nigeria is not likely to be much different from what exists in many developing countries. Nigeria must give the utmost priority to the development of telecommunications because of its multiplying effects on industrial and economic growth.

Bibliography

American Embassy, Lagos. 1994. *Nigeria—Telecommunications Overview.* Lagos: American Embassy.

Attah, M. E. 1988. Current Trends and Future Development in Telecommunications in Nigeria. Paper presented at the 1988 Annual Conference of the Nigerian Association of Radio Science, Lagos.

Ayuba, A. T. 1986. Address by the Minister of Communications on the occasion of the 18th World Telecommunication Day, Lagos, May.

Bitature, C. 1989. Development of an Effective Pan-African Telecommunications (Panaftel) Network. M.Sc thesis, Obafemi Awolowo University, Ile-Ife.

Iromantu, O. C. 1994. Models of Deregulation and Liberalization of Telecommunications for Africa and the Middle East." Paper presented at Africa Telecom '94, Cairo, April 25–29.

ITU. 1987. *PANAFTEL: The Pan-African Telecommunications Network.* Geneva: ITU.

Mark, D. A. B. 1990. Address by the Minister of Communications on the occasion of the 22nd World Telecommunication Day, Lagos, May.

Okafor, Uzo. 1993. *Nigeria—Cellular Telecommunications Equipment.* Lagos: American Embassy.

Salawu, R. I. 1991. Telecommunications in Focus. Tenth Anniversary Lecture for 2-Way Communications Ltd, Lagos, November.

10
Ghana

FRANCIS K. A. ALLOTEY AND FELIX K. AKORLI

The Republic of Ghana is a West African nation on the Gulf of Guinea with an area of about 92,000 square miles and a population of about 16 million (1995). Its topography consists of forest and savannah, and the country is divided into ten regions, with the national capital located at Accra. Ghana's six largest ethnic groups are the Akans (44 percent of the population), Ewe (13 percent), Ga-Adangbe (8 percent), Mole Dagbane (16 percent), Guans (4 percent), and Grume (3 percent). The official language is English. The Volta is Ghana's largest river and the location of the Volta Lake on which are built two hydro dams, Akosombo and Kpong, with a total capacity of about 1.12 gigawatts.

Ghana's main export products are cocoa, gold, and timber; other exports include diamonds, bauxite, and manganese. The fluctuation of world market prices and the absence of processing industries for these products have historically affected Ghana's revenues. Ghana imports oil for energy consumption, and exports hydroelectricity to Togo, Benin, and Burkina Faso.

In 1987, Ghana's per capita gross national product (GNP) was U.S.$390. The World Bank's estimate of Ghana's GNP in 1991 was about U.S.$440 per capita. Since 1984, Ghana has been implementing an Economic Recovery Program, which gives the country easier access to International Monetary Fund (IMF) and World Bank loans. In 1988, Ghana's exports, which amounted to U.S.$1.014 billion, exceeded imports (U.S.$907 million) by over U.S.$4 million.

Since 1981, Ghana has been a unitary republic ruled by decree. In November 1992, however, an election was held to select a president for the next democratically elected parliamentary government. In January 1993, following parliamentary elections, Ghana returned to a parliamentary form of government.

10.1 History

10.1.1 The Precolonial Era

The first telegraph line in Ghana (then known as the Gold Coast) was a 10-mile link installed in 1881 between the castle of the colony's then governor in Cape Coast and Elimina. The line was then extended to Christianborg Castle near

Accra, which became the seat of government, and later extended still further to Aburi, 26 miles outside Accra.

In 1882, the first public telegraph line, stretching over a distance of 2.5 miles, was erected between Christianborg and Accra. Between 1887 and 1889, these telegraph lines were extended to cover Accra, Prampram, Winneba, Saltpond, Sekondi, Ankobra, Dixcove, and Shama—all colonial castles or fort towns, as well as commercial ports and fishing centers. In 1886, telegraph lines were extended to the middle and northern parts of Ghana into the territory of the Ashantis. Between 1900 and 1901, this new communications technology was used to subdue the Ashantis in the Yaa Asantewa war.

Because Ghana's telegraph lines were often cut down by superstitious locals convinced that the cables were "magic" lines being used by the Europeans to win wars, Ghana's colonial governor entrusted the safety and security of the lines to the tribal chiefs in 1886. The colonial governors offered the chiefs handsome rewards for reporting any damage done to the lines.

In order to improve communications in the southern part of the country, the first manual telephone exchange (seventy lines) was installed in Accra in 1892. Twelve years later, in 1904, a second manual exchange consisting of thirteen lines was installed in Cape Coast.

10.1.2 Under English Colonial Rule

Ghana's telecommunications infrastructure was laid down and expanded by the colonial administration mainly to facilitate the economic, social, and political management of the colony. In 1901, for example, the Ashantis were brought under British colonial rule, and telegraph lines were accordingly extended from Accra to the capital of the Ashanti Kingdom and beyond. By the end of 1912, 1,492 miles of telegraph lines had been constructed to link forty-eight telegraph offices spread throughout the country.

Before the beginning of World War I in 1914, 170 telephone subscribers had been served in Ghana, but it was between World War I and 1920 that the backbone of the main trunk telephone routes—Accra-Takoradi, Accra-Kumasi, Kumasi-Takoradi, and Kumasi-Tamale—was built using unshielded copper wires. By 1930, the number of telephone exchange lines in Ghana had grown to 1,560, linking the coastal region with the central and northern parts of the country.

Due to the depressed global economy of the 1940s, there was little or no growth in telecommunications in Ghana during and immediately after the Second World War. Nevertheless, during that period 1+1 carrier equipment was installed on the Accra-Takoradi, Accra-Kumasi, Kumasi-Takoradi, and Kumasi-Tamale physical trunks. These were later augmented with 1+3 carrier equipment, thus increasing the trunks connecting these towns threefold.

In 1953, the first automatic telephone exchange with 200 lines was installed in Accra to replace the manual one erected sixty-three years earlier. Three years later, in 1956, the trunk lines connecting Accra, Kumasi, Takoradi, and Tamale were upgraded through the installation of a forty-eight- and twelve-channel VHF network.

10.1.3 The Postcolonial Period

The attainment of independence by Ghana in 1957 brought new dynamism to the country's telecommunications development. A seven-year development plan launched just after independence hastened the completion of a second new automatic exchange in Accra in 1957. By the end of 1963, Ghana had more than 16,000 telephone subscribers, and 32,000 rotary-type telephones were in use.

Due to the rapid growth in commercial activities in mining, timber, cocoa, shear butter, and the like in outlying parts of the country, new manual exchanges were installed at Cantoments, Accra, Swedru, Koforidua, Ho, Tamale, Sunyani, and Kumasi during the postindependence years. The installed switches were Strowger (step-by-step) and Philip UR 49 types.

The management of Ghana's telecommunications institutions was initially assigned to the Public Works Department but was transferred to the post office following the enactment of the Post Office Ordinance in 1886. Telecommunications was later administered by the government's Post and Telecommunications Department until the early 1970s.

10.2 The Present Era

10.2.1 The Post and Telecommunication Corporation

A new chapter in the development of Ghana's telecommunications system began in 1974, when the Post and Telecommunication Department was declared a public corporation by National Redemption Council Decree No. 311. The department was placed under the authority of the Ministry of Transport and Communication, which is still responsible today for policy formulation and the control of Ghana's telecommunications sector.

Under the instrument of incorporation, the Post and Telecommunication Corporation (P & T) is administered by a board of directors that functions as the corporation's governing body. There is a director general, who is the chief executive accountable to the board of directors and responsible for the organization, maintenance, and development of all the corporation's services (domestic and international), as well as the determination of financial policies. The director general also ensures that government policies on telecommunications are implemented and that rules and regulations governing the various services, as well as international conventions, are correctly interpreted and acted upon. He is assisted by two deputies—the deputy director general for engineering and the deputy director general for posts.

10.2.2 Quality of Service

In the present era, Ghana's telecommunications system has been plagued by significant service quality problems. According to *Missing Link* (the 1984 report of the International Telecommunication Union's [ITU] Independent Commission for World Telecommunication), penetration of telephones in Ghana in 1992 was only 0.32 per 100 inhabitants. Moreover, according to the published report of Ghana's 1990–92

Public Investment Programmes, the quality of service—measured by call completion rate, time required to obtain dialing tone, number of faults per line per year, availability of the link, mean time between failures (MTBF), and signal-to-noise (S/N) ratio—on a number of microwave and UHF routes in many exchange areas of the country is below the acceptable international standards for Africa as defined by the ITU.

10.2.2.1 Principal Causes of Low Service Quality

Analyses of Ghana's low quality of telecommunications service identified the following factors as the principal contributory causes:

- Extremely high exchange fill. Delays in the implementation of projects renders future service projections invalid after a project is completed. As a result, the estimated capacities of exchanges do not meet the needs of subscribers.
- Unsatisfactory state of the external plant network (including underground cables and overhead line system). Ghana's underground cables are not only old but their manufacture, predominantly lead-sheathed and paper-insulated, makes them susceptible to acts of vandalism and sensitive to moisture. Both Ghana's overhead systems and its underground cables are also susceptible to theft, with the metal contained in them highly prized by jewelry dealers. Moreover, the overhead systems are often rendered inoperative by bush fires and violent storms and rains.
- Rampant breakdown of a large proportion of power and air-conditioning equipment. Such failures have rendered most of Ghana's nontropicalized telecommunications equipment unreliable.
- Poor maintenance of service. Due to the very limited number of service vehicles and the unreliability of available vehicles, Ghana's telecommunications network has historically experienced delays in needed maintenance.

Another factor contributing to the low quality of Ghana's telecommunications service has been inadequate manpower resources. Most of Ghana's trained telecommunications manpower, for example, is attracted away from the public sector by private companies offering higher salaries and better working conditions. Other factors hampering the majority of Ghana's telecommunications links include interference from local broadcasting stations; signal fading, which occurs most often during "harmattan"—the months of Ghana's dry season (November to February); and cross talk and signal noise.

10.2.3 Ghana Frequency Registration and Control Board (GFRCB)

10.2.3.1 Establishment and Function of GFRCB

Up until the late 1970s, the P & T was responsible for assigning and issuing radio frequencies to private, public, and government institutions for the operation of radio equipment. In 1977, however, the Ghana Frequency Registration and Control Board (GFRCB) was established by the Frequency Registration Decree of 1977, and these responsibilities were transferred to it. Membership to the GFRCB is determined by Ghana's head of state with the advice of the National Security Council.

The functions of the GFRCB are as follows:

- Approve and issue licenses to commercial and amateur radio operators.
- Monitor the training of commercial and amateur radio operators.
- Perform tasks that the GFRCB deems to be incidental or conducive to the exercise of its function.

In order to approve and issue licenses for the sale, manufacture, and assembly of telecommunications equipment, a GFRCB 1984–92 Manpower Requirement Programme proposed to the government of Ghana that four interdepartmental technical committees be formed to advise the GFRCB: a frequency assignment subcommittee to be supported by the military assignment group, aeronautical assignment group, and maritime assignment group; a science and technology analysis committee consisting of a technical analysis group, a spectrum management group, and an authorization and standards group; a spectrum planning subcommittee consisting of a spectrum planning branch, a frequency liaison branch, and a spectrum utilization branch; and an international notification group.

The GFRCB proposed that the membership of these committees be composed of specialists drawn from the P & T, the Ministry of Defense, the Civil Aviation Authority, and the Maritime Assignment Group. Their duties were to respond to ITU questionnaires and other correspondence related to the notification and coordination of Ghana's frequency assignments. As of 1992, these interdepartmental advisory committees had yet to be formed.

10.2.3.2 Licensing of Telecommunications Apparatus and Stations

The GFRCB is responsible for performing the local registration and licensing of frequencies. Obtaining permits for the establishment, installation, and use of any telecommunications in Ghana is governed by the following decrees:

- No one may establish, install, or use any station apparatus for telecommunications unless he or she has obtained a license issued by the GFRCB to do so.
- No one may sell any telecommunications apparatus unless that person holds a valid dealer's license issued by the GFRCB.
- No one may begin manufacturing or assembling any telecommunications apparatus unless he or she has been registered by the GFRCB. Anyone who intends to begin manufacturing or assembling a telecommunications apparatus must notify the GFRCB of the date on which he or she intends to commence business and the name and address of the business.
- Anyone found to contravene any of these provisions is guilty of an offense and liable on summary conviction to a fine or imprisonment not exceeding twelve months, or both.

In order to enforce these decrees, the GFRCB undertook the following control measures:

- Anyone who in the course of business deals in the sale of any manufacturing and assembling telecommunications apparatus must submit to the GFRCB a monthly return stating the number of telecommunications apparatuses sold, imported, manufactured, or assembled by him or her during that month.

- Any authorized dealer or manufacturer who fails or refuses to comply with these regulations or who gives false information in any return shall be guilty of an offense and liable on summary conviction to a fine or imprisonment not exceeding six months, or both.

10.2.3.3 Operations of Telecommunications Services

By law, individuals or establishments can operate telecommunications services in Ghana only after they have been issued a license by the GFRCB. The telecommunications services covered by this regulation include aeronautical mobile services for stations on land or aboard aircraft, amateur radio services, broadcasting services, citizen radio services, experimental radio service other than broadcasting, fixed radio services including telecommunications services for diplomatic missions, land mobile radio services, and maritime mobile services for stations on land or aboard vessels twelve meters in length.

10.2.4 Manpower Training

10.2.4.1 Post and Telecommunication Engineering Training School

Programs for training manpower for the telecommunications industry in Ghana began as far back as 1948—when the then Post and Telecommunication Department established the Telecommunication Engineering School. This school was charged with the responsibility of training linesmen and technicians to operate and maintain telecommunications equipment and plants. The school's principal and teaching staff were drawn from the British Post Office, and as it was the only school of its kind in Ghana at that time, it catered to the needs of other institutions such as civil aviation, broadcasting, the police, and the military. In 1951, the first batch of twenty-one technical assistants with school certificate backgrounds was recruited to undergo a three-year training course in telephony.

In 1974, when the Post and Telecommunication Department was declared a public corporation, the Telecommunication Engineering School was renamed the Post and Telecommunication Engineering Training School and recharged with the responsibility to train personnel to operate, modify, and maintain telecommunications equipment, with an emphasis on job training.

In the 1990s, Ghana's P & T embarked on a program to modernize and expand telecommunications services, including human as well as material resources. In order to execute this commitment successfully, the ITU was approached to do the following:

- Propose and assist in the implementation of an organizational structure for the Post and Telecommunication Engineering Training School, including: handling all training functions; providing full documentation of its efforts; and promoting the awareness and importance of training within the P & T.
- Propose and assist in the implementation of appropriate training systems and procedures that apply ITU standard methodology to the management of training centers. The goal was the optimum utilization of training resources and the introduction of modern administrative and management tools—including both hardware and software where necessary.

- Train Ghana's telecommunications professionals in managing and updating the telecommunications training systems. As a result, P & T staff began to be sent for training to the United Kingdom, the United States, Japan, France, Sweden, and the Netherlands, among other countries.

In addition to the P & T Engineering Training School, the department of electrical and electronic engineering of the University of Science and Technology, Kumasi, Ghana, offers courses to students in telecommunications.

10.2.4.2 Involvement of the University of Science and Technology

The number of trained Ghanaian, high-caliber engineers produced by the United Kingdom, the United States, Japan, and other countries cannot meet the manpower needs of the P & T. Because of the sophisticated technologies used in the telecommunications field, many of the complex engineering problems encountered cannot be solved by P & T engineers alone. It thus became necessary to seek assistance from other institutions. In 1992, a collaboration agreement was signed between Ghana's P & T and the department of electrical and electronic engineering of the University of Science and Technology, Kumasi, Ghana.

The collaboration agreement was expected not only to benefit the P & T in fulfilling their need for high-caliber manpower, but also to provide university lecturers with interaction with Ghana's telecommunications industry, thus helping them solve their problems while saving the country foreign exchange funds used for foreign expatriate consultants.

10.3 Telecommunications Projects

10.3.1 The First Telecommunication Project

In 1975, the P & T began negotiating loans from many multilateral and bilateral financial institutions in order to undertake a number of development projects to modernize and expand both national and international telecommunications services in Ghana. The objective of these projects, known collectively as the First Telecommunication Project (FTP), was the rehabilitation, modernization, and expansion of Ghana's national telecommunications network. The project, which was planned to last from 1975 to 1979, involved financial commitments totaling U.S.$76 million, which came from the government of Ghana, the World Bank, Japan, the African Development Bank, and Canada.

The specific accomplishments of the FTP include:

- The installation of twelve new electronic exchanges to replace old and obsolete automatic and manual telephone exchanges, which increased Ghana's telephone line capacity by about 50 percent.
- Increasing the number of subscriber trunk dialing centers from eighteen to twenty-four.
- The construction of tertiary exchange, a telex, a message switch in the capital, and an earth station in the country.

- The installation of microwave radio links for telephone and television transmission from the capital to the northern part of Ghana and initiation of a second new microwave radio link.
- The construction of a third 500-kilometer microwave radio link joining Ghana to the two bordering nations of Togo and the Ivory Coast. This part of the FTP was intended to eliminate the transit of African telecommunications traffic through Europe and was popularly known as the Panaftel (Pan-African Telecommunications) project.

Although the FTP was delayed as a result of changes in government, economic recession, and other social factors, it was eventually completed in 1985.

10.3.2 The Second Telecommunication Project

As part of Ghana's long-term telecommunications development program, a Second Telecommunication Project (STP) with an eight-year time frame was initiated in 1987. This project was intended to provide further rehabilitation, modernization, and expansion of the telecommunications facilities not covered by the FTP, as well as the restructuring of the P & T.

The components of the STP were as follows:

- Expansion of the microwave transmission network and provision of coast station facilities at Tema for maritime telecommunications services.
- Rehabilitation and expansion of Ghana's switching network in thirteen urban centers and twenty-six rural communities.
- Rehabilitation and expansion of the external cable network to match the switching component just described above.
- Rehabilitation of Ghana's satellite earth station.
- Provision of a 330-line international telephone switch and a 1,000-line telex switch.
- Procurement of subscriber terminal equipment, spare parts, vehicles, and personal computers.
- Provision of in-house electronic data processing systems for improving billing systems, payroll, inventory, and the like.
- Acquisition of materials to provide 100 housing units located throughout the country for telecommunications staff.
- Updating and upgrading of P & T training school facilities.
- Creation of fellowship and training programs for manpower development.
- Separation and restructuring of the P & T into two entities.

By 1992, about U.S.$50 million, representing 30 percent of the loan, had been disbursed, and, as a result, the following elements of the STP had been completed: the international telephone switch, the rehabilitation of the satellite earth station, the rehabilitation of cable networks in parts of Accra, the rural radiotelephone facilities at thirty-six rural and isolated locations, the 1000-line telex switch, the Telecommunications Master Plan, and the Telecommunications Corporate Plan. Table 10.1 shows the projects that had been completed in the 1980s.

Table 10.1. Telecommunications Projects/Tasks Completed

Project/Task Description	Funding	Amount (m = millions)	Date Started	Date Completed	Contractor
NTC consultancy/master plan study	IDA	Y 40m	1986	1986	NTC
300 electronic teleprinters	CCCE (France)	FF 5,340,145	Aug. 1987	May 1988	Sagem
Accra international telephone switch	CCCE (France)	FF 33m	Dec. 1986	Oct. 1988	Alcatel-CIT (France)
29 facsimile and accessories	CCCE (France)	FF 604,500	Mar. 1988	Feb. 1989	Sagem
JICA 1	JICA (Japan)	Y 683m	Nov. 1987	Feb. 1989	
a. Provision of 17,400 cable pairs within castle ministries, state house, Korlebu, Ridge, Rangoon, Accra Central Exchange					
b. PABX at ministries			July 1988	Feb. 1989	NEC
c. HDX 10 telephone training equipment					Hitachi
Rehabilitation of Nkuntunse earth station	Exim Bank (Japan)	Y 979,477m	May 1987	Mar. 1989	Marubeni/NEC
Supply of cables and accessories	CCCE (France)	FF 25m	Feb. 1989	Oct. 1989	Acome
Supplementary supply of materials	CCCE (France)	FF 11m			Sagem
Accra North Cable Works	Dutch Govt./NKF	Del 37m	June 1987	Dec. 1987	NKF Holland
34 rural telephone systems	Ireland	2	June 1986	Jan. 1987	Telectron

Source: GPT Publication, Accra.

Note: See the section on abbreviations at the end of this chapter for definitions of the abbreviations used here.

Table 10.2. STP Financing Sources

Source	Foreign (millions of U.S.$)	Local (millions of U.S.$)	Total (millions of U.S.$)
Government of Ghana	1.5	9.8	11.3
CCCE (France)	21.7	—	21.7
Holland (NKF)	18.8	—	18.8
Japanese grant/JICA	9.2	—	9.2
Japan (EXIM)	7.0	—	7.0
Japan (OECF)	69.5	6.7	76.2
Ireland	1.7	—	1.7
IDA	18.3	0.7	19.0
Post and Telecommunication	2.3	5.5	7.8
Total	**150.0**	**22.7**	**172.7**

Source: GPT Publication, Accra.
Note: See the section on abbreviations at the end of this chapter for definitions of the abbreviations used here.

As a result of Ghana's economic growth, the cost of the STP, which was originally estimated at U.S.$140.7 million, was adjusted upward to U.S.$173 million. Table 10.2 shows the details of the project's funding sources.

The following factors hampered the P & T's ability to achieve the stated goals of the STP: delays in approving projects, due to administrative controls; poor conditions of employee service, resulting in high staff turnover and the inability of the P & T's to attract and retain qualified personnel; shortage of skilled and qualified managerial, professional, and technical staff; lack of funds to reduce excess unproductive workforce; and the absence of sufficient in-house facilities to process bills promptly.

10.3.2.1 Impact of the STP

The impact of the completed portion of the STP on telecommunications services in Ghana was modest but appreciable. Table 10.3 shows the changes in some telecommunications service data between 1986 and 1990. One result of the completion of the STP was that subscriber circuits in the rehabilitated areas were improved and the fault rate significantly reduced. Because the rehabilitation program involved the expansion of the cable network, more subscribers were also connected to the network, resulting in a 20 percent growth in subscriptions. In addition, the number of direct exchange lines in working order increased from 60 percent in 1987 to 89 percent in 1992. Also the availability since 1988 of international direct dial service in twelve exchange areas resulted in the promotion of international business and trade. The number of international satellite circuits also grew—from 41 satellite circuits in 1988 to 193 satellite circuits and 84 terrestrial circuits in 1992.

The STP also introduced AT&T/U.S. and U.K. direct service; airlines became capable of making reservations through SITA (Société International Télécommunication Aeronautique) facilities; a meteorological department was created to send meteorological and seismological data from various locations throughout Ghana

Table 10.3. Telecommunications Service Data

	1986	1990
DELs connected	38,046	44,834
DELs working	40%	89%
Telex lines connected	316	881
Telex lines working	62%	72%
International telephone circuits	41	254

Source: GPT Publication, Accra.
Note: DEL = direct exchange line.

to Accra; and press agencies/news houses, commodity markets, and financial institutions gained increased information and data transfer capabilities for transmission to and from the outside world.

Other benefits of the completed programs of the STP included high levels of generated revenue, significant growth in national and international telecommunications traffic, increased low-speed access to overseas data banks using telex services (particularly those in the United States), and the capability of the Ghana Broadcasting Corporation (GBC) to transmit voice cast (radio commentary) and live television telecasts via satellite and simultaneous television transmission from all GBC transmitters in the country.

The final completion of all projects in the STP in 1994 was expected to provide Ghana with sixty automatic exchanges linked by about 4,100 kilometers of microwave radio networks with 77,000 connected direct exchange lines. The P & T's turnover was projected at 25 billion in 1994 compared with 17.8 billion for 1992—a 40 percent increase—and the new business opportunities resulting from the full completion of the STP were expected to produce an economic facelift for all aspects of Ghana's commercial sector.

Despite the achievements of the STP, Ghana's telephone density, in 1994, (0.31 per 100 inhabitants), is still among the lowest in Africa. Typical telephone densities for other African nations include 9 percent for Libya, 1.3 percent for Zimbabwe, 0.5 percent for the Ivory Coast, 0.33 percent for Togo, 0.2 percent for Nigeria, and 0.1 percent for Burkina Faso. The enormity of the task facing Ghana and other African nations attempting telecommunications modernization becomes apparent when African telephone density rates are compared to those of selected nations in Europe and Asia: 62.4 telephones per 100 inhabitants in Sweden, 43 in the United Kingdom, 42 in Japan, 41 in France, and 8 in Malaysia. Massive investment in accelerated telecommunications programs is the answer to improving Ghana's telephone penetration rate.

10.4 Future Plans for Ghana's Telecommunications Infrastructure

The STP is the first phase of the Telecommunications Master Plan encompassing a twenty-year horizon (1987–2006), with each phase extended over a period of five years. It is expected that the Third and Fourth Telecommunication Projects—

involving an investment of about U.S.$400 million—will have been completed by the year 2000. This will improve Ghana's telephone density from the current level of 0.31 per 100 people to 0.534, with a corresponding increase in connected DELs of 52,000 (from 48,000).

These two projects plan to meet customer requirements through the following measures:

- Executing an accelerated program to implement district center telecommunications facilities to support the government's decentralization policy.
- Rehabilitating, expanding, and modernizing the switching, transmission, and local networks.
- Digitalizing the telecommunications network.
- Implementing a paging system.
- Instituting packet switching (data networks).
- Introducing an Integrated Service Digital Network (ISDN) and Intelligent Networks.

10.4.1 Deregulation of Telecommunications

Prior to 1980, the P & T was the sole supplier and distributor of all telecommunications terminal equipment to the public. It was also the only institution that could install and maintain telecommunications equipment and run telephone services to the public (with the exception of two-way radio telecommunications equipment, which was also being supplied and installed by a few private foreign companies). In 1987, however, the government of Ghana relaxed the regulations, and private companies began to be issued licenses and allocated frequencies enabling them to produce, install, and maintain any compatible telecommunications equipment. By 1992, about forty telephone companies were in operation, including a local cellular company and a paging company. Other companies supply, install, and maintain terminal equipment such as facsimiles, telephones, PBX, and PABX (with capacities ranging from 4 to 4,300 extensions and manufactured by such companies as Panasonic, Toshiba, AT&T, Alcatel, Philips, and British Telecom). Most of the companies now allowed to operate in Ghana are representatives of companies in Europe and the United States.

Despite the improvements that will result from the STP, telecommunications in Ghana is still extremely inadequate. In 1995, only 37 of the 110 administrative districts of the country had telephone exchange facilities, and there were only thirty-five pay phones in the entire country (with thirty-two in Accra). Also, there was an average of three faults per line per year, and the average duration of the faults was seven days. Further, the cost per line was U.S.$3,500 as compared with U.S.$1000 in developed countries. The Ministry of Communications estimated that a U.S.$450 million investment would be required between 1995 and 1999.

As a result of the dismal state of telecommunications, the high level of investment required, and the retrenchment of multilateral funding for telecommunications development, in 1995 Ghana began an ambitious telecommunications restructuring initiative. This policy has the following components:

1. The sale of the government's controlling interest in P & T through (a) the sale of a minority stake (which will have management control) to a strategic investor, such as an international telecommunications operator; (b) the sale of a stake to international financing institutions; and (c) the sale of a stake to local investors in Ghana.
2. The licensing of a second main operator, which will compete with the P & T and have the same rights and obligations as the P & T.
3. The licensing of private pay phone operators in each of the regional capitals.
4. Permission for large corporate users to develop their own networks.
5. The use of build-operate-transfer (BOT) and similar private-sector schemes as a tool for rapid expansion of service—particularly in rural communities.
6. The establishment of a regulatory body for the sector to be known as the National Communications Authority (NCA). The NCA regulates service standards and tariffs, as well as seeks to create the necessary environment to stimulate investment in the sector.

As a broader initiative to attract international investment, Ghana enacted a new Investment Code, Act 478, which has the following provisions: the unconditional transferability of dividends, loan service payments, fees and remittance; and a commitment that no enterprise shall be nationalized or expropriated by the government.

The NCA has worked hard to create more private sector participation in Ghana's telecom industry. As a result, dozens of privately owned data communications integrators and Internet service providers have been established. The NCA also authorized the licensing of the second national operator (SNO) in 1997. The SNO license was awarded to a consortium of private investors, which also includes the Ghana National Petroleum Company. The SNO's first task was the creation of a national VSAT infrastructure to offer business communications services.

Most important, the government partly privatized its telecom operations. It created Ghana Telecom and sold part of it off. The major strategic investor was Telkom Malaysia, which acquired about 30 percent. That company was also part of a consortium that bought 30 percent of Telkom South Africa.

It is clear that Ghana's telecommunications sector is at a turning point in its development. Ghana's leadership demonstrated awareness that the past structure was inadequate and that a change was needed to make the future of Ghana's telecommunications brighter than the past 100 years of inadequate and inefficient service.

10.5 Conclusion

A great deal needs to be done to improve and sustain higher quality of service in Ghana's telecommunications system. Deregulation of laws affecting telecommunications could go a long way toward encouraging individuals, universities, and research institutions to get involved in solving some of the local problems affecting telecommunications in Ghana. Deregulation would also encourage private individuals and institutions, both local and foreign, to invest in telecommunications and help to expand the telecommunications network by introducing new technologies that the government and the old state-run Post and Telecommunication Corporation could not financially support. Delays in the approval and implementation of projects, caused by institutions and bureaucracies, have cost Ghana considerable time, energy, and capital that could have been invested in the telecommunications network.

Abbreviations

ADB	African Development Bank
CCCE	Caisse Centrale de Coopération Economique
CESISC	Combined Earth Station and International Switching Center
Del	Dutch currency
ECOWAS	Economic Community of West African States
EXIM	Export-Import Bank
FF	French francs
GFRCB	Ghana Frequency Registration and Control Board
GPT	Ghana Post and Telecommunication
HDX	Hitachi Digital Exchange
IDA	International Development Association
ITU	International Telecommunication Union
JICA	Japanese International Cooperation Agency
NEC	Nippon Electric Company
NKF	Netherlands Kabel Factory
NTC	Nippon Telecommunication Consultancy
OECF	Overseas Economic Cooperation Fund
PABX	Private Automatic Branch Exchange
Panaftel	Pan-African Telecommunications
PATU	Pan-African Telecommunications Union
PBX	Private Branch Exchange
P & T	Post and Telecommunication
RASCOM	Regional African Satellite Communications System
SITA	Société International Telecommunication Aeronautique
STD	Subscriber Trunk Dialing
UHF	Ultra High Frequency
VHF	Very High Frequency
Y	Japanese Yen

Note

Part of this work was done when F. K. A. Allotey was a visiting professor at the University of Michigan, Ann Arbor.

Bibliography

Ghana Frequency Allocation (Registration and Control) Decree. 1977.

Ghana Telecoms Statistics. 1992. PORSPI/ISS, TUD.

Salia, Edward. 1995. "Telecommunications in Ghana: A Changing Horizon." *Africa Communications.* January/February.

World Bank. 1995. "Ghana–Telecommunications Sector Investment Project." In *World Bank Project Report.* Washington, D.C.: World Bank.

11
Out of South Africa: South Africa's Telecommunications Equipment Industry

DAVID KAPLAN

Alone among African countries, South Africa has a long-established and diversified telecommunications manufacturing industry, which is undergoing significant changes with the end of apartheid. From an industry almost entirely dependent on the domestic market, it has recently become far more export-oriented—with Africa emerging as a particularly important market. This chapter briefly surveys the history of the South African telecommunications equipment manufacturing industry and its prospects.

11.1 Historical Background

In the mid-1950s, the South African defense establishment was concerned that international hostility to the policies of apartheid might lead to the disruption of supplies of vital communications equipment. With military and strategic considerations predominant, the South African government decided to utilize the monopsonistic power of the postal and telecommunications authority—the South African Posts and Telecommunications (SAPT)—to establish extensive local production of telecommunications equipment.

Following the British example, in 1957 SAPT signed a series of ten-year manufacture and supply agreements with its principal suppliers of telecommunications equipment: Automatic Telephone Electric Company, Siemens (U.K.), and Standard Telephone and Cables. At the time, these companies were supplying similar equipment to the British Post Office. These agreements (known as "the long-term agreements") were extended for another ten years in 1968. Then in 1979, with the advent of digital technology, the agreements were extended for another fifteen years. The latter extension of the long-term agreements expired in 1995.

The long-term agreements stipulate that SAPT will satisfy its requirements for specified equipment over the entire period only from the agreement companies. In turn, the contracting companies agree to undertake local manufacture and to progressively increase their products' local content. Prices are established according to a complex formula of cost recovery plus an allowance for profit, the latter being dependent mainly on the quantity ordered and the return on capital invested. Profits earned on items supplied under the long-term agreements are shared with Telkom SA, the government-owned company in charge of South Africa's telecommunications (see Kaplan 1990, p. 90).

With the advent of digital technology, political and strategic considerations became critical factors in the choice of equipment suppliers to SAPT. For example, in one case the choice of three local suppliers of digital-switching equipment with two foreign technology sources was politically rather than economically motivated.

For the first two decades into the mid-1970s, South Africa's local telecommunications industry developed rapidly. The manufacture of electromechanical telecommunications equipment had two key features. First, it required considerable direct labor inputs. In South Africa, this was initially provided primarily by white male labor, but the rapid and progressive substitution of lower paid labor (that of women and especially of nonwhite racial groups) allowed local telecommunications companies to reap considerable profits and was a principal reason for the industry's expansion. Second, electromechanical technology was a mature technology, evolving slowly and predictably. Research and development (R&D) expenditures were therefore low, and local telecommunications producers could easily acquire the requisite technology from any one of a number of foreign firms.

In this context, SAPT was able to support local production through guaranteed orders for a small, select number of companies that manufactured telecommunications equipment locally. Output and employment in the telecommunications equipment manufacturing industry expanded rapidly. Moreover, this equipment was generally of high quality, "up to date," and not much more expensive than equipment available on the international market.

In 1977, however, SAPT decided henceforth to enter into new contracts only for digital telecommunications equipment. Digital technology completely revolutionized the production and R&D requirements of telecommunications equipment because the manufacture of digital telecommunications equipment is far more capital—and R&D—intensive.

South Africa's local telecommunications industry did not adjust adequately to these changes. The extent of SAPT support increased but, at the same time, this support proved to be less effective at sustaining the local telecommunications industry, now operating under the new digital dispensation. By the mid-1980s, measured in terms of its local content, its exports, or its technological capability, the telecommunications equipment industry was performing poorly. Moreover, SAPT began to reduce its capital expenditure considerably with the onset of the debt standstill, and this, combined with declining orders from the military (many of South Africa's telecommunications companies also produce electronic equipment for the military), further undermined the industry.

Furthermore, in 1991 the postal and telecommunications operations of SAPT were separated. Telkom SA was assigned responsibility for telecommunications and registered as a commercial company (with the government as sole shareholder). The pursuit of bottom-line profitability called into question the support and preferences traditionally offered by Telkom to the local manufacturers of telecommunications equipment. As a result, it was expected that the long-term agreements would be substantially modified or even eradicated altogether when they were due for renewal in 1995.

Over the last few years, however, there have been some signs that the industry has begun to respond more positively to the new dispensation. In particular, the exposure of the industry to international markets has recently increased markedly. This has been accompanied, in a number of cases, by enhanced technological capability. In addition, the industry has undergone considerable rationalization internally and has attracted important additional sources of foreign investment. The professed intent of all the major telecommunications equipment producers in South Africa in the mid-1990s was to progressively detach themselves from reliance on the domestic market, and more particularly reliance on orders from Telkom, and to continue to expand the proportion of their turnover that is exported.

11.2 Size and Composition of the Telecommunications Market and Industry

The South African telecommunications market has traditionally mirrored the structure evident in a wide range of developed countries, namely, a single dominant purchaser and, on the supply side, a small number of exclusive suppliers. While both sides of the market were undergoing change in the mid-1990s, this structure had so far been modified rather than transformed.

11.2.1 Market Size and Composition

The South African market for telecommunications equipment is quite large by international standards. South Africa was assessed to be the fifteenth largest market in the world—somewhat smaller than India and Australia and somewhat larger than Brazil.

The market for telecommunications equipment actually shrank in real terms between 1988 and 1992. Telkom's capital expenditure for 1993—a fair proxy for the size of the overall market—was only some 7 percent higher in nominal terms than for 1992, which means a 5 percent reduction in real terms (*Cape Times* [Cape Town], May 17, 1993, p. 12).

The principal reason for the falloff in the market is the cutbacks in Telkom expenditures. In 1994, Telkom purchases accounted for less than one-half of the total telecommunications market. This was down from about 70 percent in 1990 (Kaplan 1990, p. 85).

However, there are indications that the situation is changing. The new government was determined to accelerate the provision of telephony to disadvantaged communities. The number of working lines is expected to rise from 3.9 million in

1995 to 5.3 million by the year 2000. Telkom accordingly expected to increase its annual ordering level from the previous 200,000 exchange user ports to 330,000 exchange user ports per annum.

In order to meet the goals set for increasing telephony access, the South African government privatized Telkom in 1997. A joint venture between the Texas-based Bell company SBC and Telkom Malaysia purchased a 30 percent stake in Telkom SA.

In addition, the introduction of cellular telephony in 1994 has been a major shot in the arm for the South African market. By 1995, there were about 400,000 subscribers. Outside of Europe, South Africa was the largest user of the GSM (Global Standards for Mobile Systems) network.

Purchases will also increase for VSAT (very small aperture satellite terminals) equipment as satellite-based services become an attractive solution to provide telephony in rural areas that are not easily serviced by terrestrial lines.

11.2.2 Telecommunications Equipment Producers

On the supply side, the industry is dominated by five large companies. These are the exclusive suppliers to Telkom of a range of telecommunications equipment, particularly "the big-ticket items" of switching, transmission, and receiving equipment. Table 11.1 details the ownership structure of the five principal telecommunications equipment producers. The first four companies have had long-term supply agreements with Telkom. Together, the five companies constitute more than 85 percent of the industry.

The ownership structures of all the South African telecommunications equipment companies have undergone major changes. Changes in share ownership have been accompanied by a rationalization of production. Larger and more focused South African companies are now better able to compete in more open markets.

Two significant features of the new ownership structure should be noted:

- The local telecommunications companies are all integrally linked to one of the major conglomerates that dominate the South African economy. Anglo-American has a minority shareholding (20 percent) in Altron, the holding company of Altech; Sanlam (16 percent) and Gencor (16 percent) have minority holdings in Siemens SA and therefore an interest in Telephone Manufacturers of South Africa; and Plessey is a wholly owned subsidiary of Sankorp, a subsidiary of Sanlam. Reunert was part of Barlow Rand, which then "unbundled," but Old Mutual remains its principal shareholder.
- Siemens and Alcatel CIT have emerged as the key foreign firms with a strong presence in the local telecommunications equipment industry. The European orientation of the local industry is pronounced.

Production of different telecommunications items in South Africa tends to be highly concentrated, frequently with a single local producer. Alcatel Altech leads in transmission; Telephone Manufacturers in telephone sets, card phones, and public telephones; Siemens in telex and teletex terminals; and Plessey in test

Table 11.1. Profile of the Principal Telecommunications Equipment Companies

Company	Principal Local Shareholder	Principal Foreign Shareholder
Alcatel Altech Telecomms (Pty) Ltd.	Altron (50%) S.A.(50%)	Alcatel CIT
Siemens Telecommunications (SITEL)	Siemens SA (51%) Reunert (27.5%)	GEC Plc (21.5%)
Plessey Tellumat	Sankorp (100%)	—
Telephone Manufacturers of South Africa	Siemens SA (26%) Reunert (40.6%)	GEC Plc (33.33%)
SA Philips Pty (Ltd.)	Unspecified (25%)	Philips BV Eindhoven (75%)

Source: Company annual reports.

equipment and as sole supplier of small business telephone systems to Telkom. Siemens and Alcatel Altech produce digital exchanges—with Siemens having two-thirds of the market. Philips, Siemens, and Plessey produce mobile telephones. Only in the unregulated PABX (private automatic branch exchange) market is there fierce competition between a much larger number of local suppliers. Moreover, the larger companies (particularly Alcatel Altech) tend also to be significant players in the importation of components and in component production.

With the end of apartheid, there were two new significant entrants into the industry in 1994—AT&T and Ericsson. AT&T joined forces with Telkom to provide services such as a managed telecommunications network for multinational companies and explored a partnership with Afritel Systems, South Africa's only black-owned telecommunications company, possibly in the cellular market. Ericsson Radio Systems formed a joint venture with Plessey Tellumat SA to supply and install new cellular equipment.

11.3 The State of the Industry

High levels of market power make profitability a poor guide to the efficiency and competitive international position of the South African telecommunications equipment industry. This is particularly true given that, in South Africa, the principal customer determines the price paid and the telecommunications equipment companies are backwardly integrated (for example, into the manufacture and importation of components). Two more significant (and interrelated) indicators of the industry's efficiency and competitive international position are the technological capability of the industry and the industry's export performance.

11.3.1 Technological Capabilities

One index of well-being and viability is the extent to which the industry is capable of designing and developing new products, or at least of making significant adaptations to imported designs. Telecommunications is, of course, a technologically dynamic industry, and the development of technological capability is therefore likely to be indispensable to future success in internationally competitive markets. Technological capability is both an index and a symptom of well-being.

With some exceptions, for example in small PABXs, the South African telecommunications equipment industry derives its products and production processes from foreign companies—often with only minor modifications. Technology is generally secured via license agreements. An analysis of the license agreements operative in 1990 by one of the largest telecommunications equipment producers revealed the following features:[1]

1. Licenses are of long duration—a median period of ten years—and are very often extended for a further period. Long duration in a license agreement indicates that the local company is not making any real headway in genuinely learning or assimilating the licenser's technology. Local companies must therefore continue to rely on the extension of the license agreement. This nonassimilation of or inability to learn the licenser's technology is further evidenced by the fact that the license agreements very rarely stipulate any training of local personnel in the licensers' technology. In sum, the import of technology from abroad is often a substitute for local technological capabilities and not, as it has been elsewhere in some of the newly industrialized countries, for example, a facilitator of the development of local technological capability.
2. License agreements contain significant restrictive clauses. For example, the license agreements often stipulate the use of imported inputs (frequently from the licenser or an affiliate). Such "tied" purchasing clauses severely limit backward linkages and hence local content. But, most critical (and most common) are clauses in the license agreements that restrict the export activity of local companies. Out of eight license agreements surveyed, seven expressly limited exports to the immediate southern African region and in some cases only to Namibia or to Botswana, Lesotho, and Swaziland.
3. Royalty payments are high. Typically, they are 4.5 percent of ex-factory price, but, in addition, they often entail a front-end charge. These charges are heavy—the front-end charge in one case exceeded R1.7 million. Overall, a royalty of 7 percent per year is not uncommon. The sums paid by local companies in license fees are large by comparison with expenditures on R&D.[2] Indeed, for most of the large telecommunications companies, payments to import product technology exceed payments to develop local product technology by a large margin.

The long-term agreements expressly attempted to encourage Telkom's contractors to design and develop new products by ensuring that Telkom would pay for any expenditures (plus an allowance for profit) that the contractors incurred in this regard. But despite the very clear objective of the long-term supply agreements to encourage R&D on the part of the suppliers, the evidence is clear that they did not have the desired effect.

The reasons for this are complex.[3] The long-term agreements exclude the entry of new smaller companies that might be highly innovative. Furthermore, the agreements provide alternative and often less risky routes to achieving high levels of profitability. For example, contractors have often found it to be more profitable to establish a monopoly on the production or importation of certain inputs and to utilize this monopoly position to raise the price at which it supplies those inputs to the telecommunications industry. (The profits of component suppliers, unlike those of the telecommunications equipment producers, are not subject to the same profit-sharing arrangements with Telkom.) Profits earned through such a form of monopoly pricing would be akin to a rent. Finally, the agreements allow for any license fees paid abroad to be recouped as a cost item by the contractors.

Local telecommunications producers therefore have an incentive to simply adapt designs (especially where these have been acquired from a parent company) sufficiently so as to meet Telkom specifications. Moreover, where Telkom sets very high "specs" (and this was and is characteristic of SAPT/Telkom), the effect may also be to discourage local product design. Cost-plus pricing under the long-term agreements makes it advantageous for Telkom's contractors to focus on expanding the breadth of their product range as opposed to the depth of product design and development. Not only are Telkom's contractors therefore devoting fewer resources to R&D, but these resources also tend to be spread over a very wide range of products. Finally, the requirement that Telkom approve new product development prior to a project's being undertaken imposes substantial delays in the design cycle. Because Telkom is required to pay the full costs of such development up front, it may also be more cautious than the contractors with respect to new product development. As a user of telecommunications equipment, Telkom has little appreciation for the potential advantages and pitfalls on the manufacturing side. Development efforts will be slanted toward products needed by Telkom, and this may well have contributed to the past poor export performance of Telkom's contractors.

The expansion of exporting, a very recent phenomenon (see next section) holds much more promise for the enhancement of technological capabilities in the South African telecommunications equipment industry. While precise data are lacking, there does seem to be a link between expenditure on R&D and success in export markets. According to information supplied by Business and Marketing Intelligence (BMI), which monitors the industry, "There is a definite correlation between the export successes of companies (in telecommunications) . . . as well as in the broader context of the local electronics industry, to their R&D activity. . . . Own technology in product design has proven to be a winning factor in several sectors of the local industry, and communications products are but one example" (BMI 1991, p. 4).

While precise data are not available, quantitative investigations into R&D activity by local companies in the electronics and telecommunications sectors over the last three years has shown that "own design is essential to company export hopes in a global market" (BMI 1991, p. 5).

Not only is the local telecommunications industry becoming significantly more export-oriented, it seems quite possible that the South African subsidiaries of large multinational telecommunications equipment producers may undertake product development for the entire region. In particular, there are indications that both Siemens and Alcatel see their South African operations as a launching pad into Africa and their South African companies as undertaking product design and development so as to provide products appropriate for the entire African market.[4] Similarly, AT&T has stated that it sees its South African operations "as a springboard into Africa."

Additionally, a number of smaller suppliers focused on exports and services throughout the rest of Africa. Companies such as Aerial Empire, Dimension Data, and Protea Technology Ltd. marketed equipment and installation services throughout the continent.

11.3.2 Export Performance

Under apartheid, South African telecommunications equipment producers have been almost exclusively focused on the domestic market and have performed very poorly in export markets. The Board of Trade and Industry reported that in 1984 the industry exported only 1.5 percent of its product, equivalent to only 3 percent of telecommunications equipment imports.[5]

Exports grew very slowly until 1988. Thereafter, there has been a pronounced growth in exports. Measured in constant Rands, exports almost quadrupled between 1988 and 1993.

Moreover, exports have risen much more rapidly than imports so that the import/export ratio has declined from 24 in 1988 to 6 in 1993.

In 1994, with the introduction of cellular telephony, there was a massive increase in the importation of telecommunications products (imports rose from R1.4 billion in 1993 to R2.6 billion in 1994). Moreover, exports declined as local companies sought to meet burgeoning domestic needs, even as foreign markets opened to South Africa. Nevertheless, in 1994, exports were still more than three times larger than in 1988.

The major companies like Altech and Plessey substantially increased their exposure to international markets and were seeking to further increase the proportion of their product marketed abroad.

South Africa's export expansion after 1988 coincided with a significant decline in Telkom orders and a general contraction of the domestic market. Exports increased significantly only when it was clear that the domestic market was contracting and that this contraction was likely to continue.

Unfortunately, the trade data do not indicate the destination of telecommunications equipment exports. There are reports of significant export sales to a wide range of markets as far afield as Belgium and Indonesia. However, two markets seem to be particularly important—Eastern Europe and Africa.

Some major export orders to Eastern Europe were secured on the basis of product-appropriate characteristics developed in-house by South African companies. The most significant example here is that of Telkor (a Reunert company whose public pay phones operation is now part of Telephone Manufacturers of South Africa). Telkor was able to export coin- and card-operated pay phones very successfully, particularly to Eastern Europe, based partly on licensed technology but also on considerable in-house product development. Exports rose from 5.7 percent of company sales in 1991 to over 40 percent in 1993, and already the company captured more than 12 percent of the pay phone market in Europe (*Financial Mail* [Johannesburg], March 25, 1994, p. 98). In 1994, Telkor designed the world's first GSM cellular pay phone.

Also important in the movement toward exporting has been the impact of the reunification of Germany. With Siemens in Germany producing to full capacity and with the high-level "specs" required of Siemens (SA) by Telkom, Siemens (SA) was selected to become a major supplier of Siemens equipment for the rebuilding of the telecommunications infrastructure in the former East Germany. Siemens (SA) subsystems supplied to Siemens (Germany), have also been marketed throughout Europe.

With respect to Africa, post-apartheid South Africa is exceptionally well positioned for the African market. The African market is expected to grow rapidly. It is unlikely that the major international equipment companies will develop products specifically designed to serve Africa's needs and environment. With Telkom long recognized as demanding high performance levels (Kaplan 1990, pp. 132–33), South African companies can offer not merely telecommunications products developed and adapted for African conditions but also the market support, maintenance, installation, and network configuration skills that are frequently lacking in African countries.

In the early 1990s local telecommunications companies had already begun targeting the African market. This is especially true of Siemens and Altech. Announcing their new joint venture with Altech in 1993, Alcatel's chairman and chief executive officer, Pierre Cuichet, stated that this "will enable Alcatel to spread its wings into the rest of Africa" (*Business Day* [Johannesburg], May 6, 1993). Bill Venter, the executive chairman of Altron (Altech's holding company), stated in 1992 that the group had registered with world aid agencies and had "devoted much effort to penetrating selected markets on the African continent" (*Business Day* [Johannesburg], March 3, 1992). A little later it was reported that Altech had won a R7 million contract that could result in overall orders of R34 million to supply a radio-based rural telecommunications system to Burundi. The World Bank provided the funding (*Business Day* [Johannesburg], September 23, 1992). In his 1992 review, Altech's executive chairman declared, "While we will still take advantage of every opportunity that presents itself in the developed western world for some of our advanced systems, we will also continue to search for opportunities in Eastern Europe. However, our main thrust will be in Africa, providing African solutions to African problems" (Altech Annual Report, 1992, p. 9). In 1993, Altech's chairman reported that the joint venture with Alcatel CIT has meant that ". . . Altech secured access to the technology of the world's foremost telecommunications multinational, which has opened the door to exports, particularly to the southern sub-continent of Africa" (Allied Technologies Ltd. Annual Report, 1993, p. 13). Furthermore, ". . . selected markets in sub-Saharan Africa are being investigated in close cooperation with Alcatel Trade International" (Allied Technologies Ltd. Annual Report, 1993 p. 24).

Similarly, in the early 1990s, Siemens (SA) aggressively marketed the Siemens

D900 digital mobile communications system throughout Africa. Cameroon was its first customer (*Business Day* [Johannesburg], January 21, 1993).

There have also been increased exports in some related products. In telecommunications cable, for example, one of the two major local producers recently reported an annual growth of 15 percent in exports and new export contracts from Hong Kong, Singapore, and Central Europe (*Business Day* [Johannesburg], December 6, 1993).

After Telkom Malaysia and SBC partnered with Telkom SA, Telkom SA was poised to become the major African gateway for communications between the United States and Asia and a continental powerhouse in its own right. As Telkom begins to roll out such international communications networks, exports to these regions may follow.

11.4 Conclusion

The South African telecommunications industry is undergoing a significant transformation. There have been major changes in corporate ownership that have brought about a rationalization of the industry, and the industry has been far more successful in securing export orders. These changes are in part a response to adverse factors—specifically the decline in Telkom orders and the likely curtailment of the long-term agreements—and in part a response to positive factors, in particular the opening up of new possibilities in the export market, especially in Eastern Europe and Africa. There are indications that greater exposure in international markets is beginning to affect the industry's technological capabilities positively and that, in addition, some of the major international telecommunications companies—in particular, Siemens and Alcatel—now intend their South African operations to perform much of the product adaptation and product support for the entire African continent.

On the domestic front, the industry is almost certainly likely to experience increased competition, especially with the demise of Telkom's right to provide the first telephone instrument on customers' premises (which are supplied exclusively to Telkom by Telephone Manufacturers of South Africa). This would allow customers to purchase telephone instruments from any supplier (Coopers and Lybrand 1992, pp. 78–79).

While the future of the long-term agreements is currently unknown, they will almost certainly be substantially modified, for example, to cover fewer products with more competitive tendering and to be of much shorter duration. There have been a number of calls for the substantial modification of the system of support for the local telecommunications industry from outside the industry—and even from within.[7]

Despite increasing competition in the telephone equipment market, growth in the equipment market depends on regulations concerning the operation of the underlying telecom services. The Telecom Act passed by the new government in 1996 stated that primary services would continue to be under the Telkom monopoly, which would be phased out gradually over five years. If other companies can build networks to compete with Telkom (such as Eskom, the national electric company, and Transtel, the national transportation company), equipment suppliers will be able to sell to more than one large purchaser.

In addition, while the South African telecommunications producers will face far more competition in the domestic market than they have hitherto, the new government has increased the rate of growth of the telecommunications infrastructure, particularly to serve the needs of previously disadvantaged communities. A prima facie case can be made, on the grounds of economic growth (i.e., the positive impact of access to telecommunications on output and employment) and on the grounds of equity (i.e., access to telecommunications is severely skewed), that the post-apartheid government should engage in an accelerated development of the telecommunications network (see Kaplan 1992, pp. 96–97).

A more rapid rate of expansion of the domestic network, targeted particularly at the needs of poorly served communities such as those in the rural areas, might provide a further springboard for the export of similar products elsewhere in Africa.[8] The development of a common infrastructure in the southern African region would give a further impetus to the export activity of the South African telecommunications industry. These two developments would moreover significantly enhance the capacities of the South African telecommunications equipment producers to adapt, develop, maintain, and support telecommunications equipment that is appropriate for African conditions and markets. This would enable South African telecommunications producers to be increasingly active and successful in African markets in the future.

Notes

1. All license agreements that entail the payment of royalties abroad are deposited with the Department of Trade and Industry. The license agreements surveyed were those in force in 1990.

2. In 1991, R&D expenditure on the part of the large agreement companies was of the order of 1 percent of turnover (data supplied by Business and Marketing Intelligence).

3. Kaplan 1990, Ch. 7, analyzes in some detail why the long-term agreements were not successful in enhancing technological capabilities in the local telecommunications equipment industry.

4. In a recent talk, a Siemens executive stated that Siemens was utilizing its South African operations to make modifications to its switching and rural telephony products for the entire continental African market. Siemens's local managing director has said that "the company has developed technology geared to deal with specific African conditions" (*Business Day* [Johannesburg], January 21, 1993, p. 9). The chairman of Altech has stated that the company "would become the center for Alcatel's interests in Africa" (*Cape Times* [Cape Town], January 18, 1993, p. 12).

5. This was the lowest of all products of the electronics sector (BTI 1986, p. 11, par. 32 and p. 14, par. 42).

6. Data for exports and imports are from the Industrial Development Corporation trade database. The 1991 and 1992 import and export figures are unaudited. The 1984 ratio is from BTI 1986, p. 14, par. 42.

7. "With Telkom moving into a new commercialized environment, all incentives in supply agreements must be changed to keep up with the changing order," says John Temple,

managing director of Plessey Tellumat. (*Business Day* [Johannesburg], February 4, 1993). Only a small proportion of Plessey's sales are with Telkom under the long-term agreements. The long-term agreements are much more important for the other agreement companies, which understandably tend to favor a retention of the long-term agreements.

8. By way of illustration, at the time of writing, it has been announced that Plessey Tel-lumat will begin manufacturing rural telecommunications systems in an agreement with NEC. This is designed to meet the forecast needs of government to extend telephones to disadvantaged communities under the Reconstruction and Development Program and simultaneously to generate significant new export earnings (*Cape Times* [Cape Town], March 7, 1995 p. 9).

Bibliography

Board of Trade and Industry (BTI). 1986. Investigation into the South African Electronics Industry. Report 2455. Pretoria: BTI.

Business and Marketing Intelligence (BMI). 1991. Report on R&D Undertaken in the Telecommunications Industry. Commissioned by and prepared for the author. Mimeographed.

Coopers and Lybrand. 1992. Telecommunications Sector Strategy Study for the Department of Posts and Telecommunications. August.

Financial Times (London). 1987, 1989. Telecommunications Surveys. October 19, 1987, pp. xxiv and 19; July 1989, p. ii.

Kaplan, David. 1990. *The Crossed Line: The South African Telecommunications Industry in Transition.* Johannesburg: Witwatersrand University Press.

______. 1992. "The Development of Telecommunications in South Africa: The Equipment Supply Industry." *Critical Arts: A Journal for Cultural Studies,* University of Natal, vol. 6, no. 1.

12

South African Telecommunications: History and Prospects

ROBERT B. HORWITZ

Covering the broad tip of southern Africa and possessed of vast mineral deposits, South Africa is the most important industrial nation in Africa. Historically, it has been among the world's largest gold and diamond producers, and it continues to derive at least one-seventh of its gross domestic product (GDP) from mining and quarrying. At R431.5 billion in 1994, South Africa's GDP is by far the largest in Africa.[1]

Of course, what is unique about South Africa is its politics. For the most part, white domination of the country's majority black population has been the rule since Dutch settlers arrived in the Cape peninsula during the latter half of the seventeenth century. Although serious hostility has always existed between Afrikaners (descendants of mostly Dutch settlers) and the English (who began settling the territory after Britain seized the Cape in 1806 to protect its Indian sea route), they essentially made common cause when it came to the use of black labor. This was particularly true after diamonds and gold were discovered in the late 1800s. The emergence of South Africa as an independent republic in 1910 rested on policies explicitly designed to ensure that whites retained political power and control over the state. A wide range of measures ended the limited access of "Coloureds" (mixed race) and Africans to the vote and required that all senior positions in the state bureaucracy be filled by whites. Historically, the South African republic had been a democracy for whites.[2]

In 1948, South Africa's system of racial separation became dramatically explicit and manifested itself in a series of nefarious laws known as apartheid.

Although the apartheid policies succeeded in securing a low-wage industrial reserve army for South Africa, they created vast inefficiencies. Black "guest workers," for example, were needed in cities, but were forced to live far outside them, thus forcing South Africa's transport system to accommodate massive long-range daily population movements—at the cost of considerable state-provided subsidies. Outlays for police and security represented other inefficiencies: while always high, these outlays reached immense proportions in the years following the 1976 uprising in Soweto.

In 1993, the population of South Africa, including the ten black homelands, was approximately 38.5 million. Blacks made up 75.6 percent of this total, Asians (primarily people of Indian ancestry) 2.6 percent, mixed-race inhabitants 8.6 percent, and whites 13.2 percent. In the mid-1990s, some 89 percent of Asians, mixed-race South Africans, and whites were urbanized, compared to only 50 percent of the country's Africans. In 1992, more than half the African population was under nineteen years of age (South Africa Institute 1992, p. 2). Decades of legalized discrimination and violence against Africans had resulted in a remarkably unequal society. Employed Africans, for example, are situated overwhelmingly in semiskilled and unskilled labor markets, and although accurate statistics are often difficult to obtain, unemployment among Africans in the early 1990s was estimated at 45 to 46 percent of the economically active population (South Africa Institute 1992, p. xliii; "Recession Leaves 46% without Jobs" 1993). The repeal of "influx control" laws in 1986, which brought about a 28 percent increase in the number of urbanized blacks, did not end the nation's economic inequality. Moreover, a sizable percentage of Africans lives in informal settlements and squatter camps situated at the edges of black townships, without access to tap water or sewage facilities, much less electricity or telephones.[3] Even with the formal end of the apartheid state, per capita expenditure on schools finds Africans receiving just 25 percent of traditionally white school expenditures (South Africa Institute 1992, p. 195). Unemployment levels remained unchanged and proved to be one of the most intractable problems facing the popularly elected government in 1994.

It is in politics that the major dramatic changes in South Africa have occurred. The concatenation of three forces—widespread overt opposition to racial rule, efforts of the state to preserve white minority control, and international pressure—gave rise to a process of debilitating economic crisis and intensifying political conflict that placed intense pressure on the South African state (Price 1991). In 1990, the government of F. W. de Klerk legalized banned organizations, including the African National Congress (ANC) and the South African Communist Party (SACP), and began the process of rescinding the formal laws of apartheid.[4] This led to tremendous political activity on all fronts, a great deal of conflict (some violent), and a general worsening of the economy. In the period after the legalization of the ANC, representatives of the black majority engaged in difficult negotiations with the white government to establish new political ground rules and structures, including the resolution of fundamental questions over power-sharing arrangements versus one-person, one-vote democracy. Struggles within the African population itself also arose, the most prominent being the frequently violent confrontations between supporters of the ANC and those of the Inkatha Freedom Party (see Adam and Moodley 1992). Momentous elections held in April 1994 brought to power a coalition dominated by the ANC, which called itself the "Government of National Unity (GNU)."

Less noticed than these highly visible political changes was an important transformation in South Africa's political economy. Long a strongly interventionist state characterized by extensive government involvement in the economy, in the years of economic crisis, the South African state began to dramatically scale back its economic activity. Historically, much of the state's intervention in the economy involved controlling labor markets under the apartheid system, and apartheid secured a continuous flow of phenomenally cheap black labor to white farms, mines, and factories. But state intervention in South Africa also entailed the operation of monopolistic public corporations—known as "parastatals"—in many areas of economic life. In the past few years, the South African government began disengaging from its extensive historic involvement in the parastatals. Telecommunications is one of several sectors—including electricity and transport, oil, and iron and steel production—where this retrenchment of state intervention took place.

12.1 The "Ancient Regime"

Historically, the South African Posts and Telecommunications (SAPT) was in most respects a classic post, telephone, and telegraph (PTT) monopoly, legally monopolizing postal and telecommunications services and operating a system characterized by internal cross-subsidies (*Statutes of the Republic of South Africa* 1958). Perhaps the clearest cross-subsidy was that from telecommunications to posts. With the entire operation statutorily forbidden either to make a profit or to experience loss, profitable telecommunications services balanced out large operating losses on the postal side. The SAPT, often colloquially referred to simply as "the Post Office," was classified as a "state business enterprise" and was run through the office of the Ministry of Transport and Communications.[5] After the emergence of the South African Republic (most often referred to as "Union") in 1910, the finances of the Post Office were controlled by the Treasury. Post Office revenue was paid into the Exchequer and all Post Office expenditures came from the Exchequer. Because it operated directly out of a government ministry and hence was tied to Parliament's annual planning cycle, the SAPT generally experienced close financial oversight. The oversight system evolved such that the SAPT had to submit detailed annual reports on its budgets, accounts, and proposed capital expenditures to the Auditor-General (an instrument of Parliament), who then wrote a report on the SAPT.

Under this system of oversight, the SAPT as a rule experienced little outright corruption or gross overspending but generally suffered from a shortage of capital inasmuch as it had to compete for funds with other central government capital projects. Close government oversight also meant that the SAPT did not set its tariffs according to marginal costing principles. Since members of South Africa's Parliament felt that they had to serve their constituents' interests in low telephone rates, tariffs were set well below the level that would have enabled the SAPT to meet demand expediently. If upward tariffs were thought to contribute to inflation or proved politically difficult for the transport and communications minister, they were typically adjusted downward (Taylor 1992). In 1968, South Africa's system of financing changed to some degree when the Post Office's finances were separated by law from the Exchequer and placed under its own control. Yet while the SAPT controlled its revenues, it was still dependent on Treasury loans to finance capital expenditures. As in many nations, the SAPT operated a savings bank that provided for some of its loan requirements. The operation of the bank itself, however, was an expensive undertaking. It was only in 1972 that the SAPT finally received permission to seek its own local or offshore financing ("Posts and Telecommunications" 1986, pp. 47–48).

Judging by comparative productivity indexes, one could reasonably conclude that the SAPT functioned as a repository for public employment. For instance, on the telecommunications side access lines in service per employee measured a relatively low 45.4 as late as 1989. By way of contrast, South Korea measured 226.3, the United States 130.6, and Mexico 95.6. Even Turkey outperformed South Africa at 65.9 access lines in service per employee, according to 1989 data (Coopers & Lybrand 1992, p. 16).[6] This performance, however, must be understood in the broad context of South Africa's apartheid variant of the interventionist state: South African parastatals historically have been the focus of the "job reservation" system, whereby unemployable whites (primarily poor Afrikaners who had left farming) were given jobs by the state or its parastatals. At the very least, the SAPT participated in the job reservation system indirectly, through the contracts the Post Office fashioned with domestic equipment manufacturing companies, which particularly in the early years were under mandate to hire whites (Kaplan 1990, pp. 31–32).[7]

12.1.1 The SAPT and Local Telecommunications Equipment Manufacturing

The relationship between the SAPT and domestic equipment providers functioned, as in other nations, in the manner of an industrial policy, with local purchasing obligations and high domestic content requirements. In the main, the policy succeeded rather well. By the mid-1990s, the manufacture of telecommunications equipment constituted the largest part of the South African electronics sector. Prior to 1958, the SAPT's equipment was supplied by the principal contractors to the British Post Office: Automatic Telephone Electric Company, Siemens U.K., and Standard Telephones and Cables. In 1958, with military and strategic considerations in mind, the government utilized the SAPT's monopsonistic power to establish extensive local production of telecommunications equipment. Long-term contracts (officially known as the Manufacture and Supply Agreements) with five principal South African telecommunications equipment companies embodied national goals on local content (Kaplan 1990, pp. 27–31, 79–85).

In the mid-1990s, South Africa's four remaining major telecommunications equipment companies were the Altech Group, Siemens, Telephone Manufacturers of South Africa (Temsa), and Plessey, but other companies, including General Electric Company (GEC) and ATC, have become significant players in local equipment provision as well. Altech was South Africa's exclusive supplier of transmission equipment while Siemens, Altech, and Temsa provided switches. Siemens also supplied telex and teletex equipment as well as telephone exchange power supply equipment, and Temsa, originally set up by the local Plessey and GEC companies, was the sole manufacturer of standard telephones. Plessey was responsible for providing small private branch exchanges (PBXs), and Aberdare and ATC were selected to produce cable for South Africa's national transmission network. In the early 1990s, ATC was the only manufacturer of fiber-optic cable in South Africa. More recently, Telkor was awarded the contract to supply South Africa's pay phones (Telkom 1991, p. 31).

All the principal local producers of South Africa's telecommunications equipment and components have a significant part of their equity held by one of South Africa's largest local corporate groups.[8] Anglo-American, for example, has a 20 percent stake in Altech, and the Sanlam Group has a 32 percent stake in Siemens. Considerable ownership changes in South Africa's telecommunications manufacturing companies occurred in the 1990s, reflecting a general rationalization and a specific regrouping in the advent of a sharp decline in orders from Telkom, South Africa's telecommunications entity since 1991 (see Kaplan, this volume, chapter 11). Currently GEC holds a two-thirds stake in Temsa with the other third held by Siemens. The Reunert Group, owned (80 percent) by Barlow Rand, is the parent of both ATC and Telkor. South Africa's principal telecommunications companies all have a shareholding in South African Micro Electronic Systems (SAMES), South Africa's sole local producer of integrated circuits (Kaplan 1990, pp. 84–85). Following a major fire in 1990, SAMES was restructured, with the Industrial Development Corporation, a state agency, taking a majority stake.

Historically, as technology became more complex, the local South African companies essentially became licensees of first-world companies. While in many respects these foreign corporate telecommunications connections follow broader historical relationships (particularly with the United Kingdom), the SAPT's decision in the late 1970s to digitize the network tightened the connections with the foreign equipment suppliers. The decision was based on the fact that demand was increasing and the British-based electromechanical switching system had begun to present the SAPT with serious technical problems. However, small nations cannot easily amass the capital and expertise to create and manufacture highly advanced, general (as opposed to niche) telecommunications devices. As a result, South African equipment companies entered into licensing arrangements to assemble foreign-designed telecommunications equipment in South Africa. For example, Siemens licensed Temsa to manufacture the German EWSD digital exchange, and Teltech (part of the Altech group) was licensed to manufacture the French Alcatel E-10 digital exchange (designated locally as the SA 128E). Inasmuch as the late 1970s represented the early days of digital technology, the choice of suppliers was limited, and only a few companies could present viable digital equipment offerings at that time.

The decision to authorize two different foreign telecommunications equipment suppliers for South Africa was a political one. Given economic sanctions and the possibility of a supplier cutoff, dependence on a single source of foreign technology was considered to entail political risks. The government thought that the difficulties of meshing two different digital exchange systems was outweighed by the political slack offered by two different foreign suppliers (Kaplan 1990, pp. 41–42). In the mid-1990s, foreign owners thus have significant percentages of shareholding in most of the major South African telecommunications companies. For years, for example, Alcatel (France) had a 12.5 percent stake in Altech. In January 1993, Alcatel boosted its stake to 50 percent and allowed Altech to buy a stake in Alcatel as well, thus more fully integrating the companies. In the mid-1990s, Siemens (Zurich) had a 52 percent ownership of Siemens South Africa, and until the recent merger of Plessey and Grinaker, Plessey (United Kingdom) had a 74 percent stake in Plessey South Africa. Following the reorganization of Plessey, GEC of the United Kingdom now holds 50 percent of GEC South Africa's stake in Temsa.

Originally, South Africa's Manufacture and Supply Agreements covered a ten-year term, but the equipment suppliers persuaded the SAPT to increase the term to fifteen years because of the high investment in plant and technological transfer from overseas principals.[9] The SAPT was obliged to purchase its equipment from the domestic companies when feasible, depending on the percentage of local content and the comparison of local with international equipment prices. According to policy, the SAPT would generally purchase South African–made equipment if there was a reasonable degree of local content so long as the price did not exceed the international price by more than 25 percent. In many instances, however, the price premium was far higher, ranging from 32 percent for telecommunications equipment to 60 percent for electronic components (Kaplan 1990, pp. 106–26).[10]

While the Manufacture and Supply Agreements worked well to ensconce viable equipment firms and establish coherent standards and specifications, the pricing structures may have acted to replicate the famous Averch-Johnson-Welliscz effect. As a rule, the SAPT would take back a percentage of high supplier margins as a kind of rebate or "profit sharing." Designed to give suppliers incentives to enhance efficiency, the rebate arrangement instead gave suppliers incentives to reinvest in their own plants (such investment being part of the overall cost structure) in order to reduce the amount that the SAPT would take back (Hartyani 1992; Schulze 1992; Averch and Johnson 1962; Welliscz 1963). Although intended to foster innovation and product development, South Africa's supply agreements paradoxically may have hindered them. According to David Kaplan (1990, pp. 99–101), the agreements function to exclude the entry of new, smaller companies that might be highly innovative. The agreements provide alternative and less risky routes to achieving high levels of profitability.[11] As if to underscore the historical weakness of South African telecommunications equipment manufacturing, the South African Board of Trade and Industry reported that telecommunications manufacturers exported only 1.5 percent of their products in 1985 (Kaplan 1992). Since then, however, local telecommunications equipment exports have improved. Exports, primarily to Eastern Europe and Africa, expanded after 1988, increasing two and one half times in constant Rands (see Kaplan, this volume, chapter 11).[12] By 1991, the telecommunications equipment industry in South Africa was worth R1.839 billion (BMI TechKnowledge 1992, p. 95).

12.2 A New Telecommunications Regime

South Africa's telecommunications structure changed in October 1991 when posts and telecommunications were separated from each other and set free from direct ministerial control (Republic of South Africa 1991c). A combination of forces, including those that have affected other African telecommunications administrations as well as those specific to South African politics, led to the regime change (see Horwitz 1992). Demands from large users—who had formed effective lobbying groups—for improved service quality, freedom of choice, and the opportunity to compete began to find a receptive audience in the government. Similarly, on the technological front, rapid advances, particularly in the areas of fiber-optic and mobile communication, had begun to erode the telecommunications sector's natural monopoly conditions. Finally, and perhaps most important, in the rapidly declining economic climate of the 1980s, the government came to the tradition-jolting conclusion that privatizing its public corporations and enterprises was a better option than operating debt-ridden ones. This historic reversal was the result of a series of official investigations of the principal South African parastatals, including the SAPT, conducted by Dr. W. J. de Villiers and known as the De Villiers Report.

12.2.1 The De Villiers Report

In the case of the SAPT, de Villiers found that South Africa's telecommunications system was characterized by very heavy debt—a consequence, he concluded, of overforecast demand and "goldplating," which he characterized as a too rapid conversion to digital technology (De Villiers 1989). The SAPT's staggering debt stemmed from the timing of its expansion. With the telephone market for whites saturated, the SAPT began in the late 1970s to extend service to Indian and "Colored" areas and, later, to a number of black areas. In the midst of this expansion, the SAPT faced the problem of whether to expand its problematic electromechanical switching system or go digital. It decided to pursue the digital option. These expansion projects required a vast infusion of capital, which the SAPT was permitted to obtain through loans on the international capital market. The annual capital expenditure on telecommunications equipment increased fivefold over the seven years from 1980–81 to 1986–87 (De Villiers 1989, concise version, p. 11). As South Africa's economic condition deteriorated in the 1980s and the value of the Rand plummeted, the SAPT's—and the government's—indebtedness increased dramatically. Similar stories characterized the state of South Africa's electricity and transportation parastatals (De Villiers 1984, 1986). Reversing its historic policies, the government began moving toward a policy of privatizing its key parastatals.

The transformation of the institutional structure of South African parastatals came about essentially at the very time that South Africa was poised to realize the transition to democracy. Hence, the South African government's privatization proposals were inherently politicized. Whether or not the government intended it, the effect of such privatization would have been to take the parastatals out of the hands of a newly empowered black majority in a democratic dispensation (Horwitz 1992). And indeed, in early 1990, the newly legalized African National Congress announced that, though it was no longer wedded to nationalization as a matter of general policy, public corporations and state business enterprises that were privatized prior to political accommodation would be prime candidates for renationalization. This political line was voiced shortly after the ANC's legalization and was adopted formally at an April 1990 conference held in Harare, Zimbabwe (Battersby 1990, p. 3). The ANC's general pronouncement and the specific opposition of the Postal and Telecommunications Workers Association (POTWA, the main union representing the SAPT black employees) to the privatization of the SAPT took the wind out of the sails of the privatization gambit.

As a result, "commercialization," which had previously been viewed merely as a preparatory stage to privatization, then assumed importance as a privatization substitute. The government began to view commercialization as a step toward ridding itself of what it had become convinced was a complacent, bureaucratic, top-heavy parastatal and as a means of pushing the Post Office toward a market-oriented corporate culture. In place of the command structures of state-owned enterprises, market incentives supposedly not only would induce the SAPT to operate on a proper business footing but would inculcate a customer-oriented organizational culture in what was widely perceived as a slothful civil service bureaucracy.

12.2.2 Telkom SA

South Africa's new telecommunications entity, Telkom SA, became a company formally registered in October 1991 under the South African Companies Act, with the state as the sole shareholder. As a commercialized entity, Telkom could generate profits and pay taxes, received no state subsidies, and was responsible for obtaining its own financing (although this remained subject to ministerial oversight inasmuch as the state was Telkom's sole shareholder). Telkom's management made much of its new status and the change in corporate culture that resulted. It formed a bona fide twelve-person board of directors, consisting largely of high-ranking businessmen (including two nonwhites) and complemented by a couple of academics. The company decided to operate regionally (reserving some powers of central management) and appointed individual client managers for important corporate clients. The old SAPT relinquished its joint role as player and regulator, and a very small Department of Posts and Telecommunications acted as interim regulator. About one hundred people remained in the department while 67,000 went to Telkom.

South Africa's telecommunications sector thus entered an era of liberalization and deregulation. In the early 1990s, however, the actual complexion of this new era was still quite vague. South Africa faced the prospect of fundamentally restructuring its telecommunications institutions, which required the resolution of difficult questions regarding the boundary between monopoly and competitive services, the adjustment of tariff structures, and the coordination of a now lively telecommunications sector that previously had very few players. The difficulties of telecommunications restructuring have affected nearly all telecommunications institutions worldwide, but they are particularly pronounced in the advanced industrialized countries (see Duch 1991). What intensifies the complexity and difficulty of the process in South Africa, of course, is that these policy decisions must be considered through the historical lens of apartheid and hence are politicized to an extraordinary degree.

Lest misunderstanding arise regarding this issue of "politicization," I submit that policies on infrastructures are never the technocratic matters they are often portrayed to be. Because infrastructures invariably involve the state, tax monies, and allocative policies that have crucial public goods consequences, policies surrounding them are inherently political. In South Africa, this fact is simply intensified. The historical operation of South Africa's telecommunications, like all state services, was inscribed within the apartheid system, and the inequitable distribution of infrastructure and access to service reflected this fact. The disparities in South Africa's telephone penetration by race are striking, and the politics of the system of separate development have meant that there is inadequate infrastructure in the black townships and virtually none in the country's African rural areas. For example, 1989 figures show telephone penetration per 100 blacks at 2.4 compared with 25 for whites.[13] This disparity is not simply a matter of skewed income levels. The infrastructure is not fully in place in many areas of black settlement, and there is evidence of a fair degree of suppressed demand among blacks.[14] Thus, whereas in the industrialized nations the PTTs or regulated telephone monopolies typically marshaled cost-averaging and value-of-service pricing mechanisms to extend service universally, the SAPT historically utilized cross-subsidies to extend service mainly to whites and business.[15] How South Africa deals with this legacy, while accommodating new business needs in the postapartheid era, was the key question facing the sector during the period of transition to majority rule in the mid-1990s.

12.3 The Politics of Transition

The stated objective of the Post Office Amendment Act of 1991 was to free South Africa's postal services and telecommunications from the perceived burdens of ministerial control. Remaking major socioeconomic institutions, however, is a difficult business. The commercialization of the SAPT was a complex political process that required a fair amount of effort to secure appropriate legislation. Several full-scale parliamentary discussions were devoted to the change.[16] It is unrealistic to expect that a technically complex sector like telecommunications can be restructured in specific ways by a general body such as South Africa's Parliament. Hence, it was not surprising that in the early 1990s the determination of many of South Africa's telecommunications issues was deferred.

Two items of enormous consequence were left undecided by the legislation: namely, how the now transformed sector would be structured and how it would be regulated. The central questions confronting any transformation of the PTT system—that is, which areas of service would remain monopolized, which would be competitive and to what degree, and how to adjust tariff structures—were also left unaddressed by the Post Office Amendment Act. Less understandable perhaps was that South Africa's Parliament also did not determine who would eventually decide these fundamental questions. South Africa's now grossly pared down Department of Posts and Telecommunications was understood to be acting in the role of interim regulator without, however, having its legal standing or the scope of its authority spelled out. The 1991 Act to Amend the Post Office Act of 1958 was itself vague and contradictory with regard to the structure of South Africa's post-SAPT telecommunications. On the one hand, section 34 replaced the words "Postmaster-General" with "telecommunications company" and stated that the successor telecommunications company (Telkom):

> shall have the exclusive privilege of constructing, maintaining or using, or of authorizing any person to construct, maintain or use, any telecommunications line for the sending, conveying, transmitting or receiving of sounds, images, signs, signals, communications or other information, and of transmitting telegrams over any such telecommunications line within the Republic or the territorial waters thereof. (Republic of South Africa 1991c, p. 36)

On the other hand, section 43 stated:

> Notwithstanding anything to the contrary contained in this Act, the Minister may, after consultation with the successor company concerned, and if it is in the public interest, by notice in the Gazette, also authorize any other person to exercise any power corresponding with any part of the exclusive power to conduct the postal service or the telecommunications service which has in terms of this Act been transferred to the postal company or the telecommunications company, respectively, on such conditions as the Minister may deem fit (Republic of South Africa 1991c, p. 44)

Thus, while Telkom seemed to have an exclusive operating privilege and the sole authority to determine the network's use, there was room in the statute for the minister to undermine that privilege. Yet the statute did not clearly give the minister the power to compel interconnection with the Telkom network in the event another carrier was authorized to provide a telecommunications service. Moreover, the authority of the Postmaster General vis-à-vis Telkom was at best unclear and at worst minimal. In short, the Post Office Amendment Act left a number of central questions unaddressed and an effective vacuum in direction.[17]

Since Telkom is a commercialized company whose sole shareholder is the state, Telkom has in effect to answer to two ministers. The "policy" portfolio, held by the minister of Transport and Communications, has been separated from the "shareholder" portfolio held by the minister of Minerals and Energy Affairs and Public Enterprises. Because "policy" and "shareholding" are not mutually exclusive categories, in the early 1990s it was not in fact clear who set policy for the sector.[18]

The South African government tolerated this lack of clarity perhaps because legislative finality was unlikely and because the interim period was expected to be brief. On the legislative and regulatory fronts, the government asked Pierre Pretorius, a Transnet advocate (a lawyer admitted to argue before the Supreme Court) who had played a leading role in the commercialization of that parastatal, to draft the legislation on the telecommunications sector. Pretorius's first order of business was to propose a regulatory structure for the sector. South Africa's minister of Transport and Communications originally favored the appointment of a single regulator in the manner of the British regulatory body Oftel but was dissuaded from taking this course, according to Pretorius, because South Africa was "too divided a society" and because a group of commissioners would provide greater representation.[19] The working model of the new regulatory body at that time was the U.S. Federal Communications Commission, which oversees American broadcasting, the U.S. spectrum, and telecommunications as a whole.

On the basic policy front, the Department of Posts and Telecommunications contracted with Coopers & Lybrand, Deloitte, the U.K.-based international accounting firm, to conduct an independent study of the South African telecommunications sector. The Coopers & Lybrand study was designed to provide an independent, expert analysis of South African telecommunications to guide the writing of the new legislation. Coopers & Lybrand made available an interim draft of the report, but the government's reaction was slowed by the then pressing need to attend to the multiparty constitutional negotiations (known as Codesa—the Convention for a Democratic South Africa) in May 1992. In the meantime, the government (probably through the Department of Posts and Telecommunications Steering Group, which oversaw the review) asked Coopers & Lybrand to address several questions regarding the interim draft. The final draft was released publicly in September 1992.

12.3.1 Assessing the Sector

In 1992, the exigencies of the politics of transition to democracy complicated South Africa's telecommunications picture. At the same time that the government was moving on its own to draft legislation on the structure of the telecommunications sector and its regulation, it was obliged to discuss at least the rudimentary outline of its proposals at Codesa. This reflected the fact that though the official, formal (National Party) government continued to function, Codesa had begun to take on the trappings of an interim government.

Notwithstanding grumbling from National Party stalwarts about the sanctity of the state and "duly constituted government," consequential state policies simply could not continue to be determined solely by the white government under the fundamentally new circumstance of the transition to a democratic dispensation. All important policies effectively had to begin going to Codesa. There, however, the key discussions on the subject of a communications regulatory agency centered, perhaps not surprisingly, on broadcasting. Because of the inordinate propagandistic power of the National Party stemming from its control of the South African Broadcast Corporation (SABC) (another state monopoly), the ANC-SACP-COSATU (Congress of South African Trade Unions) alliance focused its attention on breaking up the SABC and establishing a regulatory authority to ensure fair access to broadcasting for all political parties. The focus on broadcasting made perfect sense given the likelihood of hotly contested democratic elections. In the Codesa negotiations, the telecommunications component of communications regulation was definitely of secondary importance. In terms of its importance to the South African economy, however, telecommunications dwarfs broadcasting.[20] As if to underscore this fact, the business community voiced concern that telecommunications issues would be given short shrift by any future regulatory body ("Holding the Airwaves Hostage" 1992, p. 83). That same anxiety, albeit perhaps animated by different aims, was shared by POTWA (POTWA Communications Workshop 1992).

Notwithstanding these concerns, the Codesa negotiators had essentially reached an agreement on the terms for a combined telecommunications-broadcast-spectrum regulatory authority. But when Codesa broke down in July 1992, the regulatory authority agreements came undone (Pretorius 1993).[21]

The state of South African telecommunications policy in 1993 was characterized by conflict and probably stalemate deriving from the fact that, following the collapse of Codesa, the government went ahead and began to formulate policy without first establishing a legal framework for the sector and its regulation. Moreover, the government formulated telecommunications policy without engaging an effective consultative process with the ANC and the unions. This in itself was somewhat surprising. In housing, electricity, science and technology, and the general economy, national forums in South Africa have been fashioned to serve as negotiating arenas for all relevant parties to hammer out consensual policy recommendations. These forums were pushed by extraparliamentary groups, but the government felt compelled to participate because any long-term policy it might undertake risked being vetoed without the agreement of the forums. The National Electricity Forum, for example, proposed in late 1992 and constituted in early 1993, found the government participating alongside the ANC, the National Union of Miners (NUM), the National Union of Metalworkers South Africa (NUMSA), the South African National Civic Organization (SANCO), the Association of Municipal Electricity Undertakings (AMEU), the United Municipal Executive (UME), the Chamber of Business (SACOB), and Eskom to formulate policies to restructure South Africa's electricity sector.

Even in the seemingly more contentious area of broadcasting, the ANC alliance and the government were able to work together and hammer out negotiated agreements. This resulted in the creation of an Independent Broadcast Authority in August 1993 (see Independent Broadcasting Authority Bill, final draft, 1993). However, a parallel process did not gel in the telecommunications sector. Indeed, the absence of such a process in telecommunications led constitutional negotiators to sever the previous connection between broadcasting and telecommunications and negotiate the regulatory authority for broadcasting separately (Pretorius 1993).

In early 1992, BMI TechKnowledge, a market research firm, attempted to put together a telecommunications policy forum, but participation by Telkom and the ANC was diffident and spotty, and the government took no direct part at all. A large convening conference was scheduled for October 1992, but in the aftermath of the collapse of Codesa, the ANC alliance refused to participate. In early 1993 BMI abandoned its efforts (Smit 1993).[22] In the meantime, the government received the Coopers & Lybrand report and asked for public comments within just one month. The ANC issued a blistering attack on the report, branding it as flawed and denouncing what it perceived as the government's determination to restructure South African telecommunications without any broad consultation (ANC 1992). Moreover, the report's general observation that retrenchments would probably be necessary in Telkom greatly angered COSATU, which had not been approached by the Coopers & Lybrand consultants (Fanaroff 1993b; Naidoo 1993).

At that point, all the parties began playing out a political strategy. Efforts to get representatives of the ANC and the unions and the government together came to nothing. The South African government then moved ahead with its own agenda, using particular aspects of the Coopers & Lybrand report as its guide.

12.3.1.1 The Coopers & Lybrand Report

The Coopers & Lybrand report was a comprehensive study, addressing issues pertaining to regulatory authority and the structure of the telecommunications sector. Beginning with the fact of Telkom's commercialized status, the Coopers & Lybrand report specifically addressed the problem of institutional roles. Thus, it suggested that Telkom have no regulatory functions whatsoever and that the government's role as sector policy maker be separated from its role as Telkom shareholder. Responsibility for sector policy making would remain with the minister for Posts and Telecommunications, while the role of government-as-shareholder would be exercised by the Ministry of Finance.

In the interests of ensuring adequate representation and promoting informed public debate, the report suggested that consideration be given to the establishment of a statutory consultative committee for the telecommunications sector made up of representatives of either consumers or all those with an interest in the sector. The policy-making ministry and the regulatory agency could be required to consult the committee before any major decisions were taken. A proposed Telecommunications Regulatory Agency would itself promulgate and administer specific regulations within the context of overall telecommunications policy as set out by ministers. Referring to international trends in telecommunications, the report recommended that the regulatory agency be set up as a body independent of the ministry that sets policy (Coopers & Lybrand 1992, pp. 47–70).

In the areas of competition, deregulation, and privatization, the Coopers & Lybrand report reviewed international trends and supported the principles of competition and privatization in general and in the abstract but recognized that certain service monopolies were necessary, at least for several years. Specifically, the report recommended that Telkom retain the exclusive right to supply long distance and international voice telephony for a period of at least five years. The Coopers & Lybrand consultants recognized that opening South Africa's long distance service to competition would result in widespread "creamskimming" and thus necessitate Telkom's increasing charges for local service by a factor of between two and three. Quite properly, Coopers & Lybrand understood that this would be politically unacceptable.

In a sense, in a quid pro quo for retention of the long distance monopoly, the report recommended that the objective of increased network penetration be formalized in terms of a regulatory obligation on Telkom to meet specified targets for service expansion and to provide for the financing of such obligations. The report further recommended that a system of price caps be established for those services in which Telkom retains a monopoly and that explicit regulatory controls on Telkom's prices be instituted. At the same time, the report advised that the provision of *local* service by independent network operators (such as Transtel or Eskom) not be ruled out. Such provision, however, would not be in competition with Telkom; it should instead be a supplement to the main network (Coopers & Lybrand 1992, pp. 20–26, 81–95).

In the area of enhanced services, the Coopers & Lybrand report suggested that the current draft value-added networks (VANs) license be issued immediately, subject to negotiations with the industry for further liberalization. However, because of the danger of straight resale of network capacity, the report recommended that the resale of voice services not be permitted for three to five years. Furthermore, the self-provision of short-haul circuits should be allowed subject to defined limits (the report suggested a limit of 500 meters). Very small aperture satellite terminals (VSATs) should also be liberalized but made subject to a ban on interconnection with the public-switched network.

The report recommended that provision of the first telephone instrument be liberalized and the local content requirements relaxed over three to five years. Technical standards would be supervised by an independent body as part of a move toward more open, international standards. Finally, in what would soon become the most consequential element of the report, Coopers & Lybrand recommended that consideration be given to licensing two competitive digital cellular networks, among which network sharing should be permitted to avoid unnecessary duplication of resources. International experience, the consultants argued, shows that competition works to increase penetration and lower market prices in the cellular field. Telkom should be allowed to participate in one of these network-operating franchises, the report suggested, either on its own or as part of a consortium—but only through a fully separate subsidiary. Air time resellers, the report added, should be introduced to act as intermediaries between the networks and the end users (Coopers & Lybrand 1992, pp. 71–80).

Thus, while the Coopers & Lybrand report recommended some competition to sharpen the incentives for improved performance and held out the goal of privatization as a means to introduce private-sector capital into the industry, a close reading reveals that in the short run the report suggested protecting Telkom as the primary supplier of South Africa's telecommunications services and using a mixed set of regulatory controls to push the company to behave in an efficient, public-interested manner. The only recommendation clearly contrary to this general advice was the licensing of two competitive cellular operators. Even here, the report hedged a bit, advising a prohibition on the provision of connections from mobile to fixed terminal equipment (such as PBXs) in order to protect Telkom. However, the report clearly did not adequately analyze the consequences to Telkom and the public-switched telephone network of its cellular recommendations (Coopers & Lybrand 1992, pp. 76–78). Sensitive perhaps to the very real politicoeconomic tension between the need to expand South Africa's basic network in the postapartheid period and the need to liberalize VAN services, the Coopers & Lybrand report plotted three potential scenarios that described alternative development paths for South Africa's telecommunications sector:

- Scenario 1, the "base case," essentially followed Telkom's five-year plan for modest service expansion and investment, extended forward to 2002.
- Scenario 2, labeled "network expansion," proposed to increase the availability of South Africa's basic telephone service in an aggressive manner, increasing access lines to 19.6 per 100 urban inhabitants versus 14.1 in the

base case (scenario 1), and 6.2 access lines per 100 rural inhabitants versus 3.5 in the base case. Although there would be an increase in both local and national call charges, access would be cross-subsidized mainly by business through higher rental and connection charges. International charges would decrease, however.

In scenario 2, Telkom would not face competition in voice telephony, but its profitability would be the lowest of the three options, primarily due to the significantly higher levels of capital investment. Some financial restructuring would be necessary to carry out the aims of scenario 2. Indeed, in order for the network expansion scenario to work, the report suggested the necessity of a capital injection of R4 billion in 1993–94 to retire an equivalent amount of long-term debt. This cash could come from the government or by privatizing Telkom.

- Scenario 3, the "competition" option, rebalanced all prices over five to seven years to reflect actual costs, after which competition in voice telephony is introduced and prices presumably decline. Under this scenario, business prices decline furthest and fastest. To increase the availability of basic service, public pay phone penetration would be increased substantially to nearly 300,000 by the year 2002 versus 82,600 in Telkom's base case (scenario 1).

In scenario 3, access line penetration plots out to 13.5 per 100 urban inhabitants by 2002 and 3.5 per 100 rural inhabitants—quite a bit lower than the network expansion scenario (scenario 2). Competition in all services is assumed to be a spur to productivity gains. Telkom would be in the best financial shape under this scenario.

To the extent that the Coopers & Lybrand report offered recommendations among the scenarios, it suggested that both the network expansion (scenario 2) and competition scenarios (scenario 3) were better than Telkom's base case plan (scenario 1). Telkom's plan was judged to lead only to a limited expansion of levels of access, and even that depended on network utilization increases that approached operationally infeasible levels. It seems reasonable to conclude that the Coopers & Lybrand consultants addressed what they understood to be the two main politicoeconomic policy positions with regard to telecommunications—and in their report framed these positions within the discourse of expertise.

In this sense, the Coopers & Lybrand report does in some ways represent a kind of generic, framed debate between the ANC alliance and the National Party over the future of South African telecommunications. And this perhaps was the report's real importance. It presented a version of the basic progressive demand for network expansion and framed it within acceptable boundaries. Given that the ANC alliance had neither the wherewithal nor the capacity to generate its own study, the Coopers & Lybrand report essentially became a baseline from which to assess South Africa's telecommunications policies and the debates about them. Thus, any demand that Telkom "serve the people" had to confront the specific realities of Coopers & Lybrand's scenario 2.

At the same time, any call for a competitive telecommunications system for

South Africa must address the demands for basic service by, at the very least, vastly increasing the number of pay phones and keeping call charges within reach of the poor. In other words, the Coopers & Lybrand scenarios may not be the only options, but they likely assume the status of a gold standard.

12.3.2 The Cellular Controversy

In certain respects, the Coopers & Lybrand report reflected a wariness that resolving the conflicts in South Africa's telecommunications sector would be a difficult political process. This wariness, however, was evidently not shared by the South African government. Advocate Pretorius, charged with drafting the enabling legislation, wanted to set the legal structure in place before policies were determined. Short of a legal structure, Pretorius favored a consultative process for the determination of policy.

Though seemingly placed in a powerful position, Pretorius was essentially ignored. The government went ahead with a strategy to revamp South Africa's telecommunications sector without a legal structure in place and without a consultative process. The vehicle for the revamping was the decision to establish two cellular network operators and send these licenses out to tender. Cellular thus became the key battleground in the South African telecommunications debate.

The visible driving forces in this movement were Minister of Transport and Communications Welgemoed and Eugene van Rensburg, head of the Policy Unit on Public Enterprises of the Ministry of Minerals and Energy Affairs and Public Enterprises.[23] This route was in marked contrast to that taken in other contested sectors. For instance, in 1993 South Africa's Department of Home Affairs was sitting on eighty to ninety applications for broadcast licenses, but it was decided not to make any awards until the broadcast authority was put in place (Pretorius 1993).

South Africa's cabinet met in early 1993 to decide how the proposed cellular mobile telephone system should be organized and regulated. With expressions of investment interest from big-time foreign operators—perhaps representing the largest single foreign investment in postsanctions South Africa—dangling in front of them, cabinet ministers went ahead and authorized two cellular licenses. The parties pushing for the quick authorization of cellular service pointed to the interest by foreign investors and argued that the timing was opportune and not to be squandered (see Sergeant 1993b; also Knott-Craig 1993b). The licenses were put out to tender in April 1993, and the minister of Transport and Communications appointed Ters Oosthuizen as the "regulator." Oosthuizen, an advocate who for many years worked for Eskom, was to oversee telecommunications in general and the cellular tender process in particular. Oosthuizen subsequently asked the minister to name him postmaster general because he had no staff or legal standing as regulator. However, by becoming postmaster general, Oosthuizen unambiguously identified himself with the government and thus its old structure and nonconsultative methods (Hainebach 1993).

The ANC and COSATU, joined by a group called the Cellular Telephone Consultative Forum, immediately charged that Oosthuizen had been appointed outside of both legislation and the Negotiating Forum (the resurrected Codesa constitutional negotiations). The ANC alliance claimed that the licensing of cellular telephony represented a unilateral restructuring of the industry, a form of privatization through the back door (see Fanaroff 1993a). In fact, some of the government's own pronouncements support that interpretation. Policy Unit chief Eugene van Rensburg, for example, was quoted as saying that the cellular policy can be taken as clear indication that the government favors competition ("Dial for Fair Play" 1993).

Another explanation of the government's actions put politics at center stage. According to one report, a senior government source was quoted as saying that the National Party government saw that it had the opportunity to provide telephones to the masses quickly through cellular service, and the party had no intention of waiting and thereby permitting the ANC to reap the political benefits (Sergeant 1993a).[24] Other observers invoked a political explanation of the more traditional kind, arguing that Welgemoed was a minister "of the old school"—that is, of the imperious apartheid stripe—who would never allow "interlopers" to make policy or interfere with his authority.[25]

Whatever the full explanation for the South African government's actions, in 1993 it proceeded to set the terms of application for the cellular licenses and appoint a tender review board to review the applications. A R50,000 tender application fee was assessed in order to discourage all but the most serious bidders. The basic terms of the fifteen-year license were that each operator had to pay an initial basic fee of R100 million plus an ongoing license fee of 5 percent of its net revenue. Also, R5 million in annual radio fees had to be paid to the postmaster general along with R1 million a year for each 10 MHz channel granted to the operator. Some of these fees were designed to finance the regulator, but the R100 million basic fee would go straight to the Exchequer—it was not earmarked, that is, for expanding telephone service to those without access ("Squeezing the Cellular-Phone Industry" 1993). Among other things, applicants had to specify the extent to which their choice of technology would lead to high volumes and low costs, how they would support South African industry, and how they would provide service to poor communities.[26]

Telkom was quickly awarded half of one license. It entered into a consortium with U.K.-based Vodafone and the Rembrandt Group of South Africa to form a separate subsidiary, Vodacom. Three applicants vied for the second license. Mobile Telephone Network (MTN), considered by most accounts the favorite, was a consortium of M-Net (the South African pay television provider), U.K.-based Cable & Wireless, NAFTEL (an association of black businesspeople), and Transtel. Cellstar Cellular Networks was a venture of Anglovaal's Grintek Limited and Telecom Finland. The Reunert Group applied for the license in a venture that combined Barlow Rand with the Deutsch Bundespost Telekom. Both the Reunert and Cellstar ventures left room for black participation, but at the time the tender applications were submitted there were no takers.[27]

In 1993, the African Telecommunications Forum, a loose grouping of black organizations and entrepreneurs, made inquiries about obtaining a cellular license, though it did not submit a genuine tender bid (in part because of ANC opposition to the tender process). The parties to the African Telecommunications Forum included the ANC-aligned Thebe, Afritel, Suntel, MIT, and NIT, and the forum enjoyed the backing of the National African Federated Chambers of Commerce (NAFCOC). Afritel in particular forged links with the U.S. telecommunications giant AT&T, although the U.S. investors would not commit until all state and local authority sanctions were dropped ("Dial-a-Profit" 1993).

In September 1993, the South African government announced Mobile Telephone Network as the winner of the second cellular license. Vodacom's system would be the Global System for Mobile Communications (GSM) standard, and Mobile Telephone Network proposed GSM as well. Quite apart from the licensing controversy, a minor storm arose over the technical standard. The part of the spectrum allocated to South African cellular was the 900 MHz, eliminating the American Mobile Phone Standard system, which operated in the 800 MHz spectrum. The fact that the tender documents specify a R20,000 a year fee for each 200 KHz channel led some to question whether the South African government had already chosen a standard—because only the GSM operates in 200 KHz blocks ("Squeezing the Cellular-Phone Industry" 1993).

Several observers, including Postmaster General Oosthuizen, indicated that well prior to the tender bidding Telkom had already committed itself to GSM and had installed some base stations (Oosthuizen 1993; also Maepa 1993). Moreover, Telkom's commitment to GSM, in the judgment of the chief executive of Plessey, stemmed not from its technical characteristics per se but once again from Telkom's historically close relations to its main equipment supplier, Siemens—the company that would bring in GSM technology from Europe (Temple 1993).[28] Some faulted GSM as an expensive cellular system, with high prices for handsets that are not very lightweight and probable call charges several times those of the current phone network. Others, notably Alan Knott-Craig, chief executive of Vodacom, were forthright in their defense of GSM. The key, Knott-Craig argued, was that GSM allowed the separation of service from the actual telephone. Thus, GSM cellular call boxes could go a long way toward serving underprovided communities. Vodacom's Community Tariff Plan subscribers would use a prepaid SIM card, enabling them to make calls from any GSM telephone and access a centralized voice mail system. The voice mail feature would obviate the need for the individual ownership of a telephone. In addition, Vodacom planned to franchise the GSM Community Telephones to members of underserved communities, creating the conditions for telephone entrepreneurship (Knott-Craig 1993b).[29] Of course, the complexity to the user of the SIM card and voice mail led others to doubt the very claims put forward for GSM (Davies 1993).[30] While the ANC and COSATU did not condemn GSM per se, they bitterly opposed the existing cellular tender process. In 1993, ANC Information Systems head Andile Ngcaba indicated that he believed the cellular applicants were simply using the language of universal service to win the tender. The ANC wanted cellular to be a separate, autonomous parastatal service offered by Telkom in a sector rationally planned to take advantage of all existing communications infrastructure. Such a structure would be the best means for providing universal service (Ngcaba 1993a, 1993b, p. 17).[31]

The cellular controversy thus became a major source of conflict between the government and the ANC alliance. The government argued that it would not stop the tender process; that applicants had spent millions of Rands in a good-faith effort to apply for licenses, and suspending the process would lead to a myriad of lawsuits. The government also argued that suspending the process would jeopardize foreign investment ("Government 'Will Not Back Down' on Phone Licenses" 1993; Oosthuizen 1993).[32]

The ANC alliance then called for a moratorium on the tender process. In September 1993, the politics of the cellular tenders became quite heated, with COSATU threatening strikes and the ANC threatening to revoke the licenses when it came to power. Nelson Mandela, the ANC president, with President de Klerk over the cellular license controversy, and ANC Secretary-General Cyril Ramaphosa led a delegation to meet with Public Enterprises Minister Dawie de Villiers (Makhanya 1993, p. 18). Ramaphosa was quoted as saying that a future government would immediately review, and perhaps cancel, the cellular licenses if the government went ahead and issued them ("Government, ANC Impasse on Cellular Phone Controversy" 1993).

Interestingly, the cellular controversy cooled off because another telecommunications brouhaha intervened. Moving ahead in other areas, the government tabled a new bill before Parliament to amend the Post and Telecommunications Act. While much of this bill simply clarified the power of the postmaster general, the ANC alliance again vehemently objected to the government's lack of consultation in what appeared to be a bid to restructure South African telecommunications. And in the context of the struggle over the cellular licenses, the ANC interpreted the bill's clarification of the postmaster general's authority as a cynical maneuver to deregulate Telkom (Ngcaba 1993b, p. 17). In exchange for the government's agreement to hold off on the new Post and Telecommunications amendment bill, the ANC agreed to back down on its opposition to the granting of the cellular licenses, which went forward with Vodacom and MTN. A fluid license was being considered in 1998 ("Compromise Deal on Cellular Phones" 1993).

12.4 The State of the Network During Transition

The government popularly elected in 1994 inherited a relatively large telecommunications network consisting of 3.7 million access lines—the twenty-fifth largest network in the world (Department of Posts and Telecommunications 1992, p. 49). This total had increased from 3.315 million in 1991. South Africa had 38 percent of the telephones in Africa and the highest density of telephones on the continent (Siemens 1992, p. 16; Telkom 1991, p. 1).

South Africa's network is also relatively sophisticated. It has a high degree of digitization and consists of an increasing percentage of optical fiber. As we have seen, the SAPT began the process of digitization early, in 1978–79, in part because it ran into trouble with its electromechanical switches just at the point when large foreign manufacturers were introducing digital technology and in part because digital was judged by the SAPT management to be a better technology for meeting the huge existing demand.[33] Siemens, Temsa, and Altech were awarded long-term contracts for the supply of this equipment and began installing digital exchanges in 1980.

In 1993, there were 1,586 exchange sites in South Africa's automatic telephone network, of which 929 had digital exchange equipment—a digital penetration rate of 58.5 percent. Of the over 3.5 million access lines in operation in 1993, about 56 percent terminated on digital equipment. Moreover, this rate was planned to increase to 64 percent by 1995 and to more than 80 percent by the year 2000. The penetration of digital transmission systems is even higher, with 100 percent in South Africa's large metropolitan areas and 78 percent in the rural areas (Lachenicht 1991; Telkom 1991, 1993).

Prior to 1969, South Africa's international telephone connections were carried by some twenty high-frequency circuits and two telegraph cables. In 1969, the SAT-1 submarine coaxial cable, running from Cape Town to Portugal, began operation and in December 1975 was supplemented by a satellite earth station. A new undersea fiber-optic cable (SAT-2), due to be commissioned in 1994, would carry a 565 megabits per second transmission—providing the equivalent of 7,680 voice channels. SAT-2 would link South Africa to international telecommunications nodes in Madeira and the Canary Isles (see *Proceedings of the Fifth National Congress* 1991). In the mid-1990s, a modern telex service with over 30,000 connections linked all of South Africa's major population centers—although, as in all nations, facsimile or fax service was displacing telex, and telex investments consequently became a major loss. South Africa was one of the first countries to introduce a public videotex service (Beltel), which by 1993 included an x-400 protocol electronic message handling service. By the end of March 1992, the number of registered Beltel users was nearly 30,000, but the service has historically operated at a considerable loss (BMI TechKnowledge 1992, p. 99). In 1983, an international packet-switching network (Saponet-P) replaced an analog network and by 1993 linked South Africa to thirty-five countries. In 1986, a digital point-to-point service (Diginet) for companies needing to transfer large volumes of data at high speeds was introduced.

In 1993, leased lines, including high-capacity 2-megabits lines, became available in South Africa but could be purchased only from Telkom. Under 1993 rules, a link between two offices of the same company on either side of a public road were to be provided by Telkom. Neither third-party traffic nor bypass was permitted. South Africa's cellular phone service, based on a variant of the German C450 analog system, had been operating since 1986 but by 1993 had only about 13,000 subscribers because of extremely high handset prices and the lack of available frequencies. South Africa's liberalized private Mobile Radio and radiopaging services had also grown considerably by the early 1990s, but they too were hampered by a dearth of available radio frequencies (largely because some 62 percent of the frequencies allocated by the International Telecommunications Union for Land Mobile Radio use in the early 1990s were occupied by the South African Defence Forces, although the sharp reduction in the military over the next years would create opportunities for reallocation of frequenices) (Department of Posts and Telecommunications 1992, p. 14; Coopers & Lybrand 1992, p. 76).

By the mid-1990s, various customer premises equipment, including PBXs, modems, fax machines, and cordless telephones, had been deregulated in South

Africa for several years—subject to type approval by Telkom. In addition, Telkom provided its own small business telephone systems and data modems in competition with privately supplied PABX systems and data modems. By 1993, internal wiring within buildings had been largely liberalized for several years as well. The liberalization of customer premises equipment had limits, however. Telkom retained exclusive rights to provision of the first telephone on customer premises, and the approval of attachments involved local content requirements until 1996.

In general, the quality of South Africa's telecommunications service is good and service is relatively cheap. For example, Telkom's call completion rates (just over 95 percent) and speed of fault clearance (77 percent cleared within one working day) are near the level of other industrialized countries. In white and urban areas, the average waiting time for connection in 1990 was thirty days (down from sixty-seven days in 1989), although in black areas waiting time for connection is much higher and extremely variable (Coopers & Lybrand 1992, p. 14). Notwithstanding these statistical measures, large users reported unhappiness with Telkom's response on fault clearance and connection times and frustration with Telkom bureaucracy generally (Davies 1992; Paul 1992). Moreover, the service waiting list has been growing: between 1990 and 1991, the waiting list had grown 14.56 percent to an official total of 125,448, and by 1994 to an estimated 134,000 (Department of Posts and Telecommunications 1992a, p. 12; Plessey 1995).[34]

By the early 1990s, Telkom also offered telephone information services, but the problems that plagued it in the early months of 1992 were revealing. Telkom's response to widespread customer complaints about grossly inaccurate bills, for example, was first to deny that there was any problem at all and then to insist that complainants pay their accounts (some reading in the thousands of Rands) while Telkom investigated—or risk having telephone service cut off. Finally, amid great hue and cry, Telkom's managing director conceded that computer error and/or service theft were responsible for incorrect billing. The "087" debacle (from the number for information services) lent credence to the view that, for all the talk about a new customer orientation, Telkom's corporate culture had not yet changed. The entire episode smacked of the arrogance and unconcern of a single supplier and left Telkom with bad debts of some R77 million and contributed to write-offs of R136.9 million (Telkom 1993, p. 3), a figure that would grow to R207 million by 1994.[35]

International price comparisons of a basket of telephone services with industrialized countries (based on 1989 tariffs and exchange rates) show that Telkom's tariffs for both business and residential customers, as well as residential line rental and connection charges, were low by international standards. In addition, unlike many countries, South Africa's business customers are not charged higher prices than residential customers, and business line rental and connection charges are thus even lower by international standards than those for residential customers. On the other hand, its international long distance tariffs were comparatively high. For instance, charges for calls to the United States were some 30 to 40 percent greater than U.S.-originated calls to South Africa ("Networking Pays Off" 1993). National long distance charges were in line with those in other industrialized countries. Call charges account for approximately 50 percent of Telkom's total revenue (Coopers & Lybrand 1992, pp. 11–17). Hence, South Africa's cross-subsidies can be understood as follows: international—and to a lesser degree, national—long distance charges subsidized local service (although this is not direct business subsidization of residential customers) while calls greatly subsidized rentals. According to the De Villiers Report, the actual cost of telephone rental per month in South Africa in 1987–88 was about R28, though the actual rental charge was R15. Thus, telephone rentals have been heavily subsidized by higher call tariffs. However, according to conventional wisdom (largely based on the De Villiers Report), a large proportion of subscribers (particularly new, usually nonwhite subscribers) make very few calls per day. As a result, telephone revenues rested on a narrow base of subscribers. According to the *De Villiers Report,* 6 percent of subscribers (of whom about 78 percent are business and 22 percent residential telephone subscribers) contributed 50 percent of South Africa's total telephone revenue in the late 1980s (De Villiers 1989, pp. 25, 8).[36]

12.5 Telkom's Current Condition and Financial Health

In 1994, Telkom had a monopoly in South Africa's various service areas, the most significant of which was the running of the public-switched telephone network. Telephone instruments, telex/teletex terminals, and data network terminating units connecting into the public-switched network remained Telkom's exclusive business. Private lines that were not confined to the boundaries of a client's premises were still monopolized by Telkom. With a payroll of some 62,000 people and assets valued at nearly R15 billion in 1994, Telkom had an annual operating expenditure of more than R5 billion.[37]

Notwithstanding Telkom's commercialization and its trumpeting of a new corporate culture, several factors clouded its prospects. Over the period 1987–91, Telkom's operating revenues grew by an average of 1.9 percent per year in real terms, whereas operating expenses grew nearly twice as fast, by 3.7 percent. As a result, Telkom's gross profit margins declined significantly, from 42 percent of turnover in 1987 to 30 percent in 1991 (Coopers & Lybrand 1992, pp. 16–18), a figure that would remain unchanged through 1994. Moreover, despite expected moderate surpluses in telephone services (R790 million) for the year ending March 1992, very large losses on telex (R230 million), gentex (R94 million), and postal services (R268 million) largely wiped out the surplus (Department of Posts and Telecommunications 1992b, p. 11).

Telkom's debt to equity ratio of 1.8 made it unlikely that it could easily raise more debt financing without state guarantees, although Telkom had substantially improved its financial position since the early 1990s.[38] Indeed, Telkom's capital expenditure since its commercialization (R2.2 billion in 1994) has come from internally generated funds only (Telkom 1993, p. 4). At the same time, an outflow of funds from the Post Office Savings Bank—a net amount of R521 million from 1989–90 to 1990–91—represented a loss of access to cheap capital, 64 percent of which had historically been invested in telecommunications assets (Department of Posts and Telecommunications 1992a, p. 6). In 1991–92, Telkom's total financing costs were expected to be R1.7 billion compared with a net operating profit of R2.4 billion (Coopers & Lybrand 1992, pp. 16–18). Approximately 20 percent of Telkom's budget was allocated to service its borrowings. According to Telkom's 1994 report, its official net interest bearing debt stood at R8.8 billion. The terms of its commercialization entailed government-imposed constraints on Telkom with regard to the retrenchment of staff (namely, no retrenchment as a result of commercialization), real increases in tariffs (which remain subject to the approval of the Department of Posts and Telecommunications in its role as interim regulator), recourse to external finance, and the honoring of long-term contracts with local suppliers. In addition, in 1990–91, Telkom was obliged to contribute R367 million to the postal deficit, with the expectation that it would be responsible for R500 million annually until 1995–96, to be met through taxes and dividends paid to the government. As if this were not enough, the Telkom Pension Fund was found to be short R1.59 billion and would have to be fully funded, though by the end of 1994, Telkom had succeeded in reducing the deficit to R1.1 billion.

Thus, even as revenues were declining in the period between October 1, 1991 (when it became a company), and March 1995, Telkom found itself contributing R209 million to the postal deficit, R446 million to the pension fund deficit, and R410 million in taxes (Telkom 1993, p. 56). Telkom's poor financial status translated into a significantly lower amount of capital investment—precisely at the moment when the end of apartheid dictated a vast expansion of the network. In the period 1975–86, telecommunications investments as a percentage of revenue typically averaged around 45 percent, but beginning in 1987 that figure dropped precipitously and in 1994 registered just 27 percent. At R2.227 billion, 1994 nominal values of capital investments in the network were 51 percent higher than the 1987 levels of R1.471 billion—but allowing for 15 percent average annual inflation, this was in fact negative real investment growth overall.

Moreover, according to Telkom's 1993 annual report, it spent R1.8 billion on capital programs, representing just under 17 percent of turnover (Telkom 1993). Telecommunications investments as a share of GDP, which peaked at 1.07 percent in 1986, fell to 0.64 percent in 1994. This parlous state was compounded by Telkom's depreciation rates, which, at an average of 4.7 percent between 1980–81 and 1990–91, were too low to continue the modernization of South Africa's network—particularly in those areas where it would be subject to competition (Department of Posts and Telecommunications, various years). In data and value-added services, equipment changes quickly. Without provision for replacement, Telkom cannot buy new equipment and easily retain business customers. Telkom recently went to a fifteen-year depreciation schedule for transmission and switching equipment, and ten to twenty years on cable, but some knowledgeable observers viewed even the new depreciation rate as too low (Telkom 1993; Hainebach 1993).

In short, in the late 1990s Telkom faced a difficult future. On the one hand, the end of apartheid imposed new equity-based demands on it to expand the basic telephone network to populations and areas historically denied access. The black Postal and Telecommunications Workers Association (POTWA), a mainstay of the powerful Congress of South African Trade Unions (COSATU), strongly supported the expansion of the telecommunications infrastructure both for reasons of equity and to safeguard its members' jobs.

In response to the demand to expand the network, during the 1992–93 fiscal year Telkom made a special allocation of R40 million for purchasing and commissioning "Community Telephone" facilities in remote and rural and inaccessible areas, such as squatter camps. Its objective, which was to install 250,000 phones a year between 1992–93 and 1997–98 (with emphasis on "community phones"), would be accomplished by the installation of a high-frequency radio network (Rurtel) and the establishment of base stations throughout the country. Rather than attempting to provide each household with a telephone, Telkom installed phones in all businesses and set up call boxes or card phones within "reasonable access" of everyone (Strachan 1992). Nevertheless, the R40 million earmarked for community telephony was a paltry sum when one considers the pressing needs of South Africa's disadvantaged and Telkom's R2.14 billion capital budget for 1992.[39] But the program, supplemented by offerings and subsidized rates from the cellular companies, has been a success in spreading telephony.

On the other hand, South Africa's businesses, feeling themselves restricted under a traditional, natural monopoly telecommunications structure, wanted enhanced and value-added services to be liberalized. Under the Post Office rules, it was illegal, for example, for an automobile dealer to use the data network of his or her supplying company to get information, since this constituted transferring data outside of the SAPT provision or control. The patent economic irrationality of such rules in an age of rapid information transfer, as well as the pressure brought by South Africa's National Telematics User Group (NTUG), prompted Telkom to allow companies to register a value-added network (VAN) services company in the early 1990s. Yet these agreements forced a VAN provider to become a separate company whose network and revenues were audited by Telkom in order for Telkom to ensure that the VAN was in fact a VAN and not a second telecommunications operator in disguise. The difficulties and improprieties posed by Telkom acting as regulator of competitors on the edge of its monopoly prompted an outcry among VAN providers, and in the early 1990s the VAN license recommendations again came up for negotiation (Davies 1993).

Telkom's actions surrounding the VANs issue, if blunt and heavy-handed, did, however, reflect a legitimate concern. If and when value-added services are liberalized, Telkom must rebalance its general tariffs in order to compete. In 1993, telephone traffic accounted for over 92 percent of South Africa's telecommunications income, and that traffic was growing at about 7.5 percent per year (although, with the recession of the early 1990s, telephone traffic actually declined 1.5 percent in the twelve months prior to April 1993 [Telkom 1993, p. 3]). However, it is in leased circuits and enhanced services that the growth potential for South Africa's telecommunications sector really lies. Leased circuit growth for 1991–92 topped the previous year by over 20 percent (Department of Posts and Telecommunications 1992b, p. 11; International Telecommunications Union 1992, p. 353). And it was precisely in these areas that Telkom faced potential competitors. Value-added voice and data services that allow operators to employ lines leased from Telkom for third-party traffic were a contested area in the early 1990s and represented a potential threat to Telkom. Large users want the freedom to carry third-party traffic and supply all services (including internal voice traffic) in an economically centralized way. Telkom, essentially acknowledging this as inevitable, wanted time to establish fair competition in value-added services and create a "level playing field." This could not mean anything other than moving toward cost-based pricing in all areas of service (see Knott-Craig and Hanekom 1990; National Telematic User Group 1991). Such a move, however, pointed to a fundamental tension between the need to rebalance tariffs so as to liberalize VAN services for business and the pressing need to expand the network for basic service.

In addition, because of the immense amount of data traffic they pass every day, South Africa's large users (banks in particular) were leery of total dependence on the Telkom network and thus keen to have a redundant network (or networks) available. Waiting in the wings were two and perhaps three organizations that already had the capacity to provide alternative networks. Eskom, the electricity supply monopoly, and Transtel, the now for-profit subsidiary of the commercialized transportation monopoly Transnet (formerly South African Railways and Harbours), each have large internal networks that could be made available to third parties on a point-to-point basis. Eskom operates a sophisticated microwave network and powerline carriers. Eskom and Transtel could also act as contractors for building new capacity to order, connecting their clients' private networks.[40] The South African Broadcast Corporation's signal distribution division could also, in theory, evolve into a carrier of point-to-point third-party traffic. Telkom's most likely potential competitor is Transtel, which in 1993 already operated a nationwide high-capacity digital microwave network consisting of several 34-megabit systems, a trunked radio network, and a 140-megabit fiber-optic network in the PWV and Natal (Transtel n.d.). Because South Africa's main roads and town arterials follow rail lines, Transtel's radio network extends to geographic areas not covered by the Telkom network. In addition, Transnet's connection to the PanAmSat would reduce the difficulty of reaching remote rural areas. In capacity, Transtel's network is hardly a rival to Telkom's. However, now that it has been transformed from a state enterprise into a commercialized business, Transtel is looking to maximize the utilization of its facilities in profit-making activities.[41]

Finally, due to competition from international discount carriers, in the 1990s Telkom was under pressure to offer international telephone connections at U.S. carriers' rates. Several international resellers, including Diners Club, IDT, and Word-Phone, began operations in South Africa in the early 1990s, and Telkom responded by cutting its own international rates ("Networking Pays Off" 1993). The reduction of rates by approximately 30 percent threatened to erode Telkom's earnings in overseas long distance, further undermining its ability to cross-subsidize the expansion of the basic network. An additional threat was the Internet and its use for telephony. Seizing on this potential, Telkom claimed in 1997 the right to be the country's exclusive Internet service provider. However, the regulatory authority SATRA decided to open the market to other ISPs. Undaunted, Telkom went to court and won, at least temporarily, on procedural grounds.

On the other hand, SATRA was supportive of Telkom when it outlawed the call-back operators that function worldwide as arbitageurs.

12.6 Privatization and Current Political Trends

The ANC-led Government of National Unity (GNU) inherited the issues left unresolved since the commercialization of the telecommunications industry. First, the direction of the industry had remained uncertain. Telkom was the monopoly provider of most telecommunications services, but a liberalized cellular market had claimed 400,000 subscribers by the end of 1995—10 percent of all the lines Telkom operated. Potential competitors in the equipment manufacture, cellular, and service industries demanded greater openness. And despite heightened demands for "universal access," no consensus emerged about the definition of the term.

Second, Telkom had essentially remained unregulated since its founding. Tariff setting and labor market policies, for example, had remained unchanged since the apartheid era. A new regulatory body was needed to create a framework for the growing competitiveness of the market, to determine its limits, if any, and to chart the appropriate course for the social and economic goals of the GNU.

Third, the role of Telkom itself needed to be clarified. The "commercialization" of Telkom was no more than a partial step taken as a political compromise. Issues such as its future ownership, its financing, and its obligations toward its workforce, consumers, and society remained unresolved. With its poor service record and inadequate investment in network expansion, Telkom was an unsuccessful agent of the government's goals. At the same time, its precarious financial position and inefficient operations rendered it a poor prospect for success in a competitive business environment. The Cooper's & Lybrand report framed the issue as, essentially, a choice between social responsibility and fiscal responsibility; the government could either expand its network rapidly to provide service to disadvantaged people, at a high cost, or proceed less aggressively, improve the financial performance of Telkom, and slowly establish a viable competitive environment.

Responding to political pressure from business, consumer groups, and labor, the ANC would attempt to pursue both objectives. Three events consequent to the 1994 elections began to move telecommunications policy in this direction. The long-delayed establishment of a National Telecommunications Forum (NTF) finally engaged government, business, labor, and civic organizations in the consultative processes that stakeholders in other sectors had established years earlier. A broad, Keynesian-inspired development plan for postapartheid South Africa (the Reconstruction and Development Programme or RDP) set goals for telecommunications development, including a specific goal that telephones be provided to all existing schools and health clinics within two years. And the new (ANC) minister for Posts, Telecommunications and Broadcasting, Dr. Z. Pallo Jordan, put in motion a Green Paper/White Paper process to develop telecommunications policy. This process, which would take place *outside* the old regime-identified Department of Posts and Telecommunications, tied the politics of telecommunications reform to the more open, consultative forums and mechanisms of which the NTF was the guiding spirit.

12.6.1 The Consultative Process

The Green Paper was published in July 1995 in several languages and put on the World Wide Web (Department of Posts, Telecommunications and Broadcasting 1995a). By October, when the period for submissions closed, 131 submissions had been made, amounting to over 4,000 pages of commentary. There were many points of consensus, including the need to expand the sector through new sources of investment; the need for operators, service providers, and equipment manufacturers to commit to retraining and redeployment of staff; the idea that universal *access* is a reasonable movement on the way to the goal of universal service; and the need for an independent regulator to supervise the system. On the major questions of market structure and ownership, there was a wide range of opinion, from the maintenance of state control to full competition and privatization. While most of the value-added network players, some important users, and potential competitors to Telkom (both South African and international) stressed that full-scale liberalization and the privatization of Telkom had to occur very soon, the labor unions and an important black business group (the Foundation for African Business and Consumer Services) held out strongly for minimal change and continued state control and ownership of Telkom. Telkom itself proposed a fairly long period of exclusive concession to enable it to mobilize capital and its large workforce in order to extend the network to the disadvantaged and rebalance tariffs to prepare for competition. It also advocated that it be permitted to take on strategic equity partners. The Department of Posts and Telecommunications submitted a proposal not dissimilar to Telkom's but with a much shorter period of exclusivity and a stronger endorsement of competition.

The NTF, which met in November 1995 as the National Colloquium on Telecommunications Policy, seemed unable to take a strong, coherent stand (Department of Posts, Telecommunications and Broadcasting 1995b). The colloquium's hoped-for "sufficient consensus" always meant that the key players—Telkom (which was owned by the government) and organized labor (which was allied with the government)—had to arrive at some accommodation for policy to move forward. Hence, notwithstanding the broad range of responses to the Green Paper that urged liberalization, the effective compromise had to include at least some period of exclusivity for Telkom in the public-switched voice and data networks. How much time Telkom would have exclusive reign, and when the network would open up in which market segments (and how such a process could be properly managed) were the key market structure questions. How Telkom would access sufficient capital to roll out the network while becoming more efficient in preparation for competition was the other key question.

The colloquium did and didn't arrive at sufficient consensus. A strong majority agreed that Telkom should have the primary role in universal service provision and that there would be no competition with Telkom during a three- to five-year period of exclusivity. The telecommunications infrastructure of other parastatals should be complementary, made available to Telkom under commercial conditions. Thereafter, the sector must be guided toward competition. The same majority backed Telkom's taking on strategic equity partners and/or private equity participation.

Labor continued to oppose the equity sale, supporting full state ownership and complete monopoly for Telkom indefinitely. These were "mandated" positions, laid out in the Green Paper submissions of POTWA and NUMSA, and could not be changed without further consultation. This development threatened to stymie further movement in the colloquium. The solution was to kick the problem upstairs. The question of market structure was referred to an Eminent Persons Group, a group of five "wise persons" to be nominated by the Colloquium Plenary to the Minister, whose main task, to oversee the Technical Task Team, embodied the letter and spirit of the colloquium in the writing of the White Paper. The ownership question, and whether Telkom could take on private equity, was determined to be a general government policy issue with ramifications across other state assets and was referred to Minister Jordan (Department of Posts, Telecommunications and Broadcasting 1995c). The specific question of Telkom's ownership was now recognized as just one part of a much larger debate taking place under the auspices of the Ministry for Public Enterprises on whether and how to restructure state-owned assets.

Labor's concerns were, and are, not trivial. Parastatal reform does pose dangers. Among these is the retrenchment of labor, and in a country with over 40 percent unemployment this is no small matter. Labor also harbors concerns that its collective bargaining position would be weakened consequent to any privatization. In addition, there is always the danger that a privatized monopoly, given new freedoms and having to pay dividends to shareholders, will stray from its supposed developmental mission. As an influential COSATU-associated report argued, privatization may bring greater commercial efficiency, but it is typically at the price of service delivery and developmental priorities. Because capital always attempts to maximize return on its investment, Telkom's private partner would surely push hard to increase tariffs, thus pricing the service beyond the ability of the disadvantaged to pay. Examining the privatizations in Eastern Europe, the report charged that schemes aimed at transferring ownership of state-owned corporations to disadvantaged communities have the consequence of leaving decision-making power with parastatal managers, investment funds, and stock brokers. Such schemes reduce state influence and enhance management's control. Similar criticisms were aired in the main publication of the South African Communist Party.

Notwithstanding the dangers of reform, *not* reforming poses even greater dangers. This was the understanding of at least some of the colloquium stakeholders who sided with the majority position. The abuses of monopoly providers in general, and Telkom in particular, were well understood. But politically motivated tariff schedules, poor service, and excessive labor costs were only part of the problem. The monopoly carrier was simply not equipped to supply the needs of consumers in an era of rapid technological innovation. At the time that the cellular market was liberalized, for example, Telkom provided service in a limited geographic area to 13,000 subscribers. Eighteen months later, 400,000 subscribers to the MTN and Vodacom networks could use cell phones in all but the most remote regions.

12.6.1.1 The Government Plan: The Reconstruction and Development Programme

The Reconstruction and Development Programme (RDP) dovetailed with developments that had been taking place on another front. It obliged Telkom to provide connections to schools and clinics—an obligation deceptively larger than it appears. Given Telkom's financial difficulties, the company forecast that it would need to borrow R6 billion over the next five years to fund its RDP obligations. The company put out a request for proposal (RFP) in June 1995 to provide 1 million lines in various rural areas where telephone density was below 5 percent. The government planned to work with two or more private partners, possibly chosen from the orginal equipment licensees. The government insisted on full ownership for Telkom, effectively precluding project financing schemes such as build-operate-transfer. Even so, the tender document stated that provision of the required infrastructure was to have no material effect on Telkom's debt/equity ratio.

Telkom learned some important things with the million-line RFP. First, it learned that the arrangement could not be done without some kind of equity transfer. The notion that it could gain an operating partner without this altering the company's debt/equity ratio was fantasy. Second, Telkom learned that an RFP on this scale brought much better prices than it saw previously—on the order of 30 to 40 percent lower. With these two lessons in tow, Telkom opened up the process to a much bigger affair. In November 1995, Telkom unveiled a broad expansion plan—which it called Vision 2000—of which the million-line tender was just a part. Scheduled over the next five years, the plan aimed at increasing the network by 75 percent with the addition of 2 million lines in rural and urban underserved areas and 1 million lines in areas that already have well-developed telecommunications infrastructure. Another 1.2 million lines would replace obsolete lines, aiming to complete the digitization of the network.

Telkom's normal rate of installation in the early nineties has been less than 200,000 new lines per year. The expansion of the network between 1994 and 1995 was at a rate of less than 5 percent. For the elections it was able to double that rate. (Telkom also services 800,000 transfers or disconnects per year—a very high figure.) Because some of the Vision 2000 rollout projects would be "turnkey" (that is, successful bidders would be responsible for end-to-end operation), particularly in areas where telecommunications expansion would be breaking new ground, Telkom's expectation is that it could install as many as 800,000 lines per year. This would amount to approximately a 15 percent rollout per year for five years. By way of comparison, network rollout in Mexico was contracted at a minimum of 12 percent per year, a target that was met and even slightly exceeded. Telkom's performance proved impressive. Between 1994 and 1997 the company added four million lines—an enormous increase—and cellular telephony reached one million subscribers in three years.

The financial terms of Telkom's Vision 2000 are sketchy and somewhat speculative. Telkom itself announced the cost—at today's prices and without provision for price escalation, costs of financing, and so on—at something in excess of R16 billion. This is probably too low. It means that the average cost per line would be less than R4,000, a figure never before seen in the South African context. The Department of Posts and Telecommunications estimated, conservatively, that the Vision 2000 project would cost R20 billion, and the department averaged every service to the lowest figure (Department of Posts and Telecommunications 1995). However, potential equipment suppliers confirmed that the R16 billion was a reasonable figure, in large part because of the deployment of cheaper radio-based technologies in local loops. More recent Telkom estimates put the figure at R30 billion.

Regardless as to the most accurate number, this is a very large amount of capital to be raised and deployed in a very short period of time. Where will the capital come from? Telkom's current capital investment of about R2.4 billion per year, generated from internal funds, expands the network at a rate of less than 200,000 lines per year. If the big rollout requires an average of 800,000 lines per year, it is reasonable to assume from these numbers (admittedly on a simplified basis) that nearly three-quarters of the new lines will have to be financed by new capital.

12.6.1.2 The Government Plan: The White Paper

Though Telkom's plans seemed optimistic, they were generally in line with the broad terms of the consensus reached at the colloquium, and this consonance entered into the first draft of the White Paper. The White Paper articulated a market structure scenario that gave Telkom essential exclusivity for five years, after which parts of the network would be opened up to phased entry and competition.

The specific timetable includes the following:

- In year one, fully deregulating the equipment market and ensuring that cross-subsidies flow only to noncompetitive activities (e.g., from business subscribers to low-income subscribers).
- In year three, issuing a third cellular license if appropriate.
- In year four, rebalancing Telkom's tariffs to remove cross-subsidies; allowing resale of excess capacity by private network carriers through interconnection with Telkom's network. All competitors would contribute to a Universal Service Fund that would be used to fund network expansion.
- In year six, allowing development of local loop service by local providers in cooperation with Telkom; liberalizing the long distance market; and allowing "metropolitan area networks," or networks of private networks providing service to major cities.
- In year seven, licensing a second network provider and liberalizing the international call market.

The White Paper market structure scenario reflects the particularities of the South African situation and is at the same time reasonably parallel to reforms undertaken in other countries. It gives the public telecommunications operator a period of time to expand even as it facilitates the growth of value-added and private-network entrants, and it phases in regulated competition. In this respect the plan commits the system toward a massive network rollout while also permitting the development of sophisticated value-added services through the phased regulation of competition—first at the margins, then at the core of the infrastructure. And the massive expansion of the network may alleviate the problem of potential labor retrenchments, inasmuch as network expansion will require considerable manpower. In the end, after considerable struggle, the decision was to limit the Telkom monopoly to five years (six if the terms of the license were closely met), until 2002 or 2003.

However, the plausibility of this policy is contingent upon two crucial things. First, the government must establish a competent regulatory authority very quickly, for the regulator will be expected to oversee many complicated processes and resolve a great number of technically intricate disputes. Guiding a sector into a competitive market structure is no easy task. Regulatory disputes are very likely to be dominated by Telkom, particularly at the beginning, inasmuch as it possesses most of the key information. The regulator will have the difficult task of monitoring Telkom's activities to distinguish between actions that facilitate the central goals of the sector in terms of reconstruction and development, and actions whose effect will position Telkom so powerfully as to undermine eventual competitors. The White Paper proposed the creation of a Universal Service Agency alongside the regulator to ensure that universal service remains a central focus. It also may be advisable to write as many obligations and understandings into Telkom's license as possible, particularly because it will take time to build capacity in the regulator. Performance and regulatory contracts have worked in some countries, Mexico being perhaps the best example. These would provide needed initial stability for the new market structure and at the same time afford the regulatory body time to build up its capacities. The regulatory agency created was the South African Telecommunications Regulatory Authority.

Second, assuming conservatively that Vision 2000 costs R16 billion, Telkom will need another R12 billion, or R2.4 billion per year, for the plan to work. Telkom cannot effectively expand considering its present financial state. It is the most highly "geared" (net debt as a percentage of shareholders' funds) of any telecommunications company in the world. Its capital expenditure is below other comparably situated national telecommunications operators, and this is reflected in the low rates of network growth. Without new capital, Telkom cannot expand the network in anything approaching the rapidity envisioned by the White Paper—a speed both expected by the South African majority and necessary if Telkom is to meet the challenge of competition from international operators.

12.6.2 Financing Reform

There may be several ways to infuse capital into a debt-burdened Telkom. Government funding is not an option at this time given the severe budget constraints facing the Government of National Unity. General sale of equity on the Johannesburg Stock Exchange is a possible option, but such a gambit risks replaying the consequences of the previous government's strategy because it would tend to concentrate ownership in the white elite. A giveaway of shares to favored, previously disadvantaged constituencies would increase social equity, but it would not raise capital for Telkom. Although the million-line tender has prohibited this option, build-operate-transfer (BOT) arrangements, which grant large, usually international contractors an exclusive right to a project's revenues under a time-bound concession agreement with the government, have worked in other countries, including, apparently, South Africa in the case of Water Sanitation Services South Africa's arrangement with the French company Lyonnaise des Eaux. However, governments are not always able to manage the impact of levies or tariffs on the public.

In 1996, the government enlisted an investment bank, Goldman Sachs, to help it find a strategic equity partner SEP. Such a partner was considered the most propitious way to facilitate an infusion of capital into Telkom. Not only would this raise capital quickly, but the participation of an international telecommunications operator would likely bring much-needed management and technological capabilities as well. The book value of Telkom's investments, including buildings and so on, is only about R15 billion, but the parastatal's potential market value is estimated between R31 and R35 billion. Telkom's Chairman of the Board Digang Moseneke was quoted in the press as looking for a strategic equity partner of between 20 to 30 percent. If Telkom's market value is, for purposes of argument, R30 billion, an SEP would represent a direct infusion of capital between R6 and R9 billion. Telkom would probably also borrow new money on the local and international capital markets, hoping to leverage the SEP's contribution toward a favorable interest rate. Even so, it seems that even with the SEP, Telkom will have difficulty funding its obligations under the RDP and Vision 2000 and becoming a viable competitor within five years. In 1997, the privatization process led to its cautious next step. Thirty percent of Telkom SA was sold to a private foreign consortium of a Western company (SBC, then the largest of the American Bell companies) and a non-Western firm (Telkom Malaysia). Soon, Telkom further integrated itself into the global framework of private telecom carriers by joining, in 1998, the AT&T-led global alliance World Partners. But its main focus was the development of the domestic network. Under its license, Telkom was required to install 2.8 million lines; 120,000 pay phones; 24,000 lines for schools and hospitals; and Internet access to 2,000 institutions serving the historically disadvantaged population. Furthermore, Telkom had to spend almost a half billion U.S. dollars on staff training, mostly for nonwhites, and hire at least one third of its manager from such population groups. All this was going to cost over $10 million, on top of the purchase price of $1.2 billion to SBC and Malaysia Telekom for their 30 percent of equity. To accomplish this goal, SBC brought in one of its managers, R. Geschwind, or the new chief executive of Telkom.

12.7 The Future

Continued sectoral reform requires the government and organized labor to come to a reasonable and rapid settlement of the political problem on the restructuring of state assets, a debate that pits the principal coalition partners of the ANC and COSATU against each other. The White Paper itself specifically leaves the question of ownership to future discussions, although by the end of 1995, the government had begun discussions with several international firms. In January 1996,

COSATU announced its intention to hold a general strike in response to the impending partial privatization. Fearing the economic and political consequences, the government agreed to delay the sale of equity during talks that resulted in the National Framework Agreement. This agreement permitted labor a voice in the restructuring process and some degree of control in any layoffs that would result. In exchange, COSATU dropped its unconditional opposition to privatization, though it continued to voice its preference for state control of its parastatals.

The telecommunications reform process was thrown into uncertainty due to the dismissal of Minister Pallo Jordan from the cabinet, effective April 4, 1996. The cabinet reshuffle also dismantled the RDP, distributing its functions to line ministries. Jay Naidoo, head of the RDP, was given Jordan's Posts, Telecommunications and Broadcasting portfolio. Speculation on Jordan's dismissal rested not on job performance so much as his maverick style and the fact that he had crossed President Mandela on several occasions.

Potential investors were initially alarmed at the elevation of the former labor leader and opponent of liberalization to a position responsible for the future of the telecommunications industry. Naidoo issued several statements proclaiming his commitment to the reform process, and analysts began to speculate that he might actually command a better bargaining postion from his former colleagues than the intellectual and apparently aloof Jordan. Draft legislation issued in May 1996 suggested that Naidoo would continue in the spirit of the NTF and the White Paper. The bill further delineated distinctions between policy, regulation, and operations by creating a regulator (the South African Telecommunications Regulatory Authority, or SATRA) and the Universal Service Agency and consolidating policy authority under the renamed Department of Communications.

SATRA exhibited independence from the government and from the major company, Telkom. Eventually SATRA will absorb the national regulatory body for the broadcasting industry, the IBA (Independent Broadcasting Authority). This move was designed to better reflect the convergence of data communications and broadcasting technologies that is occurring within the telecommunications industry. For example, folding SATRA and the IBA into one body, broadcasters who want to apply for a license to offer data services and Internet service providers who want to apply for a license to offer video via the Internet will not have to approach two separate regulatory bodies.

12.8 Conclusion

The Telkom debate is only a part of a greater debate over the future of the large number of state-owned enterprises in South Africa. At stake for the labor movement is not only lost jobs but also its long-term ability to control, through its traditional alliance with the ANC, large portions of South African industry. Private ownership of assets means a loss of political clout and collective bargaining strength. Investors, however, demand control in order to initiate cost-cutting measures and improve efficiency. As a financial investment, Telkom is a weak candidate (its projected rate of return is lower even than short-term interest rates), and its appeal lies in its promise of future access to the country and the rest of Africa as the market grows. To take advantage of these potential opportunities, multinationals consider some degree of strategic control to be crucial; none wishes to share control with both government and labor.

The ANC-led government, therefore, must satisfy the demands of two seemingly incompatible forces: organized labor, which has the power to bring the economy to a standstill; and the international investment community, without which none of the RDP goals for telecommunications, a viable competitive industry, or the continued financial health of Telkom would be possible. Policy makers continue to seek solutions and pursue agreements with labor and business, but in the meantime progress must be made on the promise to bring telecommunications to the disadvantaged by the year 2000. Given South Africa's moral stature and economic power, all of Africa is watching.

Notes

1. At an exchange rate average of 3.016 Rand per U.S. dollar in 1991, South Africa's GDP translated to U.S.$107.46 billion (Economist Intelligence Unit 1992, p. 3).

2. Constitutional changes in 1984 established a tricameral Parliament: a House of Assembly for whites, a House of Representatives for Coloreds, and a House of Delegates for Indians. The African majority remained excluded. A move from the former all-white, prime ministerial "Westminster" system to a strong presidency was designed to co-opt any genuine participation by Colored and Indian representatives. Among other features, a fixed 4:2:1 numerical ratio of white, Colored, and Indian representation in Parliament ensured white hegemony (Schrire 1991, pp. 59–69).

3. The Development Bank of Southern Africa estimated in 1989 that there were about 7 million people living in informal dwellings (reported in *The Star* 1989).

4. It should be noted that the processes of dismantling apartheid and exploring power-sharing options were begun under de Klerk's predecessor, P. W. Botha. Moreover, there is evidence that extensive deliberations on the end of white rule were held within the Broederbond, that secret Afrikaner fraternity, as early as 1986 (Keller 1992). The Broederbond, which began as a small Afrikaner cultural club in 1918 and grew into a formidable elite organization that sought to unite and mobilize Afrikaners against British economic power, was the closest thing to a "ruling class" one could find (see O'Meara 1983).

5. "State business enterprises" (in official government reports sometimes also referred to as "public authorities") can be considered the South African version of the more generic "state-owned enterprise," or SOE. In South Africa, "state business enterprises" were distinct from "public corporations." While both were public, as opposed to private, enterprises, public corporations were established by special legislation (hence were not chartered under the South African Companies Act) and were granted more autonomy than were state business enterprises. State business enterprises were run through ministries whereas public corporations were not. Although the SAPT was a state business enterprise, the Electricity Supply Commission (Eskom), in contrast, was a public corporation. The generic term that describes these publicly owned, utility-type enterprises, be they public corporations or state business enterprises, is "parastatals."

6. The lines in service per employee rose to about 58 in 1993 (Telkom 1993).

7. The policy of reserving certain skilled and semiskilled jobs for whites was partially ended in 1979. With the exception of the mining industry, the remaining categories were abolished in 1983 (Schrire 1991, p. 73). However, according to one source, there was still a

fair amount of salary disparity by race in the SAPT as late as the late 1980s, particularly in the lower salary grades ("Posts and Telecommunications" 1986, pp. 52–53).

8. The South African economy is characterized by extraordinarily concentrated ownership, organized into a small number of vast conglomerates. Until recently, when the Sanlam Group announced it would break up its industrial centerpiece, Gencor, just four colossal corporate groups controlled more than 80 percent of the value of all the stocks listed on the Johannesburg Stock Exchange. This concentration of economic power in English or Afrikaner conglomerates was another way in which white power historically manifested itself in South Africa—a kind of white cartel ("Gencor to Proceed with Unbundling" 1993; Keller 1993). In 1993, Gencor announced it would unbundle R6.9 billion of its R21 billion net asset value ("Gencor Stops Short of Total Unbundling" 1993).

9. The exception was the supply of underground copper wire cables, which was a competitive business. The SAPT purchased this cable on an approved list-tender basis ("Posts and Telecommunications" 1987, p. 21).

10. Moreover, according to one high-level Telkom manager (in a personal communication with the author in November 1991), local content may have been calculated fraudulently—at least in the case of one of the key suppliers. The designated company got subsidiaries to buy imported equipment, then turned around and showed this equipment as locally produced. This, of course, artificially boosted the percentage of local content. The company's high profits meant equivalent losses for the SAPT.

11. The costs of any innovation or product development carried out by companies for the SAPT were paid for by the SAPT as a separate item with an allowance for profit. Thus the companies bore no risk in the event of such innovation or development proving to be abortive. This feature of the agreements was intended to encourage innovation and product development. But because the decision whether to proceed with developing a new product was in effect removed from the contractor and placed with the SAPT, it may have actually reduced incentives for product development.

12. The growth in exports can be attributed to the following factors: (1) Telkom orders are down, pushing suppliers to become more export oriented; (2) Telkor won an important contract to supply pay phones to Hungary; (3) with the reunification of Germany, the underutilized Siemens South Africa factory could be used to service this new market; and (4) Alcatel, seeing Altech as its springboard into Africa, has taken a larger stake of Altech and pushed it to look to neighboring African markets.

13. Data on telephone penetration by race are now difficult to obtain. Telkom claims that it no longer compiles data on penetration according to race—that this kind of racial orientation is no longer acceptable in the "new South Africa." The 1989 data on penetration cited here come from Coopers & Lybrand (1992, p. 8). The following table provides a more detailed picture of penetration, showing telephone density per 100 residences by population group between 1978 and 1987.

Number of Telephones per 100 Residences

				Blacks	
	Whites	Asians	Coloreds	Metropolitan	Rural
1978	71.5	36.1	19.3	3.3	1.8
1982	83.3	61.5	46.2	24.0	8.3
1987	83.9	72.2	53.2	38.0	13.7

Source: De Villiers 1989, concise version, p. 3.

Note: Because density is measured by household, there is a built-in skew, inasmuch as household size varies by race—blacks have the largest, followed by Colored, Indian, and white, in that order.

14. As the Coopers & Lybrand report notes, while the white and black communities show levels of telephone penetration that are broadly consistent with the observed relationship between telephone penetration and per capita income, there are a number of countries that have similar per capita incomes to the black community in the Republic of South Africa but much higher telephone penetration. The Coopers & Lybrand report suggests that there is considerable suppressed demand in the black population (1992, p. 9).

15. Apartheid's legacy to infrastructure development can be seen in the fact that, while telephone service was virtually unavailable to rural Africans, it was extended to white farmers—and on a highly subsidized basis. Because of the historic economic and political power of the white farming community, the average white farmer had access to health, education, electricity, transport networks, and a telephone as a result of a commitment from the state to provide whites with such services through public-sector provision. According to one 1991 account, if tariffs for telephone service to the farming community were to be levied on a cost-related basis, the basic rental would increase from R24 to R114 per month. This compares to the calculation of an average monthly cost-related rental of R51.80 (Preiss 1991). Lest this read as a blanket statement, it should be acknowledged that some areas of black urban settlement, such as Soweto, have considerable infrastructure in place.

16. Interestingly, though, in the end the legislation was opposed only by POTWA, the black Postal and Telecommunications Workers Association, and the Conservative Party (which is well to the political right of the National Party). The union had opposed the original proposal to privatize the SAPT. It opposed the Post Office Amendment legislation because it felt it had not been properly consulted and held serious concerns over likely retrenchments and the commitment to universal service. The Conservative Party (CP), in some contrast, opposed commercialization because it believed this would jeopardize the provision of cheap communication services to its rural Afrikaner constituency. The CP also believed that commercialization would enhance the recent trend of government dismantling of the job reservation system, a system the CP heartily supported (Republic of South Africa 1991a, 1991b).

17. It should be noted, however, that notwithstanding its parliamentary form of government, South Africa is a state characterized by a strong executive. The state president and his ministers hold the bulk of political power. Hence vague legislation may have the consequence of giving the minister of Transport and Communications great discretionary authority to set basic policy.

18. In 1993, the minister of Transport and Communications administered both the policy and shareholding portfolios because the minister of Minerals and Energy Affairs and Public Enterprises, Dawie de Villiers, was too busy with his normal duties and his role as a chief negotiator in constitutional talks (Webb 1993).

19. The understanding of "representation," however, must be appreciated. Pretorius did not even like the word, because it carried with it the implication that commissioners would be representative of political parties. On the contrary, he warned, commissioners should be removed from politics. They should be "wise men" (sic)—not necessarily experts, but with access to experts on the agency staff. To ensure independence, commissioners would be appointed by the state president and could be dismissed only by Parliament, essentially only for cause. This model is similar to the protection of judges. There would be no public vetting upon appointment, due to the "political circus" this process encourages. The commission would have large powers to determine policy, though aggrieved parties could appeal decisions to court. Present grounds for judicial review are limited in South Africa, but Pretorius predicted that the South African Law Commission would liberalize and broaden the grounds of review to consider the "reasonableness" of agency decisions (Pretorius 1992).

20. And there is some evidence that the members of the progressive alliance in the Codesa Working Group had little understanding of the technical and economic issues involved in telecommunications or how broadcasting and telecommunications interact. This point was made by Michael Markowitz of the Film and Allied Workers Organization, which had been commissioned to comment on the government's communications regulation proposals for the ANC. Markowitz made this claim at a communications workshop sponsored by POTWA in Johannesburg, July 10, 1992. He voiced the opinion that with regard to telecommunications, the breakdown in Codesa talks may have been a blessing in disguise. In his estimation, the progressive alliance ran a serious risk of agreeing to things it did not understand fully and might well regret later. As for the actual content of the government's first proposals on the regulatory body at Codesa, Markowitz claims that 60 to 70 percent dealt with the problem of pensions (POTWA Communications Workshop 1992).

21. Codesa collapsed amid demands by the National Party alliance for very high majorities to pass legislation in the new government—interpreted by the ANC alliance as embedding a white veto. The second reason for the ANC-SACP-COSATU's suspension of its participation revolved around the intensity of township violence, which the progressive alliance attributed in part to the police and defense forces.

22. As more than one source explained, the ANC and POTWA were happy to meet with various players on a one-on-one basis, but they rarely attended the general meetings. The ANC alliance wanted forums to be seen as its initiative. Moreover, the ANC consistently refused to participate in a meeting or forum organized by the government or Telkom. This position has long been held by the ANC's chief player in the telecommunications debate, Information Systems head Andile Ngcaba (Ngcaba 1992, 1993a). Telkom participated in the forum, but rarely sent representatives with real managerial weight. Other participating parties had doubts about the propriety of BMI's sponsorship of the forum, citing potential conflicts of interest due to its business interests. This atmosphere of suspicion likely was an additional factor in undermining the success of the forum.

23. The evolution of the Policy Unit is revealing. During the period of the government's overt championing of privatization, it created a new ministry, the Ministry of Administration and Privatisation. Reflecting the political sensitivity of the idea of privatization following the February 1990 legalization of the ANC, the ministry's name was changed to the Ministry of Administration and Economic Coordination in March 1990. The privatization option was not dropped in toto, just made less visible. It was then embodied in a privatization policy unit of the ministry. A few additional ministerial changes later, a far less public Office of Privatisation operated under the auspices of the Ministry for Minerals and Energy Affairs and Public Enterprises. Thereafter the office's name was adjusted once again, to the Policy Unit on Public Enterprises (see Republic of South Africa, *Debates of Parliament (Hansard),* various years).

24. Projections on the number of cellular subscribers vary. Coopers & Lybrand estimates 160,000 to 220,000 subscribers by the year 2002; M-Net Communications Technologies chief executive Ian Wilkinson calculates a potential base of 300,000 to 500,000 subscribers; Vodacom managing director Allan Knott-Craig projects a potential cellular market size of 1 million subscribers over the next five years ("Dial-a-Profit" 1993, p. 15). Another figure that has been floated in the press is that 70 percent of South Africa's population will have access to cellular phones within five years. Because worldwide industry figures show more than 70 percent of all cellular calls pass through the national carrier's network, Telkom's revenues would not suffer. Vodacom and MTN say that the cellular industry could create some 70,000 jobs indirectly and 4,500 directly within their companies ("Cellular Phones" 1993, pp. 10–12).

25. Whether this observation was a comment on Welgemoed's racial politics or rather a

comment on his sense of his duly constituted authority as minister of Transport and Communications is hard to say. But Postmaster General Oosthuizen clearly evinced at least the latter in an interview. He bristled when I suggested that the Negotiating Forum at Kempton Park was now in effect the interim government and that making consequential decisions for the telecommunications sector outside the evolving constitutional framework may not be a good idea (Oosthuizen 1993).

26. The Congress of South African Trade Unions claims that because the cellular tenders are being considered on their own, outside a strategy for associated economic or industrial development, they represent only the expression of an abstract hope that privatization will provide for universal service. The tender document's requirement for views on how applicants will provide service for disadvantaged communities thus is not a strategy but an afterthought (Fanaroff 1993b; Naidoo 1993).

27. Composition of the cellular ventures is as follows.

- Vodacom: Telkom (50 percent), Vodafone (35 percent), Rembrandt (15 percent).
- MTN: M-Net (30 percent), Cable & Wireless (30 percent), Naftel (30 percent), Transtel (10 percent).
- Cellstar: Grintek (Anglovaal) (51 percent), Telecom Finland (11 percent), Finnfund (11 percent), black business (unallocated) (27 percent).
- Reunert: Barlow Rand (34 percent), Deutsche Bundespost Telekom (34 percent), black business (unallocated) (32 percent).

28. Temple, a strong advocate of an open tender system on all telecommunications equipment, told me that Telkom asked Ericsson about its cellular base stations, then showed the figures to Siemens and Altech. The traditional suppliers then undercut Ericsson's price by 2 percent. But, according to Temple, the Ericsson base station is much larger and more powerful—so the comparison is apples and oranges. Ericsson is by far the largest cellular provider worldwide. Temple adduces that notwithstanding the commercialization of Telkom, the cozy and nontransparent relationship between Telkom and its traditional suppliers continues.

29. According to Knott-Craig, Vodacom's community service comprises the provision of at least 22,000 subsidized community phones to more than sixty townships that have already been identified for this service over the next five years ("Cellular Phones" 1993).

30. Knott-Craig's rejection of the claim that GSM call charges would be too expensive inadvertently points to another problem. Claiming that calls made from one cellular community telephone to another could cost less than 20 cents a minute—independent of distance—Knott-Craig ignored the effect on Telkom's public-switched telephone network (PSTN). The common riposte is that a successful cellular service would cause telephone traffic to increase in absolute terms and hence Telkom would benefit from interface charges. But Knott-Craig made an interesting comment (perhaps a slip) in an interview. He talked about Vodacom being 50 percent owned by the "government," not by Telkom. And, good entrepreneur that he is, Knott-Craig looked forward to competing with Telkom, expressing confidence that Vodacom could be bigger than Telkom in several years (Knott-Craig 1993b). Two things should be noted. First, interface charges between cellular services and the PSTN could undermine visions of a cheap cellular service. Second, in a competitive environment the threat of cellular bypass to the PSTN may be quite real.

31. The effect of cellular on retrenchments at Telkom greatly concerns COSATU. After all, cellular could in principle take revenues away from Telkom. Also, COSATU is more concerned than is the ANC about requiring foreign contractors to source some components within South Africa (Naidoo 1993; "COSATU Calls for Delay in Cellular Telephone Licenses" 1993). Here some of the fissures in general economic policy between the ANC

and COSATU are apparent. The ANC's Ngcaba is not convinced that the old Telkom Supply Agreements shouldn't be scrapped altogether—the protection of a few jobs in South Africa may not be worth the savings that could be gained from buying cheap foreign-made equipment. This, Ngcaba averred, required study (Ngcaba 1993a). The ANC's economists have been more sanguine about basic capitalist macroeconomics, some going so far as to embrace versions of trickle down theory. In contrast, COSATU was more forthright about the need to spur manufacturing in South Africa and create domestic employment. Its policy is captured by its slogan of "growth through redistribution." In return for COSATU's support at the polls, the ANC has agreed in principle to adopt COSATU's reconstruction and development program ("ANC to Back COSATU's Reconstruction Plan" 1993).

32. The latter argument is a red herring, however, for several potential investors from the United States had made noises about an extension to the tender deadline. They were enjoined from putting forward a bid until sanctions were dropped. Indeed, there was some speculation in the industry that the ANC pushed hard to stop the tender process because it wanted a way to reward its foreign backers during the sanctions period (Makhanya 1993). On the other side, there were rumors that the government pushed ahead with the cellular tenders because it was under strong pressure from Britain on behalf of Cable & Wireless, Mobile Telephone Network's British partner.

33. This point raises the question as to whether the conclusions of the De Villiers Report were warranted. The SAPT was the last of de Villiers's investigations. According to W. J. Taylor, former deputy postmaster general for telecommunications, after finding real problems with the electricity and transportation parastatals, de Villiers came to the Post Office investigation with preconceived notions and, by then, a set agenda. Grossly ignorant of telecommunications, de Villiers insisted that the SAPT's decision to go digital had been a mistake and was a gross instance of goldplating. Yet nearly universal expert opinion argues that digital is more flexible and in the long run cheaper than electromechanical systems (Taylor 1992).

34. The waiting list reflects actual applications, hence it mainly reflects demand in areas where there is some infrastructure. The list does not adequately take underdeveloped areas into account.

35. The press minced no words in attacking Telkom. Typical was this editorial from the *Cape Times*. "The telephone service was far from perfect when run by the Post Office, earning its fair share of complaints from the public. But never did it arouse so enraged and sustained an outcry as Telkom has within a short time of taking over the service. . . . If this is the result of what is laughably called privatisation or commercialisation, then the telephone service should be re-nationalised without delay. Any private company which abused its clients in this way would quickly go out of business. But Telkom has no competition. Whatever similarity exists between its monopolistic stranglehold on telephonic communication and private enterprise is in the imagination only of the government which foisted it on the public" (*Cape Times* 1992, p. 8).

36. The uncertainty with regard to this conventional analysis stems from the difficulty of assessing black practices in the use of telephones. Telkom argued, for instance, that Soweto telephones do not generate much call traffic or revenue. But according to Alan Paul of BMI TechKnowledge, many Soweto telephones are routinely and continuously utilized for receiving calls. The revenue is generated elsewhere, but the Soweto phones—hence, by implication, black users—generate more call revenue than is ordinarily reported. Paul quoted a figure of 7 percent underreported (Paul 1992). The general point comports with the excellent ethnographic/sociological study of the rural-urban link in African telephone use by Mike Morris and S. E. Stavrou of the University of Natal. Morris and Stavrou argue that the strong rural-urban link between African family members leads to significant telephone use.

But because the rural dwellers do not have private telephones, the calls between city and village originate from pay phones located in rural areas (Morris and Stavrou 1991, 1990).

37. Until 1993, contrary to most accepted accounting practices, Telkom included the value of fully depreciated assets in its gross book value.

38. This ratio apparently is dropping. The net debt to equity ratio was down from 1.8 to 1 for 1994.

39. Some knowledgeable observers even suspect that much of the R40 million in actuality went to expand telecommunications facilities to white farmers and tourist-oriented game parks. Indeed, Telkom's annual report states that the company provided "about 300 community telephone services and purchased infrastructure equipment which will allow for the provision of another 1000 services in the first half of the 1993/94 financial year" (Telkom 1993, p. 25). This is hardly an awe-inspiring pace. Notwithstanding Telkom's difficult financial situation, its slowness in expanding service to the black population is disturbing. Concrete technology and sociological research have been available to Telkom to put in place a modest "agency" telephone system in townships for several years. The research of Mike Morris and S. E. Stavrou establishes the workability of authorizing a township telephone customer to operate an agency telephone out of his or her house. Temsa constructed a special meter that indicates the real-time cash charges of any call, thereby making charges transparent to the customer. According to Karl Hartyani of Temsa, Telkom made no commitment to the agency telephone (Morris and Stavrou 1991; Hartyani 1993).

40. Though the Post Office initially questioned the policy, Eskom and Transnet were permitted to provide their own communications links in a 1993 decision. In the case of Transnet, the cabinet determined that it was better for the railroads to take care of their own signaling because of accidents and related possible third-party problems.

41. In addition to the provision of redundant networks for large users, Transtel seeks to integrate Transnet's transport value chain, bringing Transnet's major clients directly into the company's information-communication-logistics network. And where it has infrastructure in place, Transtel would like to provide telephone service to those without. This not only would provide a public service but would make more efficient use of Transtel's plant (Webb 1993). Much of Transtel's commercial aggressiveness can be attributed to the fact that because of the terms of Transnet's commercialization, Transtel must compete with Telkom for Transnet's business; it cannot compete with Telkom outside of Transnet. Lindo Carvel Webb claims that Transtel has had a "galvanic" effect on Telkom with regard to Telkom's service. Telkom now services customers much more quickly because of the competitive pressure from Transtel (Webb 1993).

Bibliography

Adam, Heribert, and Kogila Moodley. 1992. "Political Violence, 'Tribalism,' and Inkatha," *The Journal of Modern African Studies* 30: 485–510.

ANC. 1992. "ANC Position on Telecommunications Sector Strategy Study Report for the Department of Posts and Telecommunications." Press release, August.

"ANC to Back COSATU's Reconstruction Plan." 1993. *Business Day,* September 8. Reported in *This Week in South Africa,* September 7–13.

Averch, Harvey, and Leland L. Johnson. 1962. "Behavior of the Firm under Regulatory Constraint." *American Economics Review* 52 (December): 1052–1069.

Battersby, John. 1990. "ANC Tempers Hard-Line Rhetoric on Economic Policy." *The Christian Science Monitor,* May 10.

BMI TechKnowledge. 1992. "Prospects for the Electronics Industry." 1992 ed. Johannesburg: BMI TechKnowledge.

Cape Times (Cape Town). 1992. June 15.
"Cellular Phones." 1993. *Business Day* Survey, October 6.
"Compromise Deal on Cellular Phones." 1993. *Business Day,* September 30. Reported in *This Week in South Africa,* September 28–October 4.
Coopers & Lybrand. 1992. "Telecommunications Sector Strategy Study for the Department of Posts and Telecommunications." Pretoria: Government Printer.
"COSATU Calls for Delay in Cellular Telephone Licenses." 1993. Press release, August 31.
Davies, Peter. 1992. Computer network and telecommunications consultant, interview with author, June 10.
______. 1993. Interview with author, August 31.
Department of Posts and Telecommunications. 1992a. "Annual Report of the Postmaster General of the Republic of South Africa, 1990/91." Pretoria: Department of Posts and Telecommunications.
______. 1992b. "Estimates of Revenue and Expenditure for Year Ending March 31, 1992." Pretoria: Department of Posts and Telecommunications.
______. Various years. "Annual Report of the Postmaster General of the Republic of South Africa." Pretoria: Department of Posts and Telecommunications.
De Villiers, W. J. 1984. "Report of the Commission of Inquiry into the Supply of Electricity in the Republic of South Africa." Pretoria: Government Printer.
______. 1986. "Condensed Report of the Study by Dr. W. J. de Villiers Regarding the Strategic Planning, Management Practices and Systems of the South African Transport Services." Pretoria: Government Printer.
______. 1989. "Summarized Report on the Study by Dr W. J. de Villiers Concerning the Strategy, Policy, Control Structure and Organisation of Posts and Telecommunications" (also released in "Concise Version"). Pretoria: Government Printer.
"Dial-a-Profit." 1993. *Finance Week,* September 23–29.
"Dial for Fair Play." 1993. *Finance Week,* January 21–27.
Duch, Raymond M. 1991. *Privatizing the Economy: Telecommunications Policy in Comparative Perspective.* Ann Arbor: University of Michigan Press.
Economist Intelligence Unit. 1992. "EIU Country Report: South Africa." London: Business International.
Fanaroff, Bernie. 1993a. National Secretary of Organizing Department, National Union of Metalworkers South Africa (NUMSA), interview with author, August 31.
______. 1993b. The Role of Telecommunications in Economic Growth: Projects for Southern Africa. Paper to the Telecommunications Regulatory Symposium sponsored by the Centre for the Development of Information and Telecommunications Policy (CDITP), Johannesburg, July 28–30.
"Gencor to Proceed with Unbundling." 1993. *Business Day* (Johannesburg), May 12. Excerpted in South African Consulate General, *This Week in South Africa* 4, no. 6 (May 11–17).
"Gencor Stops Short of Total Unbundling." 1993. *Business Day,* August 5. Reported in *This Week in South Africa,* August 3–9.
"Government, ANC Impasse on Cellular Phone Controversy." 1993. *Citizen,* September 15. Reported in *This Week in South Africa,* September 14–20.
"Government 'Will Not Back Down' on Phone Licenses." 1993. *Business Day,* September 13. Reported in *This Week in South Africa,* September 7–13.
Hainebach, Geoff. 1993. Joint Managing Director, Siemens Ltd, interview with author, September 6.
Hartyani, Karl. 1992. Technical Executive, Temsa, interview with author, July 23.

______. 1993. Interview with author, August 31.

"Holding the Airwaves Hostage." 1992. *Financial Mail* (Johannesburg), June 12.

Horwitz, Robert B. 1992. "The Politics of Telecommunications Reform in South Africa." *Telecommunications Policy* 16 (May/June): 291–306.

Independent Broadcasting Authority Bill (final draft). 1993. Prepared by the Technical Committee on the Independent Media Commission and the Independent Telecommunications Authority, Multi-Party Negotiation Process, September 2.

International Telecommunications Union. 1992. *Yearbook of Common Carrier Telecommunication Statistics.* 19th ed., Chronological series 1981–1990. Geneva: International Telecommunications Union.

Kaplan, David. 1990. *The Crossed Line: South African Telecommunications Industry in Transition.* Johannesburg: Witwatersrand University Press.

______. 1992. "The Development of Telecommunications in South Africa: The Equipment Supply Industry." *Critical Arts* 6: 96–107.

Keller, Bill. 1992. "Will Apartheid's Inner Circle Open Up to Blacks?" *The New York Times,* December 1.

______. 1993. "Cracks in South Africa's White Monopolies." *The New York Times,* June 17.

Knott-Craig, Alan. 1993a. "Cellular Phones the Affordable Solution." *Business Day,* September 1.

______. 1993b. Chief Executive of Vodacom, interview with author, September 3.

Knott-Craig, Alan, and Alwyn Hanekom. 1990. "Report on the Deregulation of Telematics in South Africa." Pretoria: SAPT.

Lachenicht, M. K. 1991. "ISDN and Centrex for South Africa." In *Proceedings of the Fifth National Conference on Telecommunications in South Africa,* Republic of South Africa, November 4–6.

Maepa, L. H. Nape. 1993. Non-Telecom Services and the Use of the Frequency Spectrum: Prioritisation, Management, and Regulation. Paper to the Telecommunications Regulatory Symposium sponsored by the Centre for the Development of Information and Telecommunications Policy (CDITP), Johannesburg, July 28–30.

Makhanya, Mondli. 1993. "Playing Broken Telephones." *The Weekly Mail and Guardian,* September 17–23.

Morris, Mike, and S. E. Stavrou. 1990. "The Socio-Economic Need for a Rural-Urban Telecommunication Link: A Case Study of Kwazulu/Natal, South Africa." Paper presented at the Second International Conference on Rural Telecommunications, The Institute of Electrical Engineers, London, October 29–31.

______. 1991. Rural Communications. In *Proceedings of the Fifth National Conference on Telecommunications in South Africa,* Republic of South Africa, November 4–6.

Naidoo, Jay. 1993. "Cellular Telephone Proposals Ring Impractical." *Business Day,* August 31.

National Telematic User Group (NTUG). 1991. "Response to Report on the Deregulation of Telematics in South Africa." Johannesburg: NTUG.

"Networking Pays Off." 1993. *Financial Mail,* June 25.

Ngcaba, Andile. 1992. Department of Information Systems, African National Congress, interview with author, June 8.

______. 1993a. Interview with author, September 1.

______. 1993b. "Hah! Government Caught in the Act." *The Weekly Mail and Guardian,* September 17-23.

O'Meara, Dan. 1983. *Volkskapitalisme: Class, Capital and Ideology in the Development of Afrikaner Nationalism, 1934–1948.* Cambridge: Cambridge University Press.

Oosthuizen, Ters. 1993. Republic of South Africa Postmaster General, interview with author, September 3.

Paul, Alan. 1992. Director, BMI TechKnowledge, interview with author, July 8.

Plessey Tellumat. 1995. Data on telecommunications in South Africa.

"Posts and Telecommunications." 1986. A Survey Supplement to the *Financial Mail* (Johannesburg), October 3.

"Posts and Telecommunications." 1987. A Survey Supplement to the *Financial Mail* (Johannesburg), October 23.

POTWA Communications Workshop. 1992. Johannesburg, July 10.

Preiss, W. 1991. The Tariff Policy of Telkom SA. In *Proceedings of the Fifth National Conference on Telecommunications in South Africa,* Republic of South Africa, November 4–6.

Pretorius, Pierre. 1992. Adviser to the minister of Transport and Communications and the designated author of future legislation to restructure telecommunications and to establish a telecommunications regulatory authority, interview with author, July 23.

______. 1993. Interview with author, September 6.

Price, Robert M. 1991. *The Apartheid State in Crisis: Political Transformation in South Africa, 1975–1990.* New York: Oxford University Press.

Proceedings of the Fifth National Conference on Telecommunications in South Africa. 1991. Republic of South Africa, November 4–6.

"Recession Leaves 46 percent without Jobs." 1993. *Business Day,* August 23. Reported in *This Week in South Africa,* August 17–23.

______. 1991a. *Debates of Parliament (Hansard).* "Post Office Amendment Bill," March 15.

______. 1991b. *Debates of Parliament (Hansard).* "Post Office Appropriation Bill," February 28, and March 1.

______. 1991c. *Government Gazette.* Act No. 85, Post Office Amendment Act of 1991, June 19.

Rustomjee, Zav. 1992. The Boundaries of the South African Mineral Energy Complex: The Implications for Manufacturing Led Growth Strategies." Paper presented to the Economic Trends Research Group, University of Cape Town, June.

Schrire, Robert. 1991. *Adapt or Die: The End of White Politics in South Africa.* New York: Ford Foundation and Foreign Policy Association.

Schulze, Alf M.E. 1993. Executive Manager, Transtel, interview with author, September 6.

Schulze, Peter. 1992. General Manager of Group Strategic Planning, Siemens Ltd., interview with author, July 24.

Segal, Simon A. 1994. "The R135-bn RDP Shock." *The Weekly Mail and Guardian,* June 3–9.

Sergeant, Melanie. 1993a. "Are Cellular Phones Just Another Party Line?" *Business Day,* September 2.

______. 1993b. "War of Words Over Telephones." *Computers and Telecommunications in Africa* 7, no. 4 (July).

Siemens. 1992. "1992 International Telecom Statistics." Munich: Siemens Aktiengessellschaft.

Smit, Dennis. 1993. Director of BMI TechKnowledge, interview with author, September 1.

South Africa Institute of Race Relations. 1992. *Race Relations Survey 1991/92.* Johannesburg: South Africa Institute of Race Relations.

"Squeezing the Cellular-Phone Industry." 1993. *Financial Mail,* April 16.

The Star (Johannesburg). 1989. October 26.

Statutes of the Republic of South Africa. 1958. Post Office Act, No. 44 of 1958, "Act to consolidate the laws relating to the Post Office and matters incidental thereto." Pretoria: Government Printer.

Strachan, Kathryn. 1992. "Telkom Rings the Changes in Phone System." *Business Day* (Johannesburg), September 11.

Taylor, W. J. 1992. Former Deputy Postmaster General for telecommunications, interview with author, July 7.

Telkom. 1991. "Profile." Pretoria: Telkom.

______. 1993. "The First Annual Report, 1991-10-01 to 1993-10-31." Pretoria: Telkom.

Temple, John. 1993. Managing Director of Plessey Tellumat, interview with author, August 26.

Transtel. n.d. "Profile of Transtel."

Viljoen, Christo. 1991. "Report of the Task Group on Broadcasting in South and Southern Africa." Pretoria: Government Printer.

Webb, Lindo Carvel. 1992. Deputy Executive Manager, Transtel, interview with author, July 23.

______. 1993. Interview with author, September 6.

Welliscz, Stanislaw H. 1963. "Regulation of Natural Gas Pipeline Companies: An Economic Analysis." *Journal of Political Economy* 71 (February): 30–43.

Young, Len. 1993. AT&T International Public Affairs Director for Sub-Saharan Africa, telephone interview with author, October 5.

13

Namibia's Telecommunications: The Link to Africa

KLAUS DIERKS

Communications links are the lifelines of Africa. The Republic of Namibia has developed a strong commitment to optimized communications policies in the institutional and technical fields in order to secure sustainable developments in the economic sector.

Namibia still has a structurally heterogeneous, de-integrated economy. It is characterized by the extraction and export of the country's rich mineral, agricultural, and fishing resources, but there is an impoverished, subsistence economy in many areas that serves the interests of the "modern-sector" economy. Northern Namibia, in fact, is a "residual" labor reserve with little cash income or commercialized production, so that it has little effective demand for goods and services, including telecommunications facilities. This situation is characteristic of an important aspect of Namibian communications: its imbalance between the "modern" and the former "homeland" sector. One of the principal objectives of the Namibian Ministry of Works, Transport, and Communication is to overcome this situation and develop an optimal and well-balanced communications system. Such a system will help create a structurally sound and integrated economy and help balance the "first-world" and the "third-world" Namibias.

At the time of Namibia's independence on March 21, 1990, the country had approximately 4.00 telecommunications lines per 100 inhabitants, compared with 0.72 for Kenya, 0.55 for Ghana, 1.26 for Zambia, and 0.20 for Ethiopia. Namibia also had automatic links to more than 150 countries worldwide. This situation was improved even further after independence, making Namibia one of the leaders in telecommunications in Africa. The progress in the telecommunications field was reinforced by the bridging of the "two Namibias," the modern-sector Namibia and the neglected areas, as well as the creation of links between Namibia and her landlocked neighbors in the east. Namibia's telecommunications institutional and technical system can serve as an example to other African countries, thus proving that there is less justification for Afro-pessimism today than there was in the 1980s.

13.1 The Past

13.1.1 A Geographic Country Profile

Namibia is a semiarid to arid country with a size of 824,292 square kilometers. It is located in the southwestern corner of Africa and has a total population of a little more than 1,500,000 inhabitants (1994), one of the most thinly populated countries in the world. The country is confined between two deserts, the Kalahari desert in the east and the Namib desert in the west, with an African savannah plateau in the middle (1,000 to 2,579 meters high). The capital is Windhoek, with approximately 150,000 inhabitants and an average altitude of 1,600 meters.

13.1.2 A Brief Overview of Namibia's History

The early period of Namibia's history is characterized by the complete absence of recorded data. The exploration of the country by Europeans began along the Atlantic coast as early as 1485, although access to the interior was barred by the inhospitable Namib desert. The harsh interior of Namibia aroused very little interest until the 1700s. The advent of European adventurers, explorers, and traders from the mid-eighteenth century onward, who entered from South Africa, had a disruptive impact on the existing patterns of organized society in Namibia (Dierks 1992b, p. 113).

Both the British authorities in South Africa during the 1870s and the German colonial interests in Namibia during the 1880s were able to take advantage of these developments: in 1878 the British annexed Walvis Bay, and in 1884 the formal colonization of Namibia by the German Empire began.

The outcome of the German colonial era was a divided, pluralistic society with a strong emphasis on separate cultural identities and a relative integration of the inhabitants into a common economic system. This economic system, however, remained profoundly dualistic, with a property-owning settler minority and a largely propertyless majority. This situation was coupled with the political fragmentation of Namibia's indigenous communities and the political and economic domination of the colonizing minority. Another result was the development of a remarkable modern physical infrastructure system, especially in the telecommunications and railway sectors. However, these infrastructures were not geared to promote the well-being of the indigenous Namibians; they were developed solely to foster the interest of the colonial society. This policy of group cultural separateness and fragmentation was to be expanded by the South Africans after 1915. For the indigenous inhabitants the change proved to be only from one colonial master to another, and it did not result in any improvement to their quality of life (Dierks 1992b, p. 119).

In the Peace Treaty of Versailles, which formally ended the First World War, the Union of South Africa was entrusted with several duties and responsibilities in Namibia, including, for instance, to "promote to the utmost the material and moral well-being and the social progress of the inhabitants of the territory." This "sacred trust of civilisation" was not honored by South Africa, which pursued a deliberate policy of racial domination (apartheid) and exploitation of the economic resources during its period of colonial control up to 1990.

Resistance intensified after the Second World War and under the leadership of SWAPO (South West Africa People's Organization). With the support of the United Nations and after a long liberation struggle against the South Africans, the illegal occupation ended on March 21, 1990, when Namibia gained her freedom and independence (Dierks 1992b, p. 120). Independence resulted in the advent of a stable, peaceful, multiparty democracy based on the rule of law and a deregulated, strong, free-market economy.

13.1.3 The History of Telecommunications in Namibia

Namibia is one of the last vast wilderness areas in Africa, with an extremely rich African fauna and flora. But it also possesses, compared to other African countries and seen in relation to the small population and the large size of the country, an exceptionally well-developed but unbalanced physical infrastructure.

The pace of progress was not achieved easily or cheaply. Namibia is a land of many faces: from inhospitable deserts and hard rocky outcrops to rugged mountains and undulating plains. Each of these presented different problems to the transport and communications engineers.

Namibia's excellent but unbalanced telecommunications infrastructure must be seen in the mirror of its colonial heritage. Before Namibia's date of independence in 1990, infrastructures like telecommunications were solely developed in the interest of the colonial powers. The result was a marked imbalance between the "modern, first-world" sector and the impoverished "third-world" sector of Namibia where the majority of Namibians live.

Namibia's telecommunications age commenced on January 16, 1899, when the German colonial administration contracted an agreement with the Eastern and South African Telegraph Company in London to participate in the sea cable from Mossamedes (Namibe in Angola) to Cape Town with a link to Swakopmund, the port town of "German South West Africa." Further telegraph stations followed in Karibib (1901), Okahandja (1902), and Windhoek (1902). Swakopmund was the first town to get a telephone network with twenty-eight lines in 1901. Windhoek, Okahandja, and Karibib followed in 1902 (I. G. Deutschsprachiger Südwester 1985, pp. 477, 478). The trunk connection between Swakopmund and Windhoek was constructed between 1901 and 1906.

Military needs during the Namibian resistance wars against the German forces necessitated the construction of a 495-kilometer telegraph line from Windhoek into the Namibian south via Rehoboth, Tsumis, Gibeon, and Keetmanshoop. This system was modernized in 1911–12, after a telecommunications station at Keetmanshoop was inaugurated in 1906. This system was completed by a 364-kilometer double railway telegraph line between Keetmanshoop and the southern port town of Lüderitz (completed 1907–8), with a branch line between Brakwater (south) and Chamis via Bethany (32 kilometers in 1908). The total Namibian trunk system of 1,300 kilometers was interlinked at the end of 1907, with direct telephone services between Swakopmund, Windhoek, Keetmanshoop, and

Lüderitz. Further additions were telegraph lines between Windhoek and Gobabis in the east (252 kilometers in 1905) and between Usakos and Otavi in the north (1906), as well as from Otavi to Grootfontein (91 kilometers in 1908). A trunk route from Keetmanshoop via Warmbad and Karasburg with a link into the South African system at Ramansdrift was inaugurated in 1910 (260 kilometers). This trunk line, which was extended to Cape Town in 1912, was the most expensive one so far. Between 1901 and 1907 a telegraph system of 3,616 kilometers with thirty-four stations and twelve local telephone networks was completed (I. G. Deutschsprachiger Südwester 1985, pp. 478, 479).

Between 1908 and 1914 this system grew steadily, with extensions to the diamond fields at Angras Juntas, Elizabeth Bay, Pomona, and Bogenfels.

The only missing links in the settlers' colony were telecommunications lines to the lonely and far-off farms. The first "farm party lines" came into operation in 1909 between Gibeon and Maltahöhe and in 1912 between Okahandja and Ombirisu, with a total of thirty-two farm telecommunications stations in 1913 and a cost participation by the individual farmers. Furthermore, the first wireless telegraph links over a distance of 8,000 kilometers between Germany (Nauen) and Namibia (Windhoek) via Kamina (Togo) came into being in 1913. The first official communication via this link was Germany's declaration of war, which set the stage for the First World War and led to the end of Germany's occupation of Namibia in 1915 (I. G. Deutschsprachiger Südwester 1985, pp. 479–81).

Initially the South African colonial era brought slow progress in the further development of Namibia's telecommunications system. Until 1926–27, no new projects were added. At that time it is reported that some new lines were laid and the trunk line between Windhoek and Keetmanshoop was renewed. In 1929 Windhoek received its first automatic telephone exchange (Siemens). During 1931–32, trunk lines between Otjiwarongo and Otavi as well as Grootfontein and Tsumeb were extended. Walvis Bay, Swakopmund, Usakos, Grootfontein, Keetmanshoop, Okahandja, and Mariental received new telephone exchanges. Farm party lines were erected between Otjiwarongo and Erundu as well as Okaputa and Warlencourt. The Second World War interrupted all further progress in telecommunications (I. G. Deutschsprachiger Südwester 1985, p. 482).

In 1949 the Windhoek automatic exchange was expanded to 2,000 lines. In 1950 the country possessed sixty-two exchanges with 1,033 private and 2,267 business lines, 134 public telephones, and 451 farm party lines. During the late 1950s the demand for telecommunications services still remained insatiable, and 1,066 new services and 820 supplementary services were provided in 1957, while 1,587 changes to existing services were effected. In this year the total number of hired services increased to 11,024, and 423 kilometers of trunk lines and 1,622 kilometers of farm lines were erected (South West Africa: A Report for the Calendar Year 1957 and the Activities of the Department of Posts and Telegraphs, Windhoek, pp. 2, 3).

In 1960 this system grew to 11,163 lines, with 2,413 farm lines. During the early 1960s the direct telephone links were increased from 597 to 985 with a total distance of 202,518 kilometers. Between 1965 and 1975 the investments for telecommunications increased from U.S.$20 million to U.S.$75 million. In 1972, Namibia's telecommunications system was interlinked with the automatic system in South Africa and then later replenished by an analog microwave system between Windhoek and Upington (South Africa) via Keetmanshoop. The exchanges grew from 99 to 467 at a total cost of U.S.$7 million. Direct telecommunications links with thirty-three countries were possible as of 1980. During the 1980s telecommunications links with the Owambo regions in the north, with Opuuo (Kunene region), and with Katima Mulilo (Caprivi region) were developed. In 1983 Namibia had 60,737 telephone lines (14,752 manual, 45,982 automatic, and 5,890 farm) for approximately 1.3 million inhabitants living on more than double the surface of the united Germany (I. G. Deutschsprachiger Südwester 1985, pp. 482, 483).

The year 1986 saw the introduction of the first fully digital electronic exchange in Namibia. This was the combined trunk and local exchange installed in Windhoek. The exchange increased the telephone service potential in Windhoek tenfold. Before independence on March 21, 1990, Namibia had an integrated network of analog and electronical transmission mediums of approximately 1,970,005 kilometers. However, the dominant technology for transmission links was based on analog microwave channels from the north to the south interconnected with copper-based open-wire carrier routes to the eastern and western parts of the country. These have to be extended and balanced to link the "two Namibias."

13.2 The Present

13.2.1 Assessment of the New Policies for Telecom Namibia

It is helpful to examine the commercialization of telecom services in Namibia, not only to assess the experiences, but also to understand the actors and variables that played such an important role in this process. The major players in this crucial policy decision were not different interest groups in Namibia such as business pressure groups, the trade unions, or any international institutions and investors. Instead, the policy came about due to a principal decision made by the first independent government of Namibia. This government was determined to optimize the country's scarce resources in order to initiate national development, eradicate poverty, and create a new "postapartheid Namibia" by bridging the "first-world" and the "third-world" Namibias.

13.2.1.1 Background

In line with the government's policy to embark on a new economic approach by deregulating certain functions of the state and opening up the markets to encourage international exposure and competition, the Communications Division under the Ministry of Works, Transport, and Communication was divided into four separate entities in 1992.

The four entities are as follows:

- The Ministry of Works, Transport, and Communication remained in the government fold and is, inter alia, accountable for the national policy of the postal and telecommunications sector, the appointment and removal of the board of directors, the monitoring of the performance contract, and the annual report to the Namibian Parliament on the affairs of the corporations.

- The Namibia "Communications Commission" is an independent body responsible for frequency management, the issuing of licenses, and other regulatory functions.
- Telecom Namibia Limited, the officially designated telecommunications company, is responsible for providing and maintaining the telecommunications infrastructure and services.
- Namibia Post Limited is responsible for postal and savings bank services.

Telecom Namibia Limited and Namibia Post Limited, incorporated under the Posts and Telecommunications Act of 1992, are both subsidiaries of the parent company, the Namibia Post and Telecom Holdings Limited, and are operating under the laws and regulations pertaining to a company registered in terms of the Namibian Companies Act. The accountability for the corporations' financial matters is resting with the management and its board of directors.

13.2.3 The Prospects for Telecom Namibia

13.2.3.1 Telephone Services

The Namibian telecommunications market is, compared to African standards, a well-developed market. The number of telephone lines are approximately 5 per 100 inhabitants, which is more than five times the average of the African continent. The demand for service is presently higher than the supply. The usage per line is also high on average, at U.S.$60 per line and month for automatic service.

The average growth for automatic exchanges over the 1990s has been 6 percent per year in new services. Estimated near-term growth will be 10 percent per year, bringing the average number of telephone lines to 8 per 100 inhabitants, the same as in South Africa in 1997.

To meet this demand, investments have to be approximately U.S.$20 million per year. To open up an independent international gateway, another U.S.$13 million will have to be invested annually. The expected growth in revenue will be U.S.$7 million per year, assuming 10 percent growth in new services and 5 percent growth in traffic.

Telecom Namibia covers approximately 98 percent of the market for telephone services. Competition for local traffic comes only from one other operator, the Consolidated Diamond Mines (CDM) in Oranjemund. For international traffic the only competition is presently "hubbing" via South Africa. However, in April 1995 competition arrived from a local operator of cellular telephony. More competition will come in the future, in line with government policy, which is based on competition and deregulation. For instance, banks will set up their own satellite communications links primarily for data communication, but experience from both Europe and the United States shows that these facilities also will be used for voice communication.

13.2.3.2 Data Communications

The market for data communications is looking for higher speed and more reliable services. The present service uses leased analog lines and an x-25 packet-switched network. Today the major banks are asking for digital 64-kilobit services.

The competition in data communications comes mainly from the potential customers themselves, who establish their own data communications networks (i.e., the Municipality of Windhoek, the breweries, etc.), first at their own premises but later extended and connected to other sites, both domestic and international. It is expected that Telecom Namibia's share of the market will not exceed 70 percent, including leased lines. The total turnover is approximately U.S.$3 million per year, thus representing a relatively small market for Telecom Namibia. It is, however, essential to be present in the data market, in order to protect the more viable telephony market, especially for the business customers.

The supply of modems for DATA (dedicated data channels where customers establish their own computer system) and TELNET (packet switching) services on the customer premises is left completely to the private suppliers. Telecom Namibia's DATA customer growth is above 10 percent per annum, and that of TELNET is above 20 percent.

13.3 Conclusion

As far as the telecommunications sector is concerned, Namibia is technically one of the most developed countries in Africa. It is the first priority for the government to maintain and upgrade the telecommunications systems in an optimized way. For historical reasons, physical infrastructures were always unbalanced to the disadvantage of indigenous Namibians. This imbalance was consequently addressed after the Namibian independence on March 21, 1990, and visible change has taken place since then.

The technical progress has to be supplemented by institutional reform. A process of commercializing the former Department of Posts and Telecommunications within the framework of the Ministry of Works, Transport, and Communication was initiated after independence and completed in August 1992. Telecom Namibia is one of the success stories of the country and has made great strides toward an efficient and modern telecommunications network without any cross-subsidization and without failing to fulfil the social commitments that are a heritage of the colonial era.

Bibliography

Dierks, Klaus. 1989a. *The Development of a Roads Model for Independent Namibia.* Vienna.

______. 1989b. *Namibia's Railway System—Future Link to Africa.* Harare.

______. 1991. *Second United Nations Transportation and Communication Decade in Africa.* Addis Ababa.

______. 1992a. *Growing to Nationhood—||Khauxa!nas.* Windhoek.

______. 1992b. *||Khauxa!nas—The Great Namibian Settlement.* Windhoek.

______. 1992c. *Namibian Roads in History.* Frankfurt.

______. 1992d. *Technical Aspects for Appropriate Low-Volume Roads in Namibia.* Ph.D. diss. Berlin/Windhoek.

______. 1993a. *Namibia—Key to Communication in the Southern African Region.* Washington D.C., September.

______. 1993b. *Namibia—Key to Co-operation in the Southern African Region.* Paper presented at conference, Partner of Progress, in Berlin.

Interessengemeinschaft (I. G.) Deutschsprachiger Sü dwester. 1985. *Vom Schutzgebiet bis Namibia.* Windhoek.

National Planning Commission. 1993. *National Development Plan.* Windhoek.

Telecom Namibia. *Business Plan 1994/95.* Prepared by Business and Network Development.

14

Models for the Development of Regional Telecommunications Networks in Africa

MANSUR M. NURUDDIN

The development of telecommunications in Africa began during the colonial era.[1] Telecommunications was a means to maintain control over the colonial possession[2] and was primarily limited to communications between the capital and its administrative centers within Africa. The links that were extended out from the administrative capital were primarily used to solidify control.[3] Establishment of widespread service was therefore inconceivable. Intra-African telecommunications was precluded by the colonial economic structure, which destroyed all pre-existing intra-African trading systems (Young 1986).

As African nations became independent, some saw telecommunications as a tool for empowerment and development. Yet full realization of the possible benefits of telecommunications was impeded by significant obstacles, such as the lack of indigenous expertise in telecommunications manufacturing and research. During this period, regional cooperation was articulated as a model for African telecommunications development. It was hypothesized that through the pooling of limited resources, African telecommunications carriers could overcome obstacles to telecommunications development. This model continues to dominate debate in Africa (e.g., ITU 1994b; OAU, Draft Protocol).

Although regional cooperation is often presented as a panacea, its ability to satisfy Africa's telecommunications needs must be assessed in terms of each country's individual need. This chapter seeks to find the most appropriate model/s for developing regional telecommunications networks. It addresses the essential question: Is regional cooperation the most appropriate model for developing sustainable regional telecommunications networks? In assessing the suitability of regional cooperation to regional network development, this chapter provides a comprehensive analysis of the Pan-African Telecommunications Network (Panaftel). The experience of Panaftel raises doubts about the use of the regional cooperation model to develop sustainable regional telecommunications networks. It is argued that a new model will have to emerge if sustainable regional telecommunications networks are to be developed in Africa. Three possible models are examined: the Appropriate Environment Model, the Regional Satellite Model, and the International Aggregation Model. After an explanation of each model, the future of regional telecommunications networks is discussed. The conclusion seeks to provide broad recommendations concerning the possible use of all three models. Finally, other applications of the regional cooperation model in telecommunications are broadly evaluated.

14.1 Panaftel

Since independence, African nations have viewed intra-African telecommunications capacity as essential to their initiatives toward economic, social, and political integration. Regional telecommunications networks have been considered necessary for establishing trade and government cooperation between African nations. Regional networks were also seen as a method for redefining the communications structure that had been established during the colonial period. The regional cooperation approach for the development that evolved during the 1960s was applied to the desire for intra-African telecommunications links; this spawned the most ambitious African telecommunications project to date—the Pan-African Telecommunications Network.

14.1.1 The Historical Development of Panaftel

The Pan-African Telecommunications Network was conceptualized at the first meeting of the Regional Plan Committee (the Committee) for Africa in Dakar, Senegal, in 1962.[4] At this meeting, 115 delegates sought to redefine the communications structure that had been established during the colonial period, in which no attention or resources were directed toward intra-African communications capabilities. As former African colonies achieved independence, they wanted to move toward establishing direct intra-African telecommunications links.

In 1962, there were only fifteen HF radio systems operating between African countries. There was one UHF radio system operating in East Africa (Kenya, Tanzania, and Uganda); and there were thirteen land cable systems carrying about 100 circuits between African countries (Tedros 1987). The Committee delegates envisioned a continental network that would alleviate the need to transmit intra-African communications through non-African transit centers and would facilitate the achievement of African political, economic, and social unity. They saw regional cooperation as the most effective method for establishing this network.

The second meeting of the African Plan Committee occurred in 1967 in Addis Ababa, Ethiopia. Panaftel's structure was further elaborated, and concrete steps were taken toward establishing the network. In 1968, the International Telecommunications Union (ITU) requested funding from the United Nations Development Program (UNDP) for Panaftel's preinvestment surveys. Two consultant teams began a preliminary study of thirty-eight African countries, which was completed in 1969.

In 1969, the UNDP granted U.S.$2 million to the ITU for detailed technical, economic, and financial studies. These studies concluded that a network combining coaxial cable and microwave radio-relay links would be the best system for Panaftel. These original specifications envisioned a network with 20,000 kilometers of transmission arteries and eighteen international switching centers (ITU 1974). Routes were selected that could serve as national backbones connecting major population centers in one country and providing international links between neighboring countries.

In 1972, the Meeting on the Implementation of the Pan-African Telecommunications Network was held in Addis Ababa, Ethiopia. At this meeting, the preinvestment survey recommendations were accepted. However, the surveys of the West African region were not complete. Another meeting was held in Lome, Togo, in April 1973 to discuss the results of the West African surveys. After the West African region surveys were incorporated, the estimated cost of Panaftel was U.S.$115 million (ITU 1974).

The improvement of intra-African communications links began as early as 1963—primarily through the use of VHF and UHF circuits. However, Panaftel's implementation, as a coordinated and continentwide project, began in 1975 with the UNDP/ITU project RAF/73/023: Implementation of Panaftel Network Phase I (1972–82). The major goals of this phase were: (1) installation of transmission systems and switching centers; (2) organization of traffic routing and circuit forecasting; (3) specification of radio-relay systems; and (4) training of African P&T staff. During this initial implementation phase, it became clear that coordinating Panaftel on a continental basis was unrealistic. By establishing a subregional structure, the Committee hoped to achieve better coordination and easier ratification of tariff structures and routing matrices. Subsequently, subregional coordinating committees were established for the northern, western, central, southern, and eastern African subregions.[5]

By 1983, Panaftel had progressed significantly; a considerable portion of the network was in use, and thirty-nine countries had at least one satellite earth station. The ITU and UNDP began the second phase of Panaftel: RAF/82/060: Implementation of the Panaftel Network Phase II (1983–86). The major goals of this phase were: (1) tender evaluation; (2) acceptance testing; and (3) traffic management and routing training.

By 1987, the size of the network was more than twice the original specifications: there were 35,000 kilometers of microwave relay links; 8,000 kilometers of submarine links; forty-three international switching centers; and forty-one of the member countries had at least one satellite earth station.[6] A third phase—RAF/87/011: Operation and Extension of the Panaftel Network—was initiated in 1988. The major goals of this phase were: (1) to supplement the network with new links; (2) to reactivate nonoperational links; and (3) to improve mechanisms for operation, modernization, and sustained development of the network.

By the late 1980s, Panaftel's progress had slowed considerably. This was caused by economic, political, and operational difficulties. Emphasis shifted from implementation to improving the operation and maximizing the level of traffic transmitted through the network. This shift was necessitated by operational difficulties that undermined Panaftel's overall viability.

In 1989, the status of subregional networks were as follows (ECA 1993):

1. The northern zone was relatively complete.
2. The West African zone represented a problem area for the network, due primarily to maintenance and operational difficulties.[7]
3. The East African zone had a working system. However, political differences between many of the countries hindered the system.
4. The central African zone had the least developed network. The major obstacle appeared to be lack of political will rather than technical difficulties.
5. The southern African region was almost completed.

By 1989, the existing analog transmission and switching systems were gradually being replaced by modern digital equipment.[8] In 1989, the future plans for Panaftel were as follows: (1) tap the potential of fiber-optic links; (2) improve coordination between African PTTs; (3) help establish a uniform tariff structure; and (4) complete the final links in this expansive network (ECA 1993).

As of 1990, when the last full-scale inventory of Panaftel was taken, the system consisted of 39,000 kilometers of radio-relay links, 8,000 kilometers of submarine cable, and forty-three international switching centers. Moreover, forty-two of the forty-five member countries had international satellite stations. In 1990, twenty-nine links or approximately 4,000 kilometers of links remained to be installed (ITU 1991; Riverson 1991).

14.1.2 Panaftel Problems

Many of the problems encountered during the implementation of Panaftel were identical to the perennial problems faced by African PTTs: lack of skilled personnel, inadequate funding, and poor organization. There were also significant environmental impediments: torrential rains, extreme temperatures, and a lack of accessible roads. Theft of equipment also slowed Panaftel's progress (Crecchio 1980).

Maintenance of the network was a major problem. Many of the same environmental impediments that hindered the implementation of the network also impeded its effective maintenance (e.g., torrential rains and the lack of adequate roads). The inability to obtain spare parts and fuel also led to prolonged outages. The most significant maintenance issue was the lack of well-defined maintenance procedures (Crecchio 1980). In fact, the need for such procedures was not articulated until the advanced stages of Panaftel's implementation. In 1980, the ITU initiated UNDP/ITU Project RAF/80/018: Telecommunications Maintenance—Pan-African Network. This project established the National Program for Improved Maintenance (NPIM) in fourteen countries. The NPIMs concentrated on preventive maintenance and establishing a well-defined "fault signaling system" and well-organized maintenance procedures (N'Zengou 1990).

A major challenge in implementing Panaftel was assessing the demand for intra-African communications. As was noted earlier, prior to Panaftel, there was virtually no intra-African telecommunications capacity, and no means by which to forecast the demand for intra-African communications. In 1963, the ITU and UNDP established an HF link between Addis Ababa, Ethiopia, and Abidjan, Ivory

Coast, to assess the demand for intra-African communications. Optimistically, the ITU felt ". . . that the need for telecommunication circuits between African countries was so high that it could not possibly be met with the inherently low capacity HF radio system" (ITU 1974). In assessing the demand for intra-African communications, the ITU applied an input jump of 50 percent to 1975 telephone forecasts for Panaftel. The ITU also estimated that there would be a yearly increase of 20 percent until 1980, and a yearly increase of 15 percent thereafter. These forecasts were justified on the basis of perceived "latent demand" (Okundi 1976). The actual demand for intra-African communications has never reached the optimistic levels assumed by the ITU's preinvestment studies. In 1992, only 14 percent of Africa's international telecommunications traffic was intra-regional (ITU 1994a).[9]

Due to its expansive nature, Panaftel faced numerous coordination problems. For example, it took over a decade for the African countries to standardize signaling systems (Sy 1985), and signaling problems were still present as late as 1990 (Girmaw 1990). In the southern African subregion, many countries were unable to connect through satellite communications because of the use of various incompatible carrier techniques (Dymond 1987). The most serious coordination problems were less technical and more bureaucratic. These difficulties primarily involved establishing appropriate tariff structures and routing matrices.

Political difficulties existed throughout Panaftel's development and either hindered or even precluded its implementation in particular subregions. There was little user demand for intra-African communications; instead, the impetus for the network was political—stemming from perceived economic and political benefits. In regions where there was no political will to establish Panaftel, the network floundered, as in the central African subregion. Furthermore, Panaftel's political origins made it completely reliant upon the maintenance of cordial diplomatic relations. In several instances, cooperation in Panaftel was used as leverage for achieving unrelated political goals. In at least one case, the end of cordial relations led to the disconnection of Panaftel links, namely, between Kenya and Uganda.

14.1.3 The Current Status of Panaftel

As of 1992, international funding for Panaftel[10] was discontinued. The major reasons were:

- The relatively low revenue from existing links could not justify continued investment.
- The UNDP felt that political inertia in those regions in which links were still missing—specifically, the central African subregion—could not be overcome.

Despite the termination of funding, in 1995, the ITU continued to sponsor seminars and training sessions for the African operators of the Panaftel links.

Although Panaftel's funding was discontinued, attempts to establish intra-African communications links have continued on a bilateral and subregional basis. For example, the completion of Ethiopia's Panaftel links to Sudan and Somalia was planned as part of the Second United Nations Transport and Com-

Table 14.1. Telecommunications Traffic: Tunisian Panaftel Links to North Africa, 1990

Destination Country	Minutes × 1,000
Algeria	1,500
Morocco	1,550
Libya	2,500
Egypt	700

Source: ITU, 1991.

munications Decade for Africa (Second UNTACDA) (Conference of African Ministers 1993).[11] Attempts to establish connectivity between central African countries have continued; for example, a project to establish connection between the four countries belonging to the Kagera River Basin Development Organization was initiated in 1992 (ARB 1992).

Despite ongoing attempts to establish intra-African connectivity, the future of Panaftel is in doubt. As of the mid-1990s, links were deteriorating and much of the system was being exclusively used as backbones for domestic networks. Certain subregions within the overall system are currently sustainable—primarily in the northern and the southern African subregion (e.g., see tables 14.1 and 14.2).[12] The stability of the links in the northern and southern subregions is correlated to numerous factors: the level of trade; the cost of intra-African calls between countries in these subregions; cultural/linguistic similarity; and so on. In the future, the level of traffic transmitted over these links is likely to grow. However, the thinnest routes in Panaftel will probably be allowed to deteriorate (see table 14.3), since diverting resources into these links would be an unsound policy—considering other more pressing telecommunications needs.[13]

14.1.4 An Assessment

Although Panaftel grew from the initial specifications of 20,000 kilometers of communications links and nineteen international switching centers to a network of more than 47,000 kilometers of links and thirty-nine international centers, the final assessment of Panaftel should not be based on its size. Instead, an assessment must be based on whether the network is sustainable, since this will determine whether it

Table 14.2. Telecommunications Traffic: Zimbabwean Panaftel Links to Southern African Countries, 1990

Destination Country	Minutes × 1,000
South Africa	15,265
Botswana*	1,988
Malawi*	693
Zambia	1,110

Source: ITU, 1991.
*Routed through South Africa.

Table 14.3. Selected Panaftel Thin Routes, 1990

Originating Country	Destination Country	Minutes × 1,000
Burkino Faso	Gabon	10.00
Kenya	Nigeria	2.00
Benin	Ghana	9.5
Ivory Coast	Ethiopia	21.00
Chad	Benin	2.4
Chad	Malawi	1.9
Tunisia	Ivory Coast	11.00

Source: ITU, 1991.
*Routed through South Africa.

will be able to assist in the advancement of Africa. These same ideas were articulated at the Meeting of the African Traffic Managers (1987); however, these concerns were articulated at the wrong stage of Panaftel's life cycle.

If the revenues generated are insufficient to make the network self-sustaining, then the energy and resources invested have been wasted. A look at the amount of traffic being transmitted over some links explicates the economic inviability of much of the network (see table 14.3). The future of these links and many others like them is questionable. Although certain subregions are more stable than others, Panaftel has not become a sustainable continentwide regional telecommunications network.

14.2 The Need for a New Model

The model employed for Panaftel can be most accurately termed a Multilateral, Large-Scale Regional Cooperation Model. This model is fraught with obstacles and pitfalls. For example, regional networks developed under this model are too dependent on political initiatives. As Panaftel demonstrates, politically motivated networks are not sustainable if and when political will dissipates. Further, the lack of demand for the network beyond political desires means that there is unlikely to be sufficient revenue to sustain it.

Despite the apparent inefficacy of this model, it continues to be employed by African telecommunications carriers in the development of regional networks—for example, the Kagera Basin Project. However, if African nations want to establish sustainable regional telecommunications networks, a new model must be adopted. There are at least three possible models: Appropriate Environment Model, the Regional Satellite Model, and the International Aggregation Model. These models differ in the following ways: the types of technology employed; the types of and method by which connectivity is established; and the potential revenue streams.

14.2.1 The Appropriate Environment Model

The Appropriate Environment Model (AEM)[14] concentrates on establishing the milieu necessary to spawn demand for regional networks. Panaftel clearly demonstrates that establishing connectivity is insufficient to spur intra-African communications. Communications capacity is merely an enabling factor; the effective utilization of this capacity is contingent on numerous factors, such as appropriate tariff structures, trade, travel, national telecommunications penetration, and so on. The AEM seeks to create the appropriate mix of factors that will stimulate the demand for regional telecommunications networks. This model asserts that since demand for regional networks is contingent on numerous factors, which are external to telecommunications capacity, the best method for developing regional networks is to concentrate on these factors. Further, within model regional telecommunications networks, demand can be induced through the convergence of these other factors. The factors that are central to the AEM are national telecommunications penetration, appropriate tariff and policy structures, and liberal government policies.

Increasing the national telecommunications penetration rates is essential to the AEM. These penetration levels are a key determinant of demand for regional capacity (Checchi 1968). The more individuals that are connected to each national network, the greater the value of each of these networks and the greater the value of connections between these networks. Investing in regional communications capacity without significant growth of national networks will only assure that the regional networks are not sustainable. Thus, resources and energy should be placed in national networks. As national networks develop, the demand for regional communications will increase.

To improve national telecommunications penetration levels, carriers and African governments will have to alleviate many of the obstacles that hinder the expansion of national networks. The poor national penetration levels of African telecommunications networks is partially due to structural problems within the telecommunications sector. First, more of the revenues generated by the telecommunications sector will have to be reinvested in expanding the network.[15] Also, the overall efficiency of African telecommunications providers must be improved, for example, by reducing the number of employees per line[16] and improving billing systems and bill collection.[17] Removing all of these barriers to network expansion is essential to improving national telecommunications penetration and creating the necessary environment for regional telecommunications networks. The restructuring of Africa's telecommunications sector—such as the privatization projects in Senegal and Ghana—will be critical to alleviating these obstacles and can be seen as the first step toward implementing the AEM.

Establishing appropriate tariff structures that enable the use of regional networks is essential within the AEM. If regional networks are to develop, telecommunications providers will have to revise current tariff structures that hinder the use of regional networks, namely, the exorbitant tariffs on international calls (Paltridge 1994). The networks within the northern and southern subregions, which are sustainable, have the lowest charges on intra-African calls; conversely, those subregions with the lowest level of intra-African communications have the highest charges for intra-African calls—the western and central African subregions (see table 14.4). Clearly, there is some correlation between intra-African telecommunications traffic and the cost of intra-African calls. Therefore, in creating the appropriate environment for regional telecommunications networks, African telecommu-

Table 14.4. Average Cost of a Three-Minute Intra-African Call for Selected Countries

Country	Subregion	Within Subregion	Outside Subregion	Comparable North American Price (U.S.– Canada)
Ghana	Western	$10.83	$13.43	$0.90
Benin	Western	$13.72	$22.96	$0.90
Namibia	Southern	$ 3.83	$11.00	$0.90
Zimbabwe	Southern	$ 2.4	$ 7.8	$0.90
Egypt	Northern	$ 3.35	$ 7.5	$0.90
Burundi	Central	$18.90	$15.70	$0.90
Rwanda	Central	$11.10	$11.18	$0.90

Source: ITU, 1994a.

nications providers will have to adopt tariffs that promote intra-African communication, in other words, they will have lower charges for intra-African calls.

The AEM asserts that in order to spur regional network creation, intra-African trade must increase. In reference to G7 countries, Aharon Kellerman (1990) has shown that telecommunications traffic is most directly correlated to trade. In 1993, only 8.1 percent of African trade was intra-African (IMF 1994). The amount of intra-African telecommunications traffic will remain low as long as the level of intra-African trade remains low. Those African nations that have achieved relatively high levels of bilateral trade have also achieved relatively high volumes of bilateral telecommunications traffic. For example, in 1992, Djibouti and Kenya were Ethiopia's two largest African trading partners; Djibouti accounted for 49 percent and Kenya accounted for 33 percent of Ethiopia's total intra-African trade (IMF 1994). In this same year, Ethiopia had the greatest volume of communications with these two countries: 0.9 million minutes of calls to Djibouti and 0.8 million minutes of calls to Kenya (ITU 1994a). African nations will have to increase the level of intra-African trade, if they are to create the appropriate environment for regional networks.

In stimulate intra-African trade, African nations will have to adopt policies that enable the free flow of economic factors, that is, money, people, and information.[18] For example, the 1994 expulsion of African expatriates from Gabon will likely lower the telecommunications tariff between Gabon and neighboring African countries. African nations will also have to adopt policies that surmount other obstacles to further economic integration, such as the lack of convertible currency and unequal levels of development (UNCTAD 1993). These policies will stimulate demand for regional communications, and regional networks will follow.

The AEM emphasizes the need to create a milieu that will stimulate demand for regional telecommunications capacity. Once the appropriate environment has been stabilized, regional networks should be allowed to develop in response to actual demand. If the necessary policies cannot be enacted or the investment in national telecommunications does not increase the demand for regional connectivity, then regional networks should not be developed, since they would be unable to generate sufficient revenues to sustain themselves and would ultimately divert resources from other more pressing national telecommunications needs.

14.2.2 Other Models

Although certain aspects of the Appropriate Environment Model (AEM) are surfacing in the debate on regional network development (ITU 1994b), this model is not the only one available. The AEM is optimized for regional telecommunications networks that are terrestrial systems; that carry only intra-African traffic; and whose primary revenue is generated by basic telecommunications services and data communications. However, none of these attributes is a requirement for viable regional networks. The Regional Satellite Model and the International Aggregation Model offer considerable possibilities for African regional networks. Yet these models are based on different underlying principles: they use different types of technology; they can offer different services; and they can carry different types of traffic.

14.2.3 The Regional Satellite Model

The Regional Satellite Model (RSM) has been utilized throughout the world for regional network development. The European Satellite (Eutelsat) was established in 1983; the Arab Satellite (Arabsat) was established in 1985; and the Asian Satellite (Asiasat) was established in 1990. The use of a regional satellite brings new variables into the process: new revenue sources, new economic considerations, and new coordination concerns. In 1995, African nations were moving forward with efforts to establish a regional satellite under the auspices of the Regional African Satellite Communications Organization (RASCOM).

14.2.3.1 RASCOM

Before Panaftel's implementation began, a dedicated African satellite system was proposed (Okundi 1979). It was hypothesized that Panaftel would not satisfy the demand for intra-African communications and would have to be supplemented by a satellite system. In 1975, the Conference of African Telecommunications Administrations requested a feasibility study for a dedicated African satellite system (Okundi 1976).

Between 1980 and 1984, numerous separate studies were conducted. In 1983, the Inter-Agency Coordination Committee was created to integrate all the previous studies (Akwule 1990). In 1987, the ITU conducted a final feasibility study.[19] This study examined engineering concerns, the financial and economic viability of such a system, staffing and training, and organization and management issues (ITU 1990). After some hesitation on the part of African nations (Hudson 1991), the feasibility study was approved in Abuja, Nigeria, in February 1991. An interim RASCOM organization was created in May 1992 (Jipguep 1993b). The organization became operational in November 1993, with its headquarters in Abidjan, Ivory Coast.

The specifications of RASCOM's call for a two-satellite system. The cost of the satellite segments was estimated at U.S.$250 million each, and the cost of ground segments for the fifty countries was estimated at U.S.$800 million (Rzepecki 1990). The organization is cooperatively owned, with each country purchasing minimum initial investment shares of U.S.$50,000 (Jipguep 1993b). As of March 1994, thirty-five countries had signed the Operational Convention and thirty countries had purchased their designated investment shares. To attract additional investment, RASCOM will implement a mechanism by which African and non-African investors can buy shares in the organization—the Non-Signatory Shareholder Agreement (Africa Communications 1994).[20]

After becoming operational, RASCOM's first initiative was to pool the transponder space leased from Intelsat by member nations. The organization has also begun providing commercial services; for example, in 1994, RASCOM began leasing transponder space to the South African broadcasting company ORICOM (Africa Communications 1994).

Officials stated that the first satellite would be launched in 1997, and would be a cooperative venture with another organization.[21] As of 1998, RASCOM has not launched any satellites. The organization has experienced delays due to the inability to raise the $250 million needed to launch the satellite. In an effort to raise the capital, RASCOM decided to seek a build, operate, and transfer (BOT) venture with a private sector company that would be willing to finance the RASCOM satellite project. RASCOM issued a request for proposals from various companies, including satellite manufacturers and equipment suppliers. It would be wise to adopt a wait and see position on whether or not RASCOM is able to find a willing BOT partner.

The major benefit of the RSM is the flexibility that a satellite system would provide. It would be a conduit for numerous types of telecommunications traffic; for example, RASCOM plans to offer television, video conferencing, and data transmission services on a commercial basis. It could also offer mobile services, location identification services, and remote sensing applications. A regional satellite system also offers African countries new options for addressing the dismal level of telecommunications in rural areas, such as, the use of VSATs.

Although a regional satellite system would offer numerous benefits, the RSM has many of the same obstacles inherent in the Panaftel model. For example, effective coordination and cooperation between African signatories is essential to RASCOM's success. Problems have already emerged in this regard. For example, during the process of establishing RASCOM as an operational organization, Nigeria withdrew from the organization because of a dispute over the location of the headquarters.[22]

Ultimately, RASCOM's success will hinge on its ability to generate sufficient revenue to justify and sustain it. Intra-African communication is unlikely to be able to generate the necessary revenue, since only 14 percent of African international telecommunications traffic is intra-Africa (ITU 1994a). It is also unlikely that revenues from integrating international traffic from countries in the interior of Africa with one of the fiber-optic proposals will be sufficient to sustain the network. In fact, these optical fiber proposals represent a threat to RASCOM's overall viability, since they will offer a superior alternative on the most lucrative intra-African trunk routes, for example, between the Ivory Coast and Nigeria.[23]

The major opportunity for RASCOM is in the transmission of television, since this is the only service that has the potential revenue necessary for RASCOM's success. This assertion is based on three observations: (1) the inherent suitability of satellites for point-to-multipoint communications (Podmore and Faguy 1986); (2) there are almost four times as many television sets as telephones in Africa (ITU 1994a); and (3) the Arabsat and Eutelsat satellite systems have demonstrated the feasibility and profitability of distributing of television via a regional satellite (Bloch 1992). If this service is to succeed, African governments will have to create an enabling environment for transmitting television via satellite. The high demand for television transmission over Eutelsat has been correlated to the liberalization of television broadcasting in Europe (Bloch 1992).

14.2.4 The International Aggregation Model

The International Aggregation Model (IAM) is linked to the emergence of international digital networks. The combination of international fiber networks, the digitization of telecommunications traffic, and advances in digital switching and transmission (ATM, SONET, etc.) have resulted in new economies for international telecommunications traffic. The IAM is based on the use of optical fiber to create regional networks that also have the capacity to aggregate international telecommunications traffic. The use of this model in Africa would create a regional network that would not depend exclusively on intra-African telecommunications traffic. This model takes Africa out of isolation and provides a global solution to Africa's desire for regional networks. The model has already been employed with the SEA-ME-WE2 (Southeast Asia, Middle East, Western Europe) fiber-optic cable,[24] and has been proposed for an ASEAN (Association of Southeast Asian Nations) regional network.[25] The extension of this model is related to the existence of SEA-ME-WE2 and other international optical fiber links.

Presently, there are three fiber-optic proposals slated for the continent: FLAG's Africa plan,[26] Alcatel's Pan-African Project,[27] and AT&T's Africa One proposal. For the purpose of this study, the discussion will be limited to AT&T's Africa One proposal.

14.2.4.1 Africa One

Africa One is the most ambitious of the fiber-optic proposals. Its originator, AT&T, estimates that it will require over 35,000 miles of fiber-optic cable and have an estimated cost of U.S.$1.9 billion (AT&T 1995). Moreover, AT&T proposes that Africa One will have forty-one landing points in Africa and will also connect to Italy and Saudi Arabia. The fiber will have a transmission capacity of 2.5 gigabits per second. The development of Africa One by AT&T has a three-tier plan.

1. Laying of an optical fiber ring around the continent.
2. Connecting of interior African countries through alternative means, such as satellite.
3. Providing global and transoceanic connections to the rest of the world.

According to AT&T's proposal, Africa One will be owned and operated by a regional corporation. The staff for the regional corporation will be drawn from the participating African telecommunications providers. Also, AT&T has proposed that RASCOM be the majority shareholder in the regional corporation. In early 1995, RASCOM stated unequivocally that it would not invest in the network;[28] however, in May 1995, RASCOM seemed to have reevaluated this position and stated that it would work with AT&T on Africa One—although the exact nature of this relationship was not disclosed.

In addition to RASCOM, AT&T is targeting a broad range of potential investors and funding institutions: African PTTs, private investors, multinational end-users, international carriers, and multilateral and bilateral funding agencies (AT&T 1994). African telecommunications providers will have the option of purchasing either capacity or equity in the network.[29] Major multinational end-users will be able to purchase capacity outright or may make long-term commitments in return for guaranteed pricing.[30] Private investors will be able to buy equity or hold debt. International carriers will be able to purchase capacity on the network.

In trying to sell its idea, AT&T has provided a myriad of arguments concerning the benefits and economic viability of Africa One. First, it has argued that Africa One will complement other regional projects such as RASCOM and Panaftel and will significantly increase intra-African capacity. The company also argues that a regional fiber-optic network has several other benefits:

- *Economies of Scale.* Through the use of the same technology, economies of scale can be achieved.
- *Risk Management.* Due to the regional nature, the broad communities of interest, and the economies of scale that can be achieved, risks are reduced.
- *Broad Flexibility.* Due to regional structure, the network will be able to service countries with differing levels of development and telecommunications traffic.

In addition, AT&T foresees numerous potential sources of revenue for the network:

- *Domestic Revenue.* By establishing multiple land points in one country, domestic traffic can be transmitted over the network.
- *Regional Traffic.* Revenues will be generated by intra-African telecommunications traffic.
- *Satellite Connections.* Revenues will be generated by integrating traffic from the interior into Africa One.

However, these potential revenue sources will not be sufficient to justify or sustain the network. The bedrock of Africa One's economic viability is its ability to connect to AT&T's Global Undersea Fiber Optic Network (GUFON). By the end of 1995, AT&T estimates that forty countries will be connected to its GUFON; further, AT&T estimates that by the year 2000, a hundred countries will be connected to the GUFON. The company proposes that Africa One will aggregate outbound African traffic from the region with other regional fiber-optic networks such as SEA-ME-WE2 and SAT2. This proposal is in line with African telecommunications traffic patterns, namely, that 86 percent of Africa's international telecommunications traffic is destined for countries outside the region (ITU 1994a). Further, AT&T envisions the transmission of significant global traffic through Africa One through the aggregation of global traffic at interregional gateways. The aggregation of intra-African traffic and the transmission of global traffic combined with the other revenue sources discussed earlier make Africa One a viable project from AT&T's perspective.

If the proposal is to move forward, AT&T will have to gain the African nations' support. The company will have to show how the revenue generated from the network will be distributed. For example, although the ability to transmit global traffic through Africa One gives AT&T significant flexibility in traffic routing and provides a potentially significant revenue generator, AT&T's project outline (AT&T 1994) does not indicate how this revenue will be distributed. If AT&T is to gain African support, it will have to address this and other issues that relate to Africa's ability to share in the potential benefits of Africa One.

14.3 The Future of Regional Telecommunications Networks

In the future, Panaftel's model for regional network development is likely to disappear. African telecommunications carriers will be unable to find funding for these types of projects for two primary reasons: the retrenchment of multilateral aid for these types of projects, due to the difficulties encountered by Panaftel; and the inability of African telecommunications providers to compete for funding with international telecommunications providers attempting to establish regional networks in Africa, such as AT&T.

The Appropriate Environment Model (AEM) is likely to become the dominant model at the subregional level. Several factors will precipitate the shift to this model at that level. First, the demise of the multilateral, large-scale model will require that African telecommunications providers desiring to establish regional connectivity adopt another approach. Second, there has been a shift in emphasis from the development of intra-African telecommunications links to development of domestic telecommunications networks. At the end of Panaftel's implementation, the importance of investing in national networks was realized, and Panaftel was redefined to ". . . encompass the entire public telecommunications network down to the subscriber level" (Tedros 1987). This reflects the realization that there is a real telecommunications need at the national level. This emphasis on domestic networks is likely to grow, due to the current restructuring of the African telecommunications sector and the opportunities offered by new technology. Corporatized or privatized telecommunications operators will likely look to increase revenue from the domestic network through investing in its expansion; also, many of the constraints on national network expansion will be lifted once the telecommunications provider is separated from the government (ITU 1994). Further, new telecommunications technologies offer significant opportunities for the expansion of domestic telecommunications. Cellular and VSAT technologies are already being deployed throughout the continent and are aiding the expansion of national telecommunications networks (ITU 1994a). In the future, African telecommunications providers will likely look to employ new technologies to expand domestic networks.[31]

The restructuring of the African telecommunications sector will also lead African telecommunications providers to emphasize the economic viability of international links over political justifications. Thus, if African governments want to establish regional connectivity because of the associated externalities, they will have to adopt the AEM to catalyze regional network development. Finally, the success achieved in those regions that have adopted certain aspects of this model will lead other regions to attempt to implement the principles of the Appropriate Environment Model. For example, the southern African subregion, which has the healthiest Panaftel links (ITU 1994a), has been able to establish appropriate traffic structures[32] and has significantly outpaced other regions in investing in national networks (ITU 1994a).

One should also note that rivalry among countries within the same region to become the international telecommunications hub for that region has also quickened the pace of domestic telecommunications development. For example, Ghana and Nigeria have been making strides to establish regulatory policies that encourage private investment in their national telecommunications industries in an effort to enhance national services in order to become the leader for international call switching within the West African region.

The future of the Regional Satellite Model (RSM) or RASCOM, is extremely questionable. Although plans to establish a dedicated African satellite continue to move forward, the ability of RASCOM to establish a sustainable regional network will depend on numerous factors:

1. The level of network expansion within, and the amount of international telecommunications traffic originating from, landlocked countries, since this will determine the amount of revenue that will be generated by integrating landlocked countries into one of the fiber-plans.
2. The ability to overcome the coordination problems inherent in this model. For example, the dominance of the organization by large and powerful signatories—such as Nigeria—may impede coordination.
3. The establishment of a significant market for a Direct Broadcast Satellite (DBS) in Africa. This will be a difficult task, since DBS will raise troubling questions about national identity and cultural imperialism.

Further, RASCOM raises another issue, namely, whether it is congruent with the current move toward restructuring and separation of the government and the telecommunications provider. The RSM may lead to a conflict of interests: if African telecommunications carriers are eventually privatized and seek to offer intra-African links, this will present competition to RASCOM, which is government-owned. Will African telecommunications regulators attempt to impede projects that compete directly with their own system? Ultimately, considering the issues raised earlier, and that VSAT capabilities are already offered by Intelsat and PanAmSat, the establishment of a dedicated African satellite may not be the most appropriate method for regional network development.

The International Aggregation Model is likely to take hold in Africa as it is doing throughout the world. This model holds numerous benefits for Africa; however, it is unlikely that the current fiber plans will receive much support from

African administrations. It is not clear which plan will prevail, although AT&T has put forward the most comprehensive solution. The most likely scenario is that the plans will continue mainly from international motivations for establishing global fiber systems. African telecommunications carriers will probably jump on board at a later stage.

14.4 Conclusion

The African desire to establish regional connectivity and the climate of regional cooperation that pervaded the debate over African development in the 1960s combined to create the most ambitious telecommunications project to date—the Pan-African Telecommunications Network. Despite its ambitious goals, Panaftel has not become a sustainable, integrated continentwide telecommunications network. Certain links are sustainable and certain regions have advanced further than others in establishing connectivity (e.g., SADC); however, to spur further regional network development, a new model will have to be adopted. The Appropriate Environment Model seeks to establish the necessary milieu in which sustainable networks can develop. The Regional Satellite Model seeks to develop regional networks through a cooperatively owned dedicated African satellite system. The International Aggregation Model employs optical fiber technology for regional network development and incorporates African regional network development into the wider context of global networking.

Although sustainable networks can conceivably be developed with all three models, adopting the AEM would increase the chances of success for both the Regional Satellite Model and the International Aggregation Model. For example, appropriate tariffs for international calls will be essential to the success of both RASCOM and Africa One; and an expansion of national telecommunications networks would increase the traffic transmitted through all African telecommunications systems—national, subregional, or regional; optical fiber, satellite, or microwave. All parties interested in African regional networks (e.g., RASCOM, AT&T, FLAG, African Governments, ITU, etc.) should promote the implementation of the AEM at the national, subregional, and regional levels.

In reference to the RSM, RASCOM should adopt a more flexible stance, rather than bullishly moving forward with plans that originated in the 1970s. It is a different time, in which different technological, economic, and political factors must be taken into account. Development of regional telecommunications networks must be based on the underlying economic viability of the proposed network, not on political reasoning or some other motivation. Panaftel demonstrates the folly within the "If you build it, they will come" approach to network development. Moreover, RASCOM should look at the opportunities offered through other technologies: AT&T's Africa One proposal or one of the other optical fiber plans may hold some opportunities. These plans would increase intra-African capabilities and allow RASCOM to benefit economically from the large amount of extra-African communications, as well as the global traffic transmitted over the network. The question of distribution of revenue must be brought to the table. Furthermore, RASCOM should look to market and increase the penetration of Intelsat's VSAT service; this would also generate revenue for the organization. If the African VSAT market develops and no new technologies supersede VSAT, then RASCOM should consider launching a satellite. But RASCOM must move cautiously; Africa cannot afford another ill-advised attempt at establishing a regional telecommunications network.

Those advocating the IAM will have to continue to make the economic case to African telecommunications carriers and governments, since African support is essential. From the African perspective, AT&T's Africa One proposal may provide significant opportunities; for example, potential revenue, future flexibility, and the transfer of fiber-optic technology to the continent (thereby enabling African telecommunications carriers to learn how to implement and operate fiber-optic systems). If RASCOM opts to reevaluate its position toward Africa One (as it seems to have already begun to do), it should assure that it is a full participant in every aspect of the network, and that Africans take full advantage of all the opportunities and benefits offered by the network.

It is clear that regional cooperation remains a metamodel for African development. This can be seen in the continued attempts to establish economic communities and trade blocs throughout Africa—the ECOWAS, SADC, PTA, and so on. However, the fact that regional cooperation provides opportunities in related areas should not lead to its broad and unqualified application in telecommunications development. Panaftel clearly demonstrates that regional cooperation will not be successful in overcoming all of the obstacles that hinder African telecommunications development. African nations and African telecommunications carriers will have to find appropriate opportunities for cooperation in telecommunications. Certain forms of cooperation are unlikely to be successful. For example, cooperative manufacturing projects[33] will be difficult to establish and sustain because of the political difficulties that are likely to arise, such as the selection of the facility's location and the distribution of benefits like employment.

An appropriate form of regional cooperation in telecommunications is the adoption of common standards. Common standards would increase the ability for regional networks to develop and would create economies of scale within the respective economic communities. This should enable the growth of private manufacturing initiatives. As Mr. Jean Jipguep, deputy secretary-general of the ITU, has stated, " an early agreement on a common standard for mobile communications in Africa would do more to promote investment in this high growth sector than continuing fragmentation and home-grown policies" (Jipguep 1993a). Other forms of cooperation are probably also feasible, such as cooperative training (Crecchio 1980; Riverson 1991) and cooperative purchasing of telecommunications equipment (Hashimoto 1994; Hainebach 1994). Nevertheless, one must question whether these forms of cooperation will be feasible in a more competitive environment in which the government and the telecommunications carriers have been separated and carriers face competition—possibly from other African telecommunications carriers.

Notes

I would like to give thanks to all those who freely provided information, insight, criticism, and guidance for this study. First and foremost, I must thank my wife Sasha Thomas-Nuruddin for the many hours sacrificed to reading and rereading this work. I would like to thank Dr. Martin Elton, NYU; Kevin Holmes, CITI/CTR; Dr. Raymond Akwule, AFCOM International and George Mason University; and Dr. Eli Noam, CITI and Columbia Business School, for their comments and guidance. Finally, I would like to thank all the individuals who freely provided information and resources: Ms. Zeina Afif, AFCOM; Les Breiner, CIDA; Mr. Brown, ITU, Ethiopia Field Office; Mr. LeClerc, CIDA; Carlos Cordeiro, Ph.D., Stanford University; and Mr. Michael Minges, ITU/BDT.

1. One obvious exception is the development of telecommunications in Ethiopia (see Tsige, this volume).

2. Telecommunications was used to counter threats from both African resistance fighters and other colonial powers (Headrick 1991).

3. For example, in Ghana, during the Ashanti War, the British extended telecommunications capabilities from Accra to Kumasi in order to facilitate the suppression of the uprising (Allotey and Akorli, this volume).

4. The African plan subcommittee was created in 1960 by the Plenary Assembly of the International Telegraph and Telephone Consultative Committee (CCITT) and the International Radio Consultative Committee to address the needs of newly independent African nations.

5. This subregional structure was also congruent with regional grouping that had developed in Africa, for example, the Economic Community of West African States (ECOWAS) and the Southern African Development Coordination Conference (SADCC). During this phase of Panaftel's development, the Pan-African Telecommunications Union was established as the Specialized Agency of the OAU in the field of telecommunications. It was hoped that these organizations would be able to assist in the implementation of Panaftel. Recent studies have suggested that the multitude of telecommunications organizations in Africa may have hindered, rather than assisted, telecommunications development on the continent (ITU 1994a, 1994b).

6. During this period, Panaftel also benefited from the initiation of the United Nations Transport and Communications Decade for Africa (1978–88), which provided additional resources to Panaftel.

7. Although it is not explicitly stated in any material reviewed for this study (with the exception of some discussion of the different telecommunications organizations that originated for Francophone and Anglo-West Africa in Sy (1985)), the differences in language between French- and English-speaking West Africa must have hindered telecommunications traffic. (See Kellerman 1990 for an analysis of the role of language in telecommunications flow.)

8. With the exception of the link between Djibouti and Zaire, all the switching centers installed since 1986 have been digital.

9. These data exclude communication with South Africa.

10. Much of the information for this section is based on personal communications with senior ITU officials.

11. This project has an estimated cost of U.S.$5.35 million—U.S.$1.25 million was provided by Ethiopia and U.S.$4.30 million has been contributed by the European Economic Community (Conference of African Ministers 1993).

12. The countries presented in these tables were selected on the basis of the amount of data available in ITU, 1991. These countries had the most comprehensive data. This is

probably due to a failure by other countries to respond to survey requests issued by the ITU.

13. For example, in 1994, it was estimated that to increase the current average ratio of direct exchange lines (DELs) from 1.6 per 100 to 3.0 per 100 by the year 2000 would require an estimated U.S.$30.8 billion investment (ITU 1994b). Obviously, the accuracy of this estimate will depend on the cost per line and the type of technology deployed.

14. Many aspects of this model are articulated and recommended throughout the African Green Paper; however, it is not presented as a specific model for regional network development. Further, many aspects of the Panaftel model are also recommended in the document, which does not make a clear distinction between these two or any other models for regional network development in Africa (ITU 1994b).

15. The African telecommunications sector has the highest average rate of return in the world. Africa's average annual return is 26 percent, while the world average is 16 percent (ITU 1994a). However, these profits are rarely reinvested in the telecommunications network. In 1992, on average, only 36 percent of the profits that were generated by the PTTs was reinvested in the telecommunications network (ITU 1994a).

16. In 1992, sub-Saharan Africa had an average of 58 employees per 1,000 main telephone lines. The next closest regions in the world were the Arab States and Russia with 19 employees per 1,000 DELs, then the Asia-Pacific region with 17 employees per 1,000 DELs (ITU 1994a). This inefficiency can also be found in the procurement procedures and the lack of competitive tendering. This has the effect of raising the cost per new line, which was the highest of any region in the world (ITU 1994a), and lowering the possibility for telecommunications expansion.

17. A joint UNDP/ITU report showed that for ten surveyed sub-Saharan African countries, an average of only 60 percent of bills were collected, and the state was the main debtor (cited in ITU 1994b).

18. A recent study has correlated the free flow of economic factors to trade between nations and the ability to form economic communities, trade blocs, and so on (OECD 1993).

19. A total of U.S.$1.5 million in funding was provided by the African Development Bank, the UNDP, the ITU, the OAU, and the governments of Italy and the Federal Republic of Germany (ITU 1990).

20. In April 1995, RASCOM began reviewing proposals concerning the precise terms of the Non-Signatory Agreement. At that time, the estimated levels of ownership for nonsignatory members ranged between 20 percent and 49 percent of the satellite system. There was no estimate as to the amount of capital that would be raised through the Non-Signatory Agreement. The precise terms of the agreement were to be finalized at a meeting of the RASCOM signatories in Accra, Ghana, in May 1995 (personal communications).

21. In early 1995, RASCOM management declined to disclose the name of their satellite partner.

22. Nigeria wanted the headquarters to be located in Lagos, Nigeria. However, the organization decided to establish its headquarters in Abidjan, Ivory Coast (Jipguep 1993b; African Communications 1994; Conference of African Ministers 1993). Nigeria has since rejoined RASCOM (personal communications).

23. In reference to the optical fiber plans for the continent, specifically AT&T's Africa One proposal, RASCOM asserts that these plans complement its own objectives and will not endanger RASCOM's viability (personal communications).

24. This is an 18,000-kilometer fiber-optic network that stretches from Singapore, through the east coast of Africa, to the Middle East and France. The system was built by AT&T's Submarine Systems division for a consortium of over fifty telecommunications companies.

25. In 1998, ASEAN agreed on a plan—the ASEAN Optical Fiber Cable Network (AOFSCN) (Mohammed and Supaat 1992). The ability to transmit non-ASEAN traffic may become a key revenue stream for this network.

26. Fiber-optic Link Across the Globe (FLAG) is a partnership between Nynex (United States), Gulf Associates (United States), Dallah Albaraka Group (Saudi Arabia), and Marubeni Corporation (Japan). In 1995, FLAG was pushing a proposal to connect African countries to its 31,000-kilometer fiber-optic link from the United Kingdom to Japan. The FLAG proposal has an estimated cost of U.S.$800 million (Langworth 1994).

27. Alcatel's Pan-African project focuses exclusively on West Africa; it has nineteen landing points from Cape Town to Casablanca. The project has an estimated cost of U.S.$600 million. Alcatel hopes to solicit 50 percent of the necessary financing from European, Asian, and South American investors. It proposes that the remaining half be contributed by African countries—it estimated that the average cost per country will not exceed U.S.$18 million (Ayre 1994).

28. Personal communications.

29. AT&T foresees multilateral and bilateral funding agencies as the most likely source of capital for African telecommunications providers' investment in Africa One. On March 8, 1995, William B. Carter, president of AT&T Submarine Systems, Inc., testified before the Joint Hearing on Trade and Investment in Africa. He argued that Africa One was essential to Africa's development and should be supported by the United States through the bilateral aid agencies, such as the Agency for International Development (AT&T 1995).

30. The long-term economic viability of this strategy has been questioned (Langworth 1994).

31. Of course, African nations must be cautious in their selection of technology and must be certain that their choices appropriately satisfy African needs.

32. The southern African region has adopted a Sender Keeps All tariff structure. Under this model the originating country keeps all tariffs charged to the user, which eliminates the need for settlement of payments (ITU 1994b). This model may also provide an incentive to lower international tariffs, since it would theoretically increase the amount of traffic that flows over the international link and translate into increased revenue for the national telecommunications provider.

33. This application of the Regional Cooperation Model is advocated in OAU, Draft Protocol on Transport and Communications, Article 10 (rev. 2), item h.

Bibliography

Africa Communications. 1994. "RASCOM: News Maker Interview with the New Director General." *Africa Communications,* March/April.

Akwule, Raymond. 1990. African Communications in an Information Age: What Role for a Joint Soviet–U.S. Development Program? In *Agenda for Action African-Soviet-U.S. Cooperation,* edited by Gromyko & Whitaker. Boulder, Colo.: Lynne Rienner Publishers.

ARB. 1992. "Communication and Transport." *African Research Bulletin: Economic, Financial and Technical Series,* September 16–October 15.

______. "Communication and Transport." *African Research Bulletin: Economic, Financial and Technical Series,* July 16–August 15.

AT&T. 1994. *Africa One: Project Summary Description.* AT&T.

______. 1995. "AT&T Testifies Its 'Africa One' Plan Would Provide the Region's First High-Capacity Digital Access to Global Information Superhighway." Press release, March 8.

Ayre, Jim. 1994. "Wiring Africa: Four Fiber Optic Plans on the Drawing Boards for Africa." *Africa Communications*, May/June.

Bloch, Jean-Jacques. 1992. "Lessons Drawn from Two Existing Regional Satellite Networks: Arabsat and Eutelsat." In *Regional Interests and Global Issues: The Challenge of Telecommunications Integration for the Pacific*, edited by Lofstrom and Wedemeyer. Proceedings of the Pacific Telecommunications Conference, Honolulu, Hawaii, January 12–15.

Checchi and Co. 1968. *A Study of the Economic and Technical Feasibility of Improving Telecommunications in West Africa.* Washington, D.C.: Checchi and Co.

Crecchio, F. P. de. 1980. The Panaftel Transmission Network: A Review of Some Maintenance Problems Encountered. Presented at a meeting of the Pan-African Telecommunications Union, Kinshasa, Zaire.

Conference of African Ministers. 1993. *Report of the Ninth Meeting of the Conference of African Ministers of Transport and Planning.* March 12–13.

Dymond, Andrew. 1987. "Reducing the Number of Missing Links: Regional Cooperation and Telecommunications Development in Southern Africa." *Telecommunications Policy*, June: 121–34.

ECA. 1993. *Survey of the Economic and Social Conditions in Africa, 1988–89.* Annual report, Economic Commission for Africa.

Girmaw, I. 1990. "The PANAFTEL Project." in *Working Documents*, African Telecommunications Development Conference, Harare, Zimbabwe, December 6–11.

Hainebach, Geoff. 1994. Telecommunication Equipment Procurement Strategies for African Network Operators. In *Telecommunications and Development in Africa*, edited by Kiplagat and Werner. Amsterdam: IOS Press.

Hashimoto, Satoru. 1994. Possibilities for Benefits of Regional Co-operation in Procurement of Telecommunication Systems and Their Operation. Paper presented at Africa Telecom 94, Cairo, Egypt, April 25–29.

Headrick, Daniel R. 1991. *The Invisible Weapon: Telecommunications and International Politics 1851–1945.* New York: Oxford University Press.

Hudson, Heather. 1991. "Telecommunications in Africa: The Role of the ITU." *Telecommunications Policy*, August: pp. 343–50.

IMF. 1994. *Directions of Trade Statistics Yearbook.* Washington, D.C.: IMF.

ITU. 1974. *Panaftel: The Pan-African Telecommunications Network.* Geneva: ITU.

______. 1990. *Report on the Activities of the ITU in 1989.* Geneva: ITU.

______. 1991. *General Plan for the Development of Regional Telecommunications Network in Africa 1991–1995.* Dakar: ITU.

______. 1994a. *African Telecommunications Indicators.* Geneva: ITU.

______. 1994b. *Telecommunications Policies for Africa: The African Green Paper.* Geneva: ITU.

Jipguep, Jean. 1993a. "The Role of Telecommunications in the Economic Integration of Africa." Abidjan, June 17.

______. 1993b. "Speech presented at the First Meeting of the Interim Assembly of Parties of the Regional African Satellite Communications Organization." Abidjan, September.

Kellerman, Aharon. 1990. "International Telecommunications around the World: A Flow Analysis." *Telecommunications Policy*, November: 461–75.

Langworth, G. 1994. Combining Forces: The Challenge to Connect Africa. In *Telecommunications and Development in Africa*, edited by Kiplagat and Werner. Amsterdam: IOS Press.

Meeting of African Traffic Managers. 1987. *The Panaftel Network: Current Situation,*

Identification of Problems, Recommendation and Follow-up Methods. Addis Ababa, Ethiopia: ETA.

Mohammed, A. and A. Supaat. 1992. ASEAN Optical Fiber Cable Network (AOFSCN). In *Regional Interests and Global Issues: The Challenge of Telecommunications Integration for the Pacific,* edited by Lofstrom and Wedemeyer. Proceedings of the Pacific Telecommunications Conference, Honolulu, Hawaii, January 12–15.

N'Zengou, E. 1990. Contributions of Project RAF/87/085 to the Improvement of Networks and Services. In *Working Documents,* African Telecommunications Development Conference, Harare, Zimbabwe, December 6–11.

OAU. African Economic Community: Draft OAU Protocol on Transport and Communications, Article 10 (rev.2)—Telecommunications.

OECD. 1993. *Regional Integration and Developing Countries.* Paris: Organization for Economic Cooperation and Development.

Okundi, Philip O. 1976. "Pan-African Telecommunication Network: A Case of Telecommunication in the Development of Africa." *IEEE Transactions on Communications* 24, no. 7 (July).

Paltridge, Sam. 1994. A Survey of the Tariff Structures in Africa—Comparison to the Rest of the World. Paper presented at Africa Telecom 94, Cairo, Egypt, April 25–29.

Podmore, C. and D. Faguy. 1986. "The Challenge of Optical Fibres." *Telecommunications Policy,* December: 341–51.

Riverson, Lantiri Kwabena. 1991. Telecommunications Development in Africa: A Case Study of the Pan-African Telecommunications Network (PANAFTEL) from 1962 to the Present. Ph.D. diss., Howard University.

Rzepecki, Kazimir. 1990. Summary of Documents No. 15 and 16. In *Working Documents,* African Telecommunications Development Conference, Harare, Zimbabwe, December 6–11.

Sy, Demba Jacques Habib. 1986. Capitalist Mode of Communications, Telecommunications Underdevelopment and Self-reliance: An Interdisciplinary Approach to Telecommunications History and Satellite Planning on a Pan-African Scale. Ph.D. diss., Howard University.

Tedros, Gabriel. 1987. "African Telecommunications after Panaftel." *Telecommunication Journal* 54 (March): 198–201.

UNCTAD. 1993. *Regional Integration Issues in Africa.* Report by the secretary-general of UNCTAD, February 16.

Yaker, Layashi. 1994. Telecommunications Development in Africa—State of the Art and Plans. Paper presented at Africa Telecom 94, Cairo, Egypt, April 25–29.

Young, Crawford. 1986. Africa's Colonial Legacy. In *Strategies for African Development,* edited by Robert J. Berg and Jennifer Seymour Whitaker. Berkeley: University of California Press.

15

New Communications Technologies for Development: Challenges for Africa

HEATHER E. HUDSON

The importance of expanding and upgrading telecommunications in developing countries has been a major telecommunications theme throughout the 1980s and 1990s. Nowhere are the needs greater or the problems more daunting than in Africa.

Access to telecommunications in most of Africa is extremely limited. For a population of about 570 million, there are only 3.5 million exchange lines. The poorest countries have less than one telephone for every 2,000 inhabitants and virtually no reliable communications in rural and isolated areas (Hudson 1990a). Of the approximately 151,000 villages in Africa, 121,000—or 80 percent—have no telephone service although the population of the continent is at least 70 percent rural (Jipguep 1990). Teledensities (number of telephone lines per 100 population) range from 0.1 to 8.9. At least thirty countries have less than 1 line per 100 inhabitants.

Of course these averages do not reflect actual access to telecommunications for most Africans. Telecommunications facilities are concentrated in urban areas, although typically 80 percent or more of the population live in rural areas. Thus the majority of Africans have no access to telephone service.[1] Table 15.1 shows that many countries have less than 1 line per 1,000 people in areas outside the largest city.[2]

Internet access in Africa is also extremely limited. As of mid-1994, at least twenty African countries had no Internet access. Only South Africa had full Internet connectivity. The other countries had electronic mail access only, typically at a major university (*NSF Network News* 1994).

To meet the very modest goal of 1 telephone per 100 inhabitants by the end of the century, the existing 5 million lines will have to be increased by another 4 million, requiring an investment estimated at U.S.$6 billion (Westendoerpf and Odeh 1990).

Table 15.1. African Teledensity: Urban and Rural Areas (1992)

Country	Teledensity: National	Percentage of Population in Largest City	Teledensity: Largest City	Teledensity: Rest of Country
Algeria	3.7	12.1	6.0	3.54
Angola	0.5	16.9	2.2	0.18
Benin	0.3	4.4	4.3	0.21
Burkina Faso	0.2	4.6	3.0	0.23
Burundi	0.2	4.6	4.5	0.03
Cameroon	0.4	6.7	3.3	0.25
Chad	0.1	12.9	0.5	0.01
Djibouti	1.4	75.0	2.0	0.29
Egypt	3.9	17.2	8.1	3.94
Ethiopia	0.3	3.7	4.6	0.25
Ghana	0.3	7.3	2.2	0.30
Ivory Coast	0.7	17.7	3.0	0.21
Madagascar	0.3	5.8	2.9	0.29
Malawi	0.3	3.6	4.3	0.21
Mali	0.13	7.8	1.2	0.11
Mauritania	0.3	22.0	1.2	0.11
Morocco	2.5	12.8	5.9	2.13
Mozambique	0.4	10.1	2.7	0.09
Namibia	4.0	9.6	20.5	2.23
Niger	0.1	7.6	1.2	0.05
Nigeria	0.3	6.6	1.4	0.19
Senegal	0.8	20.1	2.8	0.28
Sierra Leone	0.3	16.7	1.8	0.05
Somalia	0.2	11.5	1.7	0.04
South Africa	8.9	6.4	44.6	7.42
Sudan	0.2	7.7	0.7	0.22
Swaziland	1.9	6.5	10.9	1.35
Tanzania	0.3	6.3	2.0	0.20
Togo	0.4	19.1	1.8	0.08
Tunisia	4.5	20.0	6.8	4.03
Uganda	0.2	4.0	1.6	0.10
Zambia	0.9	12.2	2.5	0.71
Zimbabwe	1.2	11.4	5.2	0.71

Source: World Telecommunications Development Report, Geneva: ITU, 1994.

Not only is the amount of infrastructure in Africa limited, but the quality of service is also poor in many regions. Local call completion rates are less than 30 percent compared with more than 70 percent for industrialized countries (Saunders, Warford, and Wellenius 1994). The World Bank estimates that the average network expansion rate in Africa is under 7 percent per year, whereas the economic requirement is not less than 12 percent. Sub-Saharan Africa has invested only 0.3 percent of gross domestic product (GDP) in telecommunications, while Europe, which is near saturation, has invested 0.7 percent of GDP and Latin America 0.6 percent (Lomax 1990). Clearly, management of the telecommunications sector must be improved if increased investment is to be obtained and utilized effectively to meet these growth targets.

15.2 Telecommunications in the Development Process

Telecommunications is a "missing link" in much of the developing world, as the Maitland Commission noted (International Commission 1984).[3] The telecommunications link is not simply a connection between people but a link in the chain of the development process itself. There is now considerable evidence that telecommunications contributes to socioeconomic development. Projects and studies conducted in the 1980s and 1990s have shown that telecommunications can facilitate many development activities, including agriculture, industry, shipping, education, and health and social services (see, e.g., Saunders, Warford, and Wellenius 1994; Hudson 1984; and International Commission 1984). These studies have been augmented by recent research on rural telecommunications in the United States (cf. Parker and Hudson 1992; Parker et al. 1989).

In an increasingly time-conscious world, distance represents time. In economies that depend heavily upon agriculture or the extraction of resources (lumber and minerals), distance from urban markets has traditionally been alleviated only by the installation of improved transportation facilities, typically roads. Yet transportation links leave industries without the access to the information that is becoming increasingly important for the production and marketing of their commodities.

Another disadvantage faced by many developing countries is economic specialization. As they strive to diversify their economies, timely access to information on market opportunities and modern production and management techniques becomes even more critical. As developing countries join the global market by attracting multinational corporations, establishing joint ventures, and developing service industries, they soon recognize the need for a reliable and modern telecommunications network.

Telecommunications is also vital to the emerging information sectors in developing regions. The great distances between the major research institutes and development centers, as well as the vagaries of postal services and expense of airfares, mean that experts are isolated from each other and from the people they are trying to help. For example, the National Research Council points out that sharing information is vital for Africa if Africans are to contribute to finding solutions to their own development problems:

> Economic development in Africa will depend heavily on the development of the information sector. Countries will need the ability to communicate efficiently with local and overseas markets to determine where they may have comparative advantages for supplying their products to consumers or to purchase essential imports, based on current prices and services. Many of the economic development problems facing African countries have scientific and technological components that will require solutions to be developed in Africa by African scientists. . . . Lack of information is a critical constraint. (National Research Council 1990)

In sum, the ability to communicate instantaneously can facilitate the development process by increasing the following:

- Efficiency, that is, the ratio of output to cost.
- Effectiveness, or the extent to which development goals are achieved.
- Equity, that is, the distribution of development benefits throughout the society.

15.3 New Technologies and Services

Telecommunications technology has changed dramatically in the past decade. Perhaps the most telling evidence of change is the cover of the Maitland Commission report itself, which shows two rotary dial telephones. This is not to say that digital switching did not exist by 1984, but that it was not considered necessary or perhaps even appropriate for developing regions. A second indicator is that the commission specifically identified only telephone service, and proposed access "in due course [to] the other services telecommunications can provide." Today, many of those services could be available as soon as telecommunications service is provided.

Recent innovations in telecommunications and other information technologies have resulted in new equipment and services that are particularly suitable for the applications of developing countries. The following sections describe examples of these technologies and services.

15.3.1 Satellite Communications

Satellites have been considered an ideal technology for Africa because of the continent's limited infrastructure, vast distances, and widely scattered population. African countries were among the major early beneficiaries of the Intelsat system, which linked African capitals to each other and to the rest of the world. A regional satellite could extend these benefits by linking towns and villages into national networks and distributing broadcasting services.

15.3.1.1 Thin-Route Satellite Earth Stations for Voice and Data

The advent of small low-cost earth stations, such as those used for rural telephony with domestic satellites or the VISTA terminals used with Intelsat satellites, bring voice and data communications to isolated regions such as deserts, jungles, mountainous areas, and offshore islands. These earth stations may be installed in any community or project site without the need for expensive terrestrial links to the national network. They may serve the surrounding territory through line-of-site radio links. Earth stations designed for mobile communications such as Inmarsat's suitcase terminals could also be used for communications with refugee camps and to coordinate emergency relief activities (Hudson 1990a).

15.3.1.2 Data Broadcasting

The flow of information within the developing world has been hampered by the cost of distribution and by the lack of access to telecommunications facilities in rural areas. Very small aperture terminals (VSATs) now make it possible for news service information to be disseminated to virtually any location. News service copy is transmitted by satellite from a hub earth station that may be shared with other data, voice, and video customers. These "micro earth stations" may be powered using photovoltaics or portable generators (Parker 1987). Reuters uses this VSAT technology for news service feeds to Latin America. In Asia, the World Broadcast Service based in Hong Kong uplinks news service feeds to Intelsat's

Indian Ocean satellite, which covers 80 percent of the world's population, including much of Africa.

15.3.1.3 VSATs for Interactive Data Communications

Microcomputers or terminals linked to mainframes via interactive VSAT technology can be used to collect and update information from the field. A VSAT network called NICNET operated by the Indian government's National Informatics Center (NIC) now links 160 locations and will be expanded in the next stage to more than 500 (Blair 1988). Similar systems may be used for electronic banking, whether linking teller machines to computers or remote bank branches to headquarters—and for other interactive applications such as reservation systems, weather and pipeline monitoring, and other field data collection. Most recently, VSATs are being employed for Internet services. From the VSAT hub, an Internet Service Provider located in African can link into a major Internet backbone in the United States via satellite.

15.3.1.4 Regional Geostationary Satellites

Until 1995, the only geostationary satellite system providing voice and data communications for Africa was the Intelsat system. Intelsat has linked African capitals to each other and to the rest of the world for more than two decades. Some countries such as Algeria, Sudan, and Nigeria have also leased Intelsat capacity for domestic communications, typically linking their provincial capitals. However, the Intelsat system is designed for heavy-route trunking rather than thin-route services. There have been many attempts to obtain a regional satellite system for Africa that would be specifically designed to meet needs for rural telecommunications and broadcasting services. The ITU, the Organization for African Unity (OAU), and other organizations sponsored the RASCOM (Regional African Satellite Communications) project to examine the feasibility of a regional African satellite. Yet the difficulties of working out ownership, management, and financing have stymied regional proposals such as RASCOM. However, other satellite capacity for Africa has become available. In 1995, PanAmSat launched its PAS-3 Atlantic Ocean Satellite, which offers coverage over Africa as well as connectivity between Europe and the United States. In the same year, PanAmSat also launched its PAS-4 Indian Ocean Satellite, which includes dedicated high-powered Ku band-spot beams over Southern Africa to support the region's first direct-to-home satellite television platform. Other global geostationary satellite systems may follow to compete with Intelsat. Meanwhile, the commercial viability of other regional satellites for either multipurpose use or specialized broadcasting applications is being evaluated by several entities. And Intelsat itself modifyied its footprints to accommodate developing country requirements.

15.3.1.5 Microsatellites

Very small satellites that provide limited but very inexpensive message communications are known as microsatellites. SatelLife of Cambridge, Massachusetts, operates two microsatellites, Healthnet I and Healthnet II, with sixteen stations licensed in Africa. These satellites provide store-and-forward data communications; when the satellite passes overhead, data can be transmitted and received by very cheap earth stations linked to personal computers. Field reports from the Gambia cited improved efficiency of collecting epidemiological data from vaccine trials using Healthnet instead of having a person from the Ministry of Health travel 500 kilometers every week to pick up the data. Similarly, in Cameroon, Healthnet is used for logistics coordination, administration, and communication, instead of someone traveling from province to province (SatelLife News 1994b).

15.3.1.6 Low Earth Orbiting Satellites

Another satellite option is likely to be low earth orbiting (LEO) systems designed for personal and mobile communications. Such systems could also be used for fixed communications in rural areas. Several global satellite systems have been proposed using LEO satellites. Frequencies were allocated for LEO systems at the ITU World Administrative Radio Conference in 1993. Low earth orbiting satellites are intended to be used for mobile communications from small handheld terminals. Proponents of LEO systems have stated that this technology could be used to provide communications in isolated regions such as much of rural Africa. While the availability of handheld units for ubiquitous communications is appealing, price estimates for this service in the range of U.S.$3 per minute are far beyond the means of most African users. Even if prices were to drop considerably, LEO communication would not be affordable for rural Africans. Major rural economic interests might use this service, for example, to provide telephone service for tourists in game parks, to reach remote mining sites and plantations, and so on.

15.3.2 New Radio Technologies

Advances in radio technology such as cellular radio and rural radio subscriber systems offer affordable means of serving rural communities. A recent news magazine article on the communications revolution showed a Masai warrior talking on a cellular phone.

Radio technologies also make it possible to reach villages without laying cable or stringing copper wire. A radio network may extend to clinics, development offices, and villages from a backbone microwave system, as is done in Zimbabwe and Rwanda (see chapter 7). Radio may also link surrounding villages to satellite earth stations, an approach used in Peru.

15.3.3 Digital Compression

With voice compression technology, digitized voice signals can now be "compressed" so that several voice signals can share a single 64-kilobit channel. Sixteen-kilobit voice has proved viable for telephony; some vendors have also introduced 8-kilobit voice with adequate quality for basic telephony. The advantage of the compression is that the cost per voice circuit can be significantly reduced if up to eight voice transmissions can share a single channel.

In video compression technology, digitized video can be sampled and compressed so that video signals require as few as 64 kilobits instead of the typical 1.5 to 2 megabits. The quality of the compressed picture has improved and the price of video codecs (coders/decoders) has dropped considerably in the past couple of years, with the result that video conferencing may be a viable alternative in Africa where travel between capitals is both expensive and time-consuming.

15.3.4 Conferencing and Messaging

A thin-route service with considerable promise for development applications is audio teleconferencing. Several sites can be linked together through a bridge at a switching point or by assigning a common frequency on a satellite audio channel. Electronic mail or communication via computer is a means of exchanging information immediately. Personal computer users worldwide may now interact using various electronic mail networks. Messages may be sent from one computer to another by communication through a host computer that is equipped with communications and message-processing software including "mailboxes" for subscribers. These services are cheaper than voice communications and overcome the time zone differences that hinder real-time communications. Users may dial into local nodes of packet-switched networks to reduce transmission costs. Specialized electronic mail networks have been established for users in developing countries (International Development Research Centre 1989).

Another application of computer communications is computer conferencing, that is, interaction of many users through a central host computer. Each conference member may share ideas with the others and respond to their comments. Participants may log on at their convenience, thus avoiding the need for scheduling to accommodate individual schedules and time zone differences.

15.3.5 Transmitting Hard Copies

Another technology with widespread development applications is the facsimile or fax machine, which enables any type of hard copy including print, graphics, handwritten messages, and the like to be transmitted over a telephone line. Also, "fax boards" may now be installed in personal computers to allow a message created on a personal computer to be sent directly to a facsimile machine.

15.3.6 Personal Computers

Information in the form of databases, full text of journals, video images, and other graphics may now be stored on compact discs and retrieved with a relatively inexpensive reader attached to a personal computer. The advantages of CD-ROMs (compact disc, read-only memory) include vast storage potential, low cost, durability, and ease of use. In addition, CD-ROMs can be used on a standalone basis, without the need for on-line access to databases. Of course, the discs must be frequently updated to keep information current. University and other research libraries in many African countries have rapidly adopted CD-ROMs because of the vast amount of information from journals and other sources that can be accessed.

The enhanced graphics capabilities of personal computers now make it possible to use desktop publishing systems to produce newsletters and other printed materials without typesetting. These features may be particularly valuable in countries where newspapers, texts, and development materials in local languages may be scarce and costly to produce. Development agencies may now produce their own materials in-house. Storefront desktop publishers may enable many small users to share the desktop publishing equipment and software. Desktop publishing may be combined with telecommunications, such as facsimile, for example, so publications may be inexpensively produced and distributed.

15.3.7 Photovoltaics

Photovoltaics are also important for the future of African telecommunications. Harnessing sunlight is an ideal solution for generating power in a continent where power grids are overtaxed in cities and often nonexistent in rural areas. Solar-powered rural call offices, repeaters, and satellite terminals are being introduced, but more affordable and appropriate designs are needed.

Electrification is an important component of any development strategy. A priest who had founded a cooperative in Rwanda that assembled solar panels explained to the author: "Development begins in the head." The package he had developed included a power supply for two lights and a radio for a small house, so that the family could read and listen to the radio after dark (Hudson 1991a).

15.4 Using New Technologies for African Development

15.4.1 Applications and Benefits

The new technologies offer many opportunities to overcome the barriers of distance that make information acquisition and dissemination so difficult in many developing regions. The following sections describe some examples.

15.4.1.1 Benefits for Commerce

A Nairobi industrial spare parts firm expanded 35 percent after the installation of additional telephone lines (Saunders, Warford, and Wellenius 1994). Farmers in the Nile delta are able to obtain better prices for their produce by contacting buyers in Alexandria directly by telephone rather than dealing through local middlemen (Hudson, unpublished research). National parks not only protect the environment but generate significant foreign exchange from tourism in many parts of Africa. Game wardens and park rangers use two-way radios to combat poaching and assist visitors.

Computers combined with telecommunications enable organizations to conduct business from virtually any location. Banks may transfer funds internationally using the SWIFT network (Hudson and York 1988). Airlines may book reservations from ticket offices, airports, and travel agencies. Brokers and traders may buy and sell coffee, soybeans, copper, petroleum, and the like, electronically. With reliable telecommunications links, these activities need not be limited to cities. Agricultural cooperatives may use computer terminals to find where to get the best prices for their crops. Tourist lodges in scenic areas may book reservations.

15.4.1.2 Logistics

Field reports from the Gambia cited improved efficiency of collecting epidemiological data from vaccine trials using Healthnet instead of having a person from the Ministry of Health travel 500 kilometers every week to pick up the data. Similarly, in Cameroon, Healthnet is used for logistics coordination, administration, and communication, instead of someone traveling from province to province (*SatelLife News* 1994b).

Unfortunately, limited access means that the benefits of improved telecommunications too often remain hypothetical. Studies by the ITU, World Bank, and other agencies have documented the benefits of telecommunications in rural and urban activities that could be obtained if reliable telecommunications were available. For example, Ugandan cotton and coffee cooperatives could eliminate unnecessary trips by managers and improve logistics if telecommunications linked the cooperatives with each other and their processing and marketing operations. It was estimated that these cost savings would offset the expense of the telecommunications investment in one to four years, depending on the technology selected for the network (Saunders, Warford, and Wellenius, 1994, pp. 185–186).

15.4.1.3 Electronic Messaging and Meetings

Facsimile transmission and electronic mail may be particularly viable alternatives to sending hard copies of correspondence and documents through the mail, where service may be slow or unreliable. Managers and researchers located in different cities may exchange information quickly. These technologies can also be used to link project staff in the field with each other and with headquarters.

Travel in Africa is both very expensive and very time-consuming. Officials may require two to three days to travel within the African continent because of inconvenient flight schedules and the limited frequency of interregional flights. Airfares are also high because of lack of competition on most routes. However, managers, development experts, or project staff may now stay in touch electronically rather than having to travel for face-to-face meetings. Audio conferencing allows participants at several sites to participate in the same meeting, while computer conferencing allows for group members to interact at their convenience by reading and contributing to a discussion stored on a host computer. Video conferencing may also be a viable option for linking major African capitals or research centers.

These electronic meetings do not offer the richness of face-to-face interaction, but they may be particularly important as a substitute for traveling to meetings where transportation costs severely strain limited travel budgets.

Researchers have also hypothesized that reducing isolation can help to reduce personnel turnover. While causal data are difficult to obtain, it appears that communication is at least an important factor. For example, better communications is cited as one of several factors encouraging reversal of the medical brain drain in Navrongo, Ghana (*SatelLife News* 1994a).

15.4.1.4 Training

Audio conferencing may be used to update field staff without bringing them to the cities for training. For example, in Peru the Rural Communication Services Project linked seven rural communities, three via satellite, and four via VHF radio and then via satellite to the national network. More than 650 audio teleconferences concerning agriculture, education, and health were carried out during the project (Mayo et al. 1987).

The ITU has piloted training courses on telecommunications that use computer conferencing and facsimile to supplement written materials.

15.4.1.5 Distance Education

Audio conferencing has been used in other parts of the world to link isolated students who are studying by correspondence. For example, the University of the South Pacific uses a satellite-based audio conferencing network to provide tutorials to correspondence students scattered in ten island nations of the South Pacific. The University of the West Indies (UWI) also offers instruction to students at extension centers throughout the Caribbean using a combination of satellite and terrestrial audio links (Hudson 1990a).

In many parts of Africa, there are well-established correspondence courses, which were typically developed to help students complete their high school education and teachers to attain certification. However, dropout rates are high. The experience at USP and UWI indicates that regular tutorials can reduce dropout rates by providing advice and feedback for students and by creating a sense of community with other correspondence students. Reliable telecommunications links in Africa would enable students to participate in audio or computer conferences from local community centers. Several distance learning projects which use satellite technology to educate students living in remote rural areas have taken off in South Africa. For example, the Africa Growth Network offers interactive educational courses to over 2,000 remote sites throughout the country on the PanAmSat satellite.

15.4.1.6 Research

Computer terminals or personal computers with modems linked to the telecommunications network can provide access to databases anywhere in the world. Agricultural researchers, for example, may access the Food and Agriculture Organization (FAO) databases in Rome. Health researchers may search the database of the National Library of Medicine in Bethesda, Maryland. Others may search specialized development databases such as those for agriculture and energy in India and for development project management in Malaysia.

Costs may be reduced further if these searches can be localized. Databases may be downloaded onto computers within the country, with updates transmitted at regular intervals using telecommunications. The search then becomes local, without the cost of connect time.

Databases in CD-ROM format are also proliferating rapidly. Several universities and research centers in Africa now search databases locally using CD-ROM drives attached to personal computers (National Research Council 1990). Many bibliographic databases contain detailed abstracts. Some CD-ROM materials now include full text, an important consideration where books and journals are expensive and difficult to obtain.

15.4.1.7 Dissemination of Information

Information for use in publications may be transmitted from the field and from regional centers to desktop publishing locations via telecommunications networks. For example, development workers and reporters in the field could send in reports by facsimile; these materials would then be edited and published in newsletters in the city. Posters and notices could be faxed to the rural communities. Newsletters could be faxed either directly to the communities or to regional centers for duplication and dispatch to schools, clinics, or government offices in their territory. Information obtained from various sources such as news services, databases, and teleconferences could also be disseminated to development workers throughout the country or region via facsimile.

These technologies can also help to foster democracy. In Rwanda, before the massacres of 1994, fledgling political parties used fax machines to organize rallies and sent audiocassettes and videotapes of the president's speeches around the country so that people would know what their leaders were saying. Reporters used newly installed rural telephones to file stories.

15.4.2 New and Changing Demands for Services

The proliferation of technologies and services has several implications for African telecommunications planning. First, there are likely to be new and changing demands for telecommunications services. It is not unrealistic to expect that project managers will want to communicate from laptop computers in the field, rural businesses will want to send messages by facsimile, and environmental agencies will want to monitor data on rainfall, soil erosion, and livestock collected in rural locations.

15.4.2.1 Voice and Data

While basic voice communication is still the first priority, many African users now have requirements for data communications as well, particularly facsimile and relatively low-speed data communications. Thus transmission channels must be reliable enough to handle data and voice traffic.

15.4.2.2 Urban and Rural

The availability of relatively low-cost radio and satellite technologies for serving rural areas makes it possible to reach even the most remote locations and to base priorities for service on need rather than proximity to the terrestrial network. But these links also need to be highly reliable if rural users are to take advantage of the applications of facsimile and data communications just described.

15.4.3 Responding to Demand for New Services

The new technologies now available are likely to create demand for new services such as packet-data networks and teleconferencing networks. African telecommunications managers must be able to respond to these demands both by providing the technical facilities and by setting realistic tariffs if users are to take advantage of the information-sharing potential now possible via telecommunications.

15.5 Bottlenecks and Bypass

15.5.1 The PTT Bottleneck

Today more than 95 percent of Africa's population has no access to reliable telecommunications, despite the availability of VSATs and other relatively low-cost technologies (VHF, UHF, rural subscriber microwave systems, cellular radio, and the like). Given the demand, the reduction in costs, and the potential benefits, why is the diffusion rate so low? There are many reasons, but a major problem is that in most cases the telecommunications administration (PTT) acts as a gatekeeper or bottleneck that prevents customers from obtaining equipment and services. Thus the government-operated utility model that was adopted to protect the public interest now acts as a constraint to retard growth of the telecommunications sector and, as a result, the economy as a whole.

Some examples illustrate the result of the bottleneck. First, the number of television sets has greatly exceeded the number of telephones in many African countries. This imbalance is generally not the result of government priorities but of either lack of awareness of the role of telecommunications, lack of coordinated planning in delivery of services, or unwillingness to give up political and economic control.

Data from forty countries show that on average there are four times as many television sets as telephone lines in Africa (see table 15.2). However, the range is even more striking. Sudan, for example, has 29.8 times as many television sets as telephone lines; Nigeria and Liberia have more than ten times as many television sets as telephone lines, while the Ivory Coast has 8.9 and Uganda has 6.9. Countries where there are approximately the same number of television sets as telephone lines are typically among the poorest, with very limited access to either.[4]

Similarly, personal computers are becoming more widely available than telephone lines in many development agencies, businesses, and universities in Africa. The reduction in cost and increase in computing power of personal computers have been accompanied by advances in telecommunications that should make affordable voice and data communications available virtually anywhere. Yet access to computers is by way of a competitive marketplace, whereas telecommunications services are provided in most countries through access to a single national network. While a personal computer is a standalone technology (despite the fact that it can be linked to other computers through a network), a telephone set and a satellite terminal are useless by themselves—they require a network. Thus, users cannot meet the need for telecommunications alone by buying a telephone set, a cellular phone, or a VSAT. They must be able to connect to a network, and the PTT controls access to the only network. If the PTT is not responsive to consumer needs, as is often the case, frustrated consumers remain without service. These same consumers can usually buy standalone technologies such as computers and videocassette recorders (VCRs) on the open market.

Table 15.2. Ratio of Television Sets to Telephone Lines in Selected African Countries (1992)

Country	Telephone Lines per 100	TV Sets per 100	Ratio of TV Sets/Tel. Lines
Algeria	3.7	3.0	0.8
Angola	0.5	0.6	1.3
Benin	0.3	0.5	1.5
Burkino Faso	0.2	0.5	2.5
Burundi	0.2	0.1	0.4
Cameroon	0.4	2.3	5.4
Central African Republic	0.2	0.4	2.4
Chad	0.1	0.1	1.7
Congo	0.8	0.6	0.8
Djibouti	1.4	5.2	3.6
Egypt	3.9	10.8	2.7
Equatorial Guinea	0.4	0.7	2.1
Ethiopia	0.3	0.2	0.9
Gabon	1.9	3.8	2.0
Ghana	0.3	1.5	5.0
Guinea	0.2	0.7	3.7
Ivory Coast	0.7	6.0	8.9
Lesotho	0.6	0.6	1.0
Liberia	0.2	1.8	10.8
Madagascar	0.3	2.1	7.1
Mali	0.1	0.1	0.9
Mauritania	0.3	1.7	5.4
Morocco	2.5	7.4	3.0
Mozambique	0.4	0.3	0.7
Namibia	4.0	2.6	0.7
Niger	0.1	0.5	3.8
Nigeria	0.3	3.0	10.6
Senegal	0.8	3.6	4.8
Sierra Leone	0.3	1.0	3.2
Somalia	0.2	1.7	10.4
South Africa	8.9	10.3	1.2
Sudan	0.2	7.2	29.8
Swaziland	1.9	2.0	1.0
Tanzania	0.3	0.2	0.5
Togo	0.4	1.1	2.6
Tunisia	4.5	8.0	1.8
Uganda	0.2	1.0	6.9
Zaire	0.1	0.1	1.2
Zambia	0.9	3.1	3.5
Zimbabwe	1.2	3.1	2.5
Unweighted Average	**1.1**	**2.5**	**4.0**

Source: World Telecommunications Development Report. Geneva: ITU, 1994.

Another bottleneck was pointed out in an article in the Harare *Sunday Herald* during the 1989 Africom Development Conference sponsored by the International Telecommunications Union. Under the headline "Telephone Blues," the article reported that the Zimbabwe Postmaster General had announced that the freeze on telephone installations imposed in 1989 would continue because there were no available telephone instruments. He confirmed that there were telephone lines available in certain areas, but no telephone sets, and that the department had applied to the government for a foreign exchange allocation to source the sets. The article stated that there were 70,000 applicants waiting for telephones ("Telephone Blues" 1990).

While it is unusual to have extra lines in most parts of Africa, the situation in Harare pointed out that some bottlenecks could be overcome simply by opening up the market, in this case for terminal equipment. Africans could probably obtain reconditioned rotary dial telephone sets for next to nothing from countries that have switched to tone dialing.

It should be noted that several African countries are opening up their markets for terminal equipment. However, this equipment is generally imported from overseas and must be paid for with hard currency. Some African countries do produce telephone sets, so it would be possible to establish a market in which they could sell sets in other African countries directly to users rather than to the PTTs.

15.5.2 Bypass: The Users' Response

When users are unable to obtain the capacity they need, or to afford to use available services, they look for alternative solutions. In the old days, they turned to high-frequency (HF) radio. There are still numerous private HF radio networks today in many parts of Africa, and although HF is frustrating in its signal quality and varying reliability, the price is right. If the users own their radios, they can use them whenever they want without paying a carrier. Now, satellites offer a more reliable bypass option.

The Flying Doctor Service operated by the African Medical and Relief Foundation (AMREF) in Kenya, Tanzania, Malawi, and other countries continues to use its own dedicated radio networks rather than the public telephone system because of the high cost and low reliability of the telephone service. As noted earlier, another nonprofit development organization has gone even further and launched its own satellite for medical communications in the developing world. SatelLife's microsatellite, launched in July 1991, provides store-and-forward data communications to small terminals in developing countries. SatelLife was founded by the International Physicians for the Prevention of Nuclear War to reflect their belief that the greatest threat to our common humanity is the gap that exists between health conditions in the developing world and those in industrialized countries.

Why did physicians feel compelled to raise funds for their own satellite? Because, despite modern technology, telecommunications facilities in the poorest regions were either unavailable or unaffordable. "The need in Africa for electronic mail not dependent on traditional communications infrastructures is desperate: In Zambia, international calls are billed at US$6 per minute. In Kenya, a fax costs $7.70 per page outgoing. In Tanzania . . . the minimal cost of a telex [is] a little more than US$25" (Clements 1991). African researchers have been able to use the satellite for free for the first three years and have gained access to medical libraries and other sources of expertise (*SatelLife News* 1993).

These applications may seem rather inconsequential in terms of usage and revenues lost to carriers. But in other parts of the world, commercial users such as banks, brokerages, news services, and oil companies are turning to bypass on a much bigger scale. In most cases, users would rather not have to develop expertise in telecommunications and set up their own networks. Physicians, educators, and bankers simply are looking for affordable and reliable service. They would invariably rather deal with the carriers and leave the technical details to them. But out of frustration, and sometimes desperation, they have turned to setting up their own networks.

15.6 Restructuring the Telecommunications Sector

Since telecommunications is critical to Africa's social and economic development, well-managed telecommunications networks are vitally important. As noted by David Lomax (1990), improved management will be needed not only to improve service quality but to manage the huge investments required to expand the networks. Increasingly, as computers and facsimile machines proliferate, African networks will be required to carry not only voice but facsimile and data communications traffic as well.

Several African countries have begun to reorganize their telecommunications sector. Nigeria has separated its domestic telecommunications from the post office and brought it together with external communications under a new organization, NITEL (Nigerian Telecommunications Ltd.), which operates as a limited liability company with the Nigerian government as majority shareholder (see chapter 9). The company has increased its rates dramatically and also instituted management practices such as aggressive bill collection and downsizing the workforce. Both of the procedures have increased revenues but have been unpopular among users, some of whom claim that service quality has not improved commensurate with local rates.

In 1985, Senegal established the National Society for Telecommunications (SONATEL), which extended the jurisdiction of the former international carrier (TELESENEGAL) to include domestic communications and did so by taking over domestic services formerly operated by the Office of Posts and Telecommunications (OPT). The state is the sole shareholder of SONATEL, which operates as a national company. It has introduced modern management techniques including bonuses and incentives for employees based on merit, as well as improved planning, engineering, and commercial services.

In 1975, Zambia turned its General Post Office (GPO) into a parastatal organization, the Post and Telecommunications Corporation (PTC). In 1987, Zambia made a further move to establish the independence of telecommunications from the government by transforming the state-controlled and state-funded PTC into a limited liability company as part of the Zambia Industrial and Mining Corporation (ZIMCO). Under ZIMCO, the government's investments will have strict limitations, and the company is to operate without subsidies (Akwule 1990).

In the late 1990s, several major countries moved toward a postprivatization of their telecom operations: South Africa, Ivory Coast, Ghana, Egypt, Moracco, and Tunisia. These are all indicators of the movement toward greater autonomy of the telecommunications sector in developing countries by setting up an entity with independent management, commercial goals, and foreign investment. In the short run, this approach is likely to lead to greater efficiency simply because the entity has an incentive to generate surplus revenues if it is allowed to retain and reinvest its own profits. However, if the government remains the majority shareholder, the government does have a stake in the operation of the company. This role may make it possible for the government to establish other goals as well, such as subsidized service for rural areas or introduction of new services. On the other hand, it also places the government in a potential conflict of interest between the goals of the company—to operate commercially—and the goals of the state. It is not clear how these conflicts would be resolved.

As the experience in some industrialized countries such as the United States, Canada, and the United Kingdom has shown, a private monopoly eventually reaches the point where it has few incentives to serve the public interest or to hold down costs. If competition is not to be introduced, a system of incentives that encourages efficiency but requires that certain public policy and service goals be met is required. There do not appear to be models of incentive regulation or other comparable mechanisms in Africa as yet.

15.7 Financing Options

While investment in developing world telecommunications has increased dramatically in the decade since the Maitland report was written, Africa still lags in attracting investment despite the immense unmet demand. The major reason for this continued underinvestment is the perceived high level of risk. Political instability in the form of coups and civil wars continues to plague numerous countries, while corruption at high levels is endemic in many regimes.

Yet Africa is not a single monolith; there are many bright spots. The end of apartheid in South Africa is the most visible reform, but there have been other significant advances such as the independence of Namibia and the reconstruction of the Ugandan economy. South Africa may become a regional economic engine; however, conditions on the continent continue to vary greatly in terms of political stability, economic organization, and existing infrastructure.

Another reason for underinvestment is the reluctance of many African governments to restructure the telecommunications sector. While there is a move toward corporatization, so that telecommunications can be run as an autonomous government entity, many governments appear to want to maintain control over the sector and the revenues it generates. Yet, as noted earlier, some have adopted the approach of allowing the private sector to provide new services such as cellular systems and value-added networks. This strategy may eventually serve as a transition to general privatization.

Another method of harnessing private initiative while maintaining public-sector control is to license telephones to private agents. For example, in Rwanda, kiosks that sell newspapers and soft drinks also offer telephone and sometimes fax services. The kiosk owner gets a percentage of the revenue, protects the telephone from vandals, and typically stays open much longer than the post office, where other pay phones are located.

Financing for telecommunications in Africa typically involves a mixture of World Bank and other development bank loans, bilateral aid, often in the form of tied aid requiring installation of equipment from the donor country, and vendor financing. Models introduced in other regions such as build-operate-yransfer (BOT) (e.g., in Thailand and Indonesia) and obligations to invest as part of franchise agreements (Philippines) may also be used as incentives in Africa if national political and economic risks can be minimized.

The most recent trend for financing telecommunications in Africa, particularly for the PTTs, is through privatization. Many major national telecommunications operators in Africa, from Nigeria to Uganda, have been privatized or are planning to be privatized. The most likely financing partner for the African PTTs are national telecommunications operators in Europe. However, Texas-based SBC and Telekom Malaysia have also participated with privatizations throughout Africa. Although most PTTs stand to gain capital infusions by being privatized, African countries also run the risk of losing control of one of their most valuable national assets.

15.8 The Role of the International Telecommunication Union: Beyond Technical Assistance

Through its Technical Cooperation Department, the International Telecommunication Union (ITU) has carried out telecommunications projects approved by the United Nations Development Program (UNDP) and the developing countries involved. As such, it has been largely a subcontractor for the UNDP (Codding and Rutkowski 1982). The Maitland Commission recommended the creation of a Centre for Telecommunications Development to take a more proactive role in responding to the needs of developing countries and to provide a vehicle for participation by the private sector. In 1989, the ITU's Advisory Group on Telecommunications Policy noted in its report *The Changing Telecommunication Environment* that "Developing countries require advice with greater emphasis on economical, financial, managerial and regulatory issues both to stimulate expansion of basic telephone service and to respond to the new telecommunications environment. They turn to ITU as the most appropriate organization for providing such advice" (Advisory Group 1989).

A new structure to respond to developing country needs was approved by the ITU Plenipotentiary Conference in 1990. A Bureau for Technical Cooperation was approved, to take over and expand upon the functions of the Technical Cooperation Department. (The Plenipotentiary extended the life of the Center for Telecommunications Development for two years, but the functions of the center have been folded into the bureau.)

What should be the role of the ITU in providing assistance to Africa? Clearly, there is a continuing need for training, both technical and managerial, and assistance in preparing technical plans. These have been the traditional roles of the ITU, through training courses and seminars and provision of on-site experts to work with the staff of developing countries.

A second role may be as a regional coordinator. Through RASCOM, the ITU helped African nations to integrate the overlapping and conflicting activities of several regional organizations, coordinated activities designed to gather data on the continent's telecommunications requirements, and examined various options for obtaining satellite services for domestic and regional use. A third role could be to provide a wider range of advisory and consulting services to assist African countries in dealing with the complex policy issues they face in deciding how to restructure the telecommunications sector, introduce new services, consider privatization and competition, revise tariffs, monitor service quality, and other activities.

Does the ITU have a further role to play? Perhaps it is now up to the countries themselves either to establish their own autonomous regional satellite organization or to use the data from the RASCOM studies to plan for their own national or regional communications networks, to develop strategies to entice carriers and equipment suppliers, and to obtain the necessary financing (see Hudson 1991b).

15.9 Implications for African Telecommunications Planning

15.9.1 Integrating Planning across Sectors

In order for new communications technologies to serve development goals, communications planning in Africa must be integrated with national planning. If a country intends to open up new areas for settlement or resource development, telecommunications facilities will be required. If a country intends to diversify its economy, it will need to ensure that adequate infrastructure is in place. It may also need to upgrade the skills of its workforce, perhaps by using instructional technologies.

Lack of coordination between communications and other sectors results in wasted resources and lost opportunities. Some countries have been unable to attract new industries because they lack the necessary telecommunications infrastructure. Too often, telecommunications planning is done in isolation without information about government development priorities or new economic activities.

New technologies allow telecommunications planners to respond to changing needs by installing radio or satellite links, for example, to serve new customers or development projects. Yet planners often do not take advantage of this new technical flexibility by authorizing modifications to existing network plans. They may also hinder development even if the equipment is in place if they enforce unrealistic technical standards or adopt tariffs that make it virtually impossible for potential customers to take advantage of the newly installed facilities.

In sum, the following is needed for coordinated communications planning to occur:

- Telecommunications administrations must be informed about national priorities and development plans.
- National planners must be made aware of the importance of telecommunications infrastructure to national development.
- Resources for extension and improvement of facilities must be allocated to the communications sector, and resources for training and utilization of facilities must be included in the sector budgets.
- Potential users must be made aware of the services available and how they could benefit from them.

15.9.2 User Involvement

Users are rarely heard from when the telecommunications plans for developing countries are being prepared. Yet the users are the most important element of any plan; without an understanding of their needs and constraints, telecommunications services may be inappropriately designed or priced. Why are users such as business managers, development agency officials, and researchers so often unrepresented? They may not have the technical expertise usually expected in planning activities, or they may be unaware of how and when to get involved.

In a sense, telecommunications planners have to act like extension agents to get out and meet with users, learn about their needs, and help them to translate their requirements into facilities and services. This is a new role for telecommunications carriers worldwide. Yet it is a particularly important function in developing countries where resources for new facilities are limited and failure to meet user needs can hinder economic development as well as limit the carriers' projected revenues.

15.10 Conclusion

The goal of using telecommunications to contribute to national development requires an active government policy to ensure that telecommunications plans and services are designed to meet national goals. It also requires flexibility and innovation in services, equipment, and pricing to respond to user needs.

The important fact for African policy makers to keep in mind is that the telecommunications network is more than simply a financial asset and a source of revenue. It is a vital strategic resource for their nations and for the continent. By meeting the challenges posed by new and converging technologies, telecommunications planners can maximize the benefits of telecommunications for African development.

Notes

1. Even in South Africa, which has the highest teledensity on the continent, telephone access in black townships, including those in urban areas, is extremely limited.
2. Data on urban versus rural areas are not available for some of the poorest countries, which tend to have extremely low telephone densities.

3. The author was a special adviser to the Maitland Commission and drafted sections of the report on the role of telecommunications in socioeconomic development.

4. South Africa is an exception, with approximately 1.2 times as many television sets as telephone lines, according to ITU data.

Bibliography

Advisory Group on Telecommunications Policy. 1989. *The Changing Telecommunication Environment.* Geneva: International Telecommunications Union, February.

Akwule, Raymond U. 1990. "Modern Telecommunications Policy and Management Trends." *Proceedings of Africa Telecom 90.* Harare, Zimbabwe, December 3–6.

AT&T. 1990. *The World's Telephones.* Bedminster, N.J.: AT&T.

Blair, Michael L. 1988. "VSAT Systems in Developing Countries." *Proceedings of the Pacific Telecommunications Conference.* Honolulu, February.

Clements, Charles. 1991. "HealthNet." Cambridge, Mass.: SatelLife, May.

Codding, George A., Jr., and Anthony M. Rutkowski. 1982. *The International Telecommunication Union in a Changing World.* Norwood, Mass.: Artech House.

Cooperman, William, Lori Mukaida, and Donald M. Topping. 1991. "The Return of PEACESAT." *Proceedings of the Pacific Telecommunications Conference.* Honolulu: January.

Goldschmidt, Douglas. 1991. On Curbing Predation in Newly Competitive Markets. Unpublished paper. Greenwich, Conn.: Alpha Lyracom (PanAmSat).

Hills, Jill. 1990. "The Telecommunications Rich and Poor." *Third World Quarterly* 12(2): 71–90.

Hudson, Heather E. 1984. *When Telephones Reach the Village: The Role of Telecommunications in Rural Development.* Norwood, N.J.: Ablex.

______. 1990a. *Communication Satellites: Their Development and Impact.* New York: Free Press.

______. 1990b. "Forging New Links: Telecommunications Management for African Development." *Proceedings of Africa Telecom 90.* Harare, Zimbabwe, December 3–6.

______. 1991a. Rural Reception of Television in Rwanda. Report prepared for the International Telecommunications Union, November.

______. 1991b. "Telecommunications in Africa: The Role of the ITU." *Telecommunications Policy* 15, no. 4 (August).

______. 1993. Maximizing Benefits from New Technologies: Policy Challenges for Developing Countries." In *Global Telecommunications Policies: The Challenge of Change,* edited by Meheroo Jussawalla. Westport, Conn.: Greenwood Press.

______, ed. 1987. "Innovative Strategies for Telecommunications Development." Special issue of *Telematics and Informatics* 4, no. 2 (Spring).

Hudson, Heather E., and Lynn York. 1988. "Generating Foreign Exchange in Developing Countries: The Potential of Telecommunications Investments." *Telecommunications Policy,* September.

Hukill, Mark A., and Meheroo Jussawalla. 1989. "Telecommunications Policies and Markets in the Asean Countries." *Columbia Journal of World Business.* 24, no. 1 (Spring).

INTELECON. 1991. "Opportunities for Distance Education in Commonwealth African Countries." Vancouver: Commonwealth of Learning.

Intelsat. 1993. *Annual Report.* Washington, D.C.: Intelsat.

International Commission for Worldwide Telecommunications Development (The Maitland Commission). 1984. *The Missing Link.* Geneva: International Telecommunications Union.

International Development Research Centre. 1989. *Sharing Knowledge for Development: IDRC's Information Strategy for Africa.* Ottawa: IDRC.

International Telecommunications Union. 1986. *Information, Telecommunications, and Development.* Geneva: ITU.

______. 1994. *World Telecommunication Development Report.* Geneva: ITU.

Jipguep, Jean. 1990. "New Dimensions: Regional African Satellite Communications System (RASCOM)." *Proceedings of Africa Telecom 90.* Harare, Zimbabwe, December 3–6.

Johnson, Tony. 1991. "Microsatellite That Turns Information into Medical Power." *The (Manchester) Guardian,* April 26.

Lomax, David. 1990. "Issues and Constraints for Telecommunications Development in Africa." *Proceedings of Africa Telecom 90.* Harare, Zimbabwe, December 3–6.

Mayo, John K., G. R. Heald, S. J. Klees, and M. Cruz de Yanes. 1987. *Peru Rural Communication Services Project Final Evaluation Report.* Washington, D.C.: Academy for International Development.

National Research Council, Board on Science and Technology for International Development. 1990. *Science and Technology Information Services and Systems in Africa.* Washington, D.C.: National Academy Press.

NSF Network News. 1994. Vol. 1, no. 4 (September/October).

Parker, Edwin B. 1987. "MicroEarth Station Satellite Networks and Economic Development." *Telematics and Informatics* 4, no. 2.

Parker, Edwin B., and Heather E. Hudson. 1992. *Electronic Byways: State Policies for Rural Development through Telecommunications.* Boulder, Colo.: Westview Press.

Parker, Edwin B., Heather E. Hudson, Don A. Dillman, and Andrew D. Roscoe. 1989. *Rural America in the Information Age: Telecommunications Policy for Rural Development.* Lanham, Md.: University Press of America.

Pierce, William B., and Nicolas Jequier. 1983. *Telecommunications for Development.* Geneva: ITU.

SatelLife News. 1993. Fifth issue, November.

______. 1994a. Sixth issue, February.

______. 1994b. Seventh issue, May.

Saunders, Robert, Jeremy Warford, and Bjorn Wellenius. 1994. *Telecommunications and Economic Development.* 2d ed. Baltimore: Johns Hopkins University Press.

Stover, William J. 1984. *Information Technology in the Third World: Can I.T. Lead to Humane International Development?* Boulder, Colo.: Westview Press.

"Telephone Blues." 1990. *The Sunday Mail,* Harare, Zimbabwe, December 9, p. 1.

Tietjen, Karen. 1989. *AID Rural Satellite Program: An Overview.* Washington, D.C.: Academy for Educational Development.

Warwick. Martyn. 1994. "Out of Africa." *Communications International* 21, no. 1 (January): 58–62.

Wellenius, Bjorn, Peter A. Stern, Timothy E. Nulty, and Richard J. Stern, eds. 1989. *Restructuring and Managing the Telecommunications Sector.* Washington, D.C.: World Bank.

Westendoerpf, Dietrich, and A. Odeh. 1990. "The Case for Investing in Telecommunications in Africa." *Proceedings of Africa Telecom 90.* Harare, Zimbabwe, December 3–6.

Index

www.ingramcontent.com/pod-product-compliance
Ingram Content Group UK Ltd.
Pitfield, Milton Keynes, MK11 3LW, UK
UKHW020849150926
475595UK00003B/80